Photothermal Investigations of Solids and Fluids

PHOTOTHERMAL INVESTIGATIONS OF SOLIDS AND FLUIDS

Edited by

Jeffrey A. Sell

PHYSICS DEPARTMENT
GENERAL MOTORS RESEARCH LABORATORIES
WARREN, MICHIGAN

ACADEMIC PRESS, INC.
Harcourt Brace Jovanovich, Publishers

Boston San Diego New York
Berkeley London Sydney
Tokyo Toronto

ACADEMIC PRESS, INC.
1250 Sixth Avenue, San Diego, CA 92101

United Kingdom Edition published by
ACADEMIC PRESS INC. (LONDON) LTD.
24-28 Oval Road, London NW1 7DX

Library of Congress Cataloging-in-Publication Data

Photothermal investigations of solids and fluids/edited by Jeffrey A. Sell.
p. cm.
Includes bibliographies and index.
ISBN 0-12-636345-5
1. Photothermal spectroscopy. I. Sell, Jeffrey A., Date
QD96.P54P46 1988
543'.0858--dc19 88-10420
CIP

Printed in the United States of America
88 89 90 91 9 8 7 6 5 4 3 2 1

Contents

List of Contributors ix
Preface xi

1. Overview of Photothermal Spectroscopy

A. C. Tam

I. Introduction 1
II. PT Detections and Applications 3
III. Conclusion 30

2. Photothermal Investigation of Solids: Basic Physical Principles

D. Fournier and A. C. Boccara

I. Physical Parameters Involved in a Photothermal Experiment 36
II. Detection of the Photothermal Signal 37
III. Photothermal Signal Generation: Theoretical Approach 43
IV. Specific Problems 63
V. Spectral and Spatial Multiplexing the Photothermal Signals 76

3. The Theory of the Photothermal Effect in Fluids

R. Gupta

I. Formation of the Thermal Image 82
II. Detection of the Thermal Image 93

4. Photothermal Spectroscopy: Applications in Chromatography and Electrophoresis

Michael D. Morris and Fotius K. Fotiou

I. General Introduction to Chromatography and Electrophoresis 127
II. Photothermal Gas–Chromatography Detectors 131

III. Photothermal Liquid–Chromatography Detectors 133
IV. Photothermal Flow–Injection Analysis Detectors 140
V. Photothermal Electrophoresis Detectors 141
VI. Photothermal Detectors for Thin–Layer Chromatography 145
VII. Conclusions 153

5. Photothermal Studies of Energy Transfer and Reaction Rates

John R. Barker and Beatriz M. Toselli

I. Introduction 155
II. Time-Resolved Optoacoustics 156
III. Time-Dependent Thermal Lensing 157
IV. Timescales 160
V. TDTL and TROA Theory 162
VI. Experimental Methods 169
VII. Observed and Calculated Results 170
VIII. Conclusions 188

6. Mirage Detection of Thermal Waves

P. K. Kuo, L. D. Favro, and R. L. Thomas

I. Imaging of Surface and Subsurface Defects in Solids 191
II. Characterization of Material Properties 198

7. Fluid Velocimetry Using the Photothermal Deflection Effect

Jeffrey A. Sell

I. Introduction 213
II. Photothermal Deflection Effect 214
III. Related Research 220
IV. Laminar Flow (Low Velocity) 221
V. Laminar Flow (High Velocity) 228
VI. Turbulent Flows 240
VII. Two–Dimensional Velocity Measurements 242
VIII. Velocities for Flowing Liquids 243
IX. Temperature and Pressure Dependence of PD Signals 245
X. Conclusions 246
XI. Appendix 246

8. Combustion Diagnostics by Photothermal Deflection Spectroscopy

R. Gupta

I. Introduction 249
II. Variation of the PTDS Signal with Temperature 253
III. Variation of Relevant Parameters with Temperatures 255
IV. Observation of PTDS Signals in Flames 257
V. Concentration Measurements 260
VI. Temperature Measurements 263
VII. Flow–Velocity Measurements 264
VIII. Flame Spectroscopy 265

9. Photothermal Characterization of Electrochemical Systems

Andreas Mandelis

I. Introduction 269
II. Out–of–Cell Characterization of Electrodes 271
III. *In situ* Characterization of Electrodes 273
IV. Electrochemical Interface Reactions 277
V. Corrosion and Thin–Film Growth Processes 285
VI. Diffusion of Electrochemical Species 290
VII. Energy–Transfer Physics at Photoelectrochemical Interfaces 293
VIII. Conclusions and Future Directions 306

10. Photothermal Spectroscopy of Aerosols

A. J. Campillo and H.-B. Lin

I. Introduction 309
II. Survey of Possible Photothermal Schemes 315
III. Early Work 320
IV. Photothermal Interferometry 321
V. Photothermal Modulation of Light Scattering 329
VI. Photophoresis 334
VII. Conclusion 340

Index 343

Contributors

Numbers in parentheses indicate the pages on which the authors' contributions begin.

JOHN R. BARKER, Department of Atmospheric and Oceanic Science, 7106 Space Research Building, University of Michigan, Ann Arbor, Michigan 48109-2143 (155)

A. C. BOCCARA, Laboratoire d'Optique Physique, Ecole Superieure de Physique et Chemie Industrielles, 10 rue Vauquelin, F-75231 Paris Cedex 05, France (35)

A. J. CAMPILLO, Optical Sciences Division, Naval Research Laboratory, Washington, DC 20375 (309)

L. D. FAVRO, Department of Physics and Astronomy and Institute for Manufacturing Research, Wayne State University, Detroit, Michigan 48202 (191)

FOTIOS K. FOTIOU, Department of Chemistry, University of Michigan, Ann Arbor, Michigan 48109-2143 (127)

D. FOURNIER, Laboratoire d'Optique Physique, Ecole Superieure de Physique et Chemie Industrielles, 10 rue Vauquelin, F-75231 Paris Cedex 05, France (35)

R. GUPTA, Department of Physics, University of Arkansas, Fayetteville, Arkansas 72701 (81, 249)

P. K. KUO, Department of Physics and Astronomy and Institute for Manufacturing Research, Wayne State University, Detroit, Michigan 48202 (191)

H.-B. LIN, Optical Sciences Division, Naval Research Laboratory, Washington, DC 20375 (309)

ANDREAS MANDELIS, Photoacoustic and Photothermal Sciences Laboratory, Department of Mechanical Engineering, University of Toronto, Toronto, Ontario M5S 1A4, Canada (269)

MICHAEL D. MORRIS, Department of Chemistry, University of Michigan, Ann Arbor, Michigan 48109 (127)

JEFFREY A. SELL, Physics Department, General Motors Research Laboratories, Warren, Michigan 48090-9055 (213)

A. C. TAM, IBM Research, Almaden Research Center, Dept. K06/803(E), 650 Harry Road, San Jose, California 95120-6099 (1)

R. L. THOMAS, Department of Physics and Astronomy and Institute for Manufacturing Research, Wayne State University, Detroit, Michigan 48202 (191)

BEATRIZ M. TOSELLI, Department of Atmospheric and Oceanic Science, Space Physics Research Laboratory, University of Michigan, Ann Arbor, Michigan 48109-2143 (155)

Preface

In the last few years there has been a surge of interest in investigations of materials (solids and fluids) using photothermal optical diagnostic techniques. The reason for the sudden popularity of this research field is the wide applicability of these techniques and the realized or potential advantages over alternative techniques. Also, the development of lasers as convenient and powerful sources of localized energy has contributed greatly to the success of the field.

To the best of this researcher's knowledge, there is a gap in the literature in coverage of this field of research—apparently there are no reference books on the subject of photothermal optical diagnostics or spectroscopy. Numerous research papers on these subjects have appeared in many different journals including, but not limited to, Applied Optics, Optics Communications, Optical Engineering, Optics Letters, Journal of Applied Physics, Applied Physics, Applied Physics Letters, Journal of the Electrochemical Society, Analytical Chemistry, and Applied Spectroscopy. In addition, there have been several short review articles on photothermal spectroscopy (Morris, 1986; Bialkowski, 1986; Harris, 1986; Fang and Swofford, 1983). The best collection of papers from the field are the conference proceedings from the Third, Fourth, and Fifth International Topical Meetings on Photoacoustics, Thermal, and Related Sciences, in Paris (1983), Ville d'Esterel, Quebec (1985), and Heidelberg (1987), respectively. However, these are collections of summaries of the papers presented at the conferences and are not intended as reference books on the subject. (Selected papers from the conferences have also been published separately in *J. de Physique*, *Supplement C6* **44**, 1983, *Canadian J. Physics* **64**, 1986, and the volume by Hess and Pelzl, 1988.) The purpose of this book is to fill the gap in the literature and serve as a reference source for photothermal investigations.

Photothermal techniques generally include the areas of photothermal spectroscopy, imaging, and velocimetry. Photothermal spectroscopy is defined for our purposes here as the field in which the nature of matter is probed using optical excitation of a medium and optical probing of the thermal energy which results from this excitation. Normally, both the excitation, or pumping, and probing are accomplished with laser beams, although not always. Detection can be in the form of deflection, diffraction, refraction, displacement, and lensing (convergence or divergence) of the probe beam. In addition, detection by interferometry is considered here to be within the realm of photothermal spectroscopy. Photothermal imaging and velocimetry are closely related to photothermal spectroscopy, but normally do not involve variation of the pump wavelength. Rather, in imaging either the sample or the pump or probe beams are moved, and in velocimetry the sample is in motion. The emphasis in this volume is on photothermal deflection since much more research has been conducted in this area recently than the other detection schemes mentioned above. The subject of photothermal spectroscopy is closely linked to photoacoustic spectroscopy, also called optoacoustic spectroscopy; the main difference is in the detection technique. Photoacoustic spectroscopy uses a microphone or piezoelectric transducer to detect acoustic waves and photothermal spectroscopy uses optics to detect thermal or acoustic waves. Optical detection of acoustic waves which are a result of optical pumping, normally called photoacoustic deflection, is considered here to be a part of photothermal spectroscopy. There are already several reference texts (Pao, 1977; Rosenscweig, 1980; Zharov and Letokhov, 1986) and review papers (Tam, 1986; Patel and Tam, 1981) on photoacoustic spectroscopy and this book will tend to minimize the discussion of photoacoustics, although the two subjects are so closely tied that some discussion in this book is necessary. In addition, photothermal radiometry is not treated in detail here since it has recently been reviewed (Tam, 1985).

This monograph is intended for graduate students and professionals (professors, and research scientists) who may or may not be familiar with the field. It is expected that the book should be very useful for researchers already active in the field, since the theory (covered in Chapters 1 to 3) is the most complete to date and the most recent experimental advances are discussed. (The emphasis is on experimental techniques, however.) Also, the applications covered here are so diverse that there is likely to be material that is new to researchers already familiar with the field. The organization of the book is as follows: Chapter 1 is an introduction and overview of the field. Chapter 2 covers further basic principles and the theory relevent to solids. Chapter 3 covers the theory for fluids (liquids and gases). Then, the bulk of the book summarizes the many diverse applications. Chapter 4 covers applications of photothermal techniques in chromatography and electrophoresis. Chapter 5 treats the measurement of energy transfer and relaxation rates. Chapter 6 summarizes mirage detection of thermal wave imaging. Chapter 7 treats the measurements of fluid velocimetry by photothermal deflection. Chapter 8 covers applications in combustion diagnostics. Chapter 9 emphasizes electrochemical studies. Finally, Chapter 10 discusses photothermal spectroscopy of aerosols and gases.

The editor would like to express his appreciation to Kay E. Sutherland for assistance in this project.

REFERENCES

Bialkowski, S. E. (1986). "Pulsed-laser photothermal spectroscopy," *Spectroscopy* **1**, 26.

Fang, H. L., and Swofford, R. L. (1983). "The thermal lens in absorption spectroscopy," in *Ultrasensitive Laser Spectroscopy*, D. S. Kliger, ed., Academic Press, New York, p. 176.

Harris, J. M. (1986). "Photothermal methods for detection of molecules in liquids," *Optics News*, 8.

Hess, P., and Pelzl, J. (1988). *Photoacoustic and Photothermal Phenomena*, Springer Series in Optical Sciences Vol. 58, Springer Verlag, Berlin.

Morris, M. D. (1986). "Photothermal effects in chemical analysis," *Analytical Chemistry* **58**, 811A.

Pao, Y.-H. (1977). *Optoacoustic Spectroscopy and Detection*, Academic Press, New York.

Patel, C. K. N., and Tam, A. C. (1981). "Pulsed optoacoustic spectroscopy of condensed matter," *Rev. Mod. Phys.* **53**, 517.

Rosenscweig, A. (1980). *Photoacoustics and Photoacoustic Spectroscopy*, Wiley and Sons, New York.

Tam, A. C. (1985). "Pulsed photothermal radiometry for noncontact spectroscopy, material testing and inspection measurements," *Infrared Physics* **25**, 305.

Tam, A. C. (1986). "Applications of photoacoustic sensing techniques," *Rev. Mod. Phys.* **58**, 381.

Zharov, V. P., and Letokhov, V. S. (1986). *Laser Optoacoustic Spectroscopy*, Springer Series in Optical Sciences Vol. 37, Springer Verlag, Berlin

CHAPTER 1

OVERVIEW OF PHOTOTHERMAL SPECTROSCOPY

A. C. TAM
IBM Research
Almaden Research Center
San Jose, CA

I. Introduction . . . 1
II. PT Detections and Applications . . . 3
A. Temperature Rise . . . 4
B. Pressure Change . . . 5
C. Refractive-Index Gradients . . . 7
D. Surface Deformation . . . 22
E. Photothermal Radiometry . . . 27
F. Other PT Changes . . . 29
III. Conclusion . . . 30
References . . . 31

I. Introduction

Photothermal (PT) generation refers to the heating of a sample due to the absorption of electromagnetic radiation. Such radiation exciting the sample can be in the optical range as well as in other ranges (X-ray, ultraviolet, infrared, microwave, and radio frequency). Furthermore, effects similar to PT generation can be produced by other types of excitation beams (electron, proton, ions, etc.) instead of an electromagnetic beam. PT generation is an example of energy conversion and has, in general, three types of applications, namely, (a) PT material probing, (b) PT material processing, and (c) PT material destruction. Thus, type (a) applications cause no sample modification, type (b) applications cause the sample to change to another useful form, and type (c) applications render the sample useless. The temperatures involved also generally increase in this order. This article is only concerned with type (a) applications of PT generation; PT material probing is based on the ideas shown in Fig. 1. Optical excitation of a

ISBN 0-12-636345-5

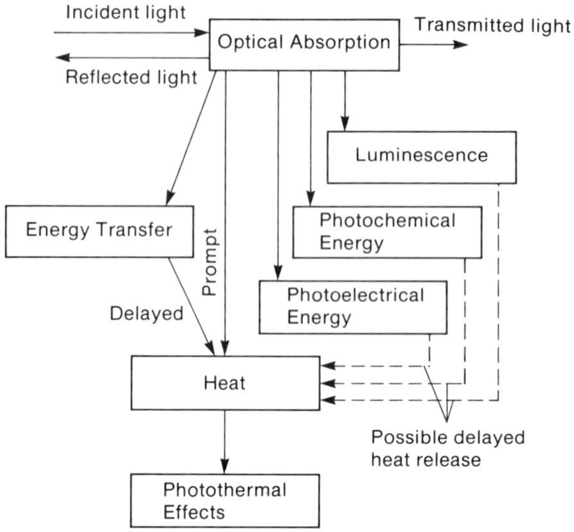

FIG. 1. Block diagram to indicate the possible consequences of optical absorption, leading to "prompt" or "delayed" heat production in competition with other de-excitation channels.

sample can result in the production of several forms of energy: heat, luminescence, chemical energy, or electrical energy. The heat can be produced promptly, or at various time delays due to energy-transfer mechanisms. All these resulting energy forms must add up to equal the absorbed optical energy; in other words, the various de-excitation branches shown in Fig. 1 are "complementary."

PT material probing or characterization techniques generally rely on the use of high-sensitivity detection methods to monitor the effects caused by PT heating of a sample. Such possible effects are indicated in Fig. 2. Many of these PT effects occur simultaneously; e.g., PT heating of a sample in air will produce temperature rise, photoacoustic waves, and refractive-index changes in the sample and in the adjacent air, infrared thermal radiation changes, etc., at the same time. Thus, the choice of a suitable PT effect for detection will depend on the nature of the sample and its environment, the light source used, and the purpose of the measurement. This article provides an overview of the experiment arrangements and detection schemes for the various PT effects and gives a summary of possible applications. Emphasis is given on the use of probe-beam refraction (PBR) schemes to detect PT refractive-index gradients, and approximate analytical solutions for some typical configurations are provided to show how the observed beam deflection signal depends on various parameters, including delayed heat release

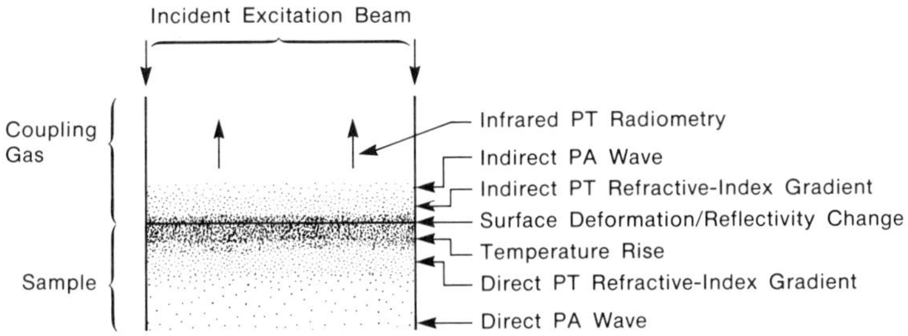

FIG. 2. Common photothermal effects produced by modulated optical absorption in a sample. Direct effects observable in the sample and indirect effects observable in the transparent coupling medium adjacent to the sample are indicated. Local temperature increase is indicated by the density of "dots."

due to a long thermal de-excitation time constant. Literature citations are not intended to be exhaustive; rather, only some of the most recent and representative work in this field is provided here. More details and earlier work are found in other chapters in this book as well as in other reviews (e.g., Tam, 1986; Hutchins and Tam, 1986; Bialkowski, 1986; Rose, Vyas, and Gupta, 1986).

II. PT Detections and Applications

Detection methods for various PT effects are summarized in Table I. These detection methods can either be applied to the sample itself or to the "coupling fluid" adjacent to a condensed sample. (We generally assume that only the sample absorbs the incident light, but not the adjacent coupling fluid, although Low, Morterra, and Khosrofian (1986) have considered a more complicated case of an absorbing coupling gas.) We shall call the former case "direct PT detection," and the latter case "indirect PT detection." All the PT detection schemes require a modulation in the excitation light (or at least a step change). Such a modulation can be in the form of short intense pulses separated by long dark periods (so-called pulsed PT detection) or continuous train of pulses at nearly 50% duty cycle (so-called continuous-modulated PT detection). The former detection scheme is typically in the "time domain" where the PT signal magnitude and shape after the pulse excitation is recorded, whereas the latter detection scheme is typically in the "frequency domain" where the PT signal magnitude and phase are measured by "lock-in detection" with respect to the excitation. Details and examples of the various PT detection methods follow.

TABLE I

THE HEATING OF A SAMPLE DUE TO OPTICAL ABSORPTION CAN RESULT IN VARIOUS PHOTOTHERMAL EFFECTS AND PROVIDE THE CORRESPONDING DETECTION TECHNIQUES.

Photothermal effects	Detection methods (applicable to sample S or to adjacent fluid F)
Temperature rise	Laser calorimetry (S or F)
Pressure change	Direct photoacoustic detection (S) Indirect photoacoustic detection (F)
Refractive-index change (Thermal or acoustic)	Probe-beam refraction (S or F) Probe-beam diffraction (S or F) Other optical probes (S or F)
Surface deformation (Thermal or acoustic)	Probe-beam deflection (S) Optical interference (S)
Thermal emission change	Photothermal radiometry (S)
Reflectivity/absorptivity change	Transient thermal reflectance (S) Transient piezo-reflectance (S) Optical transmission monitoring (S or F)

A. TEMPERATURE RISE

The most direct method for measuring PT heating is the monitoring of the rise in temperature; this is sometimes called "optical calorimetry" or "laser calorimetry," since a laser beam is frequently used for excitation. To detect the laser-induced temperature rise, thermocouples or thermisters have been used, as indicated in Fig. 3a (Brilmyer *et al.*, 1977; Bass, Van Stryland, and Steward, 1979; Bass and Liou, 1984), and pyroelectric detectors with higher sensitivity have also been exploited, as shown in Fig. 3b (Baumann, Dacol, and Melcher, 1983; Coufal, 1984). Even absorption spectroscopy can be used to detect the rise in temperature if the corresponding Boltzmann molecular population change can be analyzed (Zapka and Tam, 1982a).

There are advantages and disadvantages of using a temperature sensor to detect PT temperature rise directly rather than using other detection methods described later. The main advantage is that absolute calibration is readily available, i.e., the observed temperature rise can be directly measured and related to physical parameters like absorption coefficients. The disadvantages are that response is usually slow, and sensitivity is typically low compared to other methods; furthermore, in what is basically a "dc

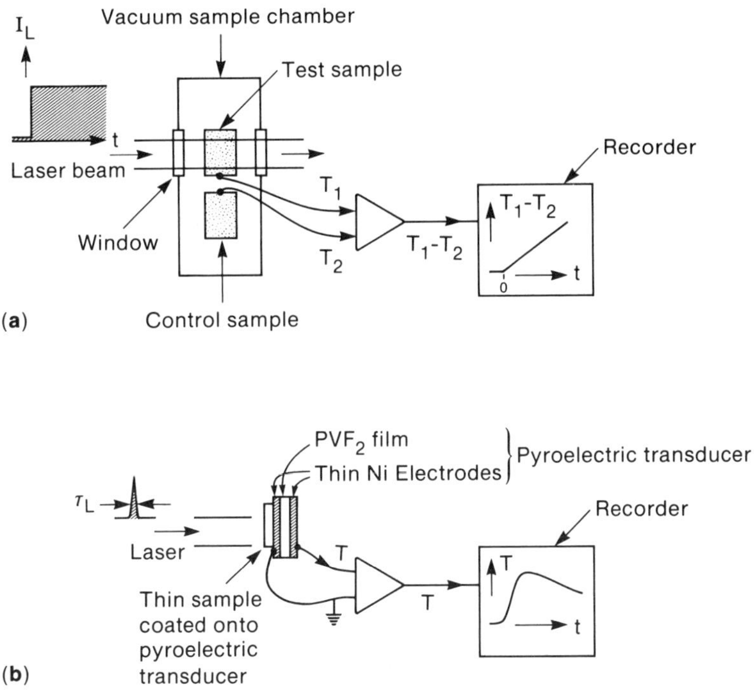

FIG. 3. Examples of laser calorimetry measurements: (a) step heating of an isolated sample; (b) pulsed heating of a thin sample in contact with a thin-film pyrometer.

method," heat leakage from the sample must be minimized by elaborate thermal isolation. However, Coufal (1984) and Coufal and Hefferle (1985) have shown that fast risetime and high sensitivity for a thin-film sample is possible if it is directly coated onto a thin-film pyroelectric detector.

B. PRESSURE CHANGE

Pressure variations or modulations resulting from the absorption of modulated light by a sample is usually referred to as "photoacoustic" (PA) or "optoacoustic" (OA) generation. PA generation mechanisms include electrostriction, thermoelastic expansion, volume changes due to photochemistry, gas evolution, boiling or ablation, and dielectric breakdown, with the PA generation efficiency η (i.e., acoustic-energy generated/light-energy absorbed) generally increasing in this order. For electrostriction and for

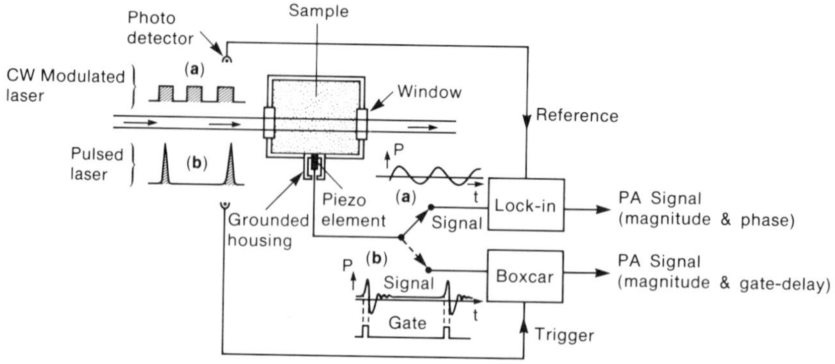

FIG. 4. Schematics of direct PA measurement using (a) continuous modulated excitation, and (b) pulsed excitation.

thermal expansion (also called thermoelastic) mechanisms, η is small, often on the order of 10^{-12} to 10^{-8}, whereas for breakdown mechanisms, η can be as large as 30% (Teslenko, 1977). PT generation via thermoelastic expansion, where η is small, is the most common mechanism used in PA spectroscopy. To get a qualitative spectrum of a sample, the wavelength of the excitation beam is scanned, and the corresponding magnitude of the acoustic signal normalized by the excitation pulse energy is measured to

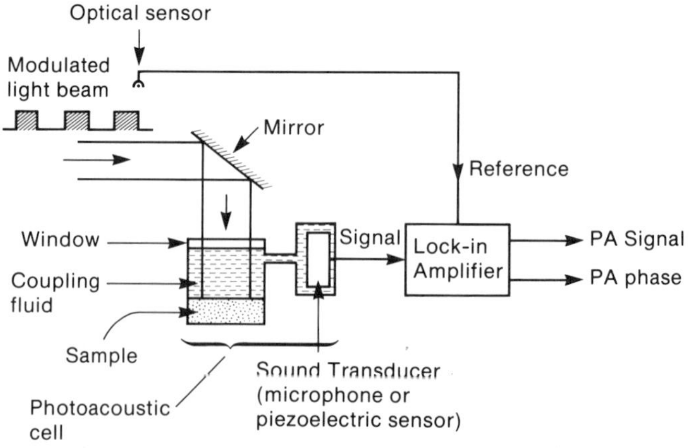

FIG. 5. Schematic of indirect PA measurement using a continuous modulated excitation beam.

provide an "excitation spectrum," called a PA spectrum. To quantify this PA spectrum, the various other "branching ratios" for the de-excitation in the excitation wavelength range (see Fig. 1) must be accounted for.

As discussed for the general case of PT generation, PA generation can be classified as either direct or indirect. In direct PA generation (Fig. 4), the acoustic wave is produced in the sample where the excitation beam is adsorbed. In indirect PA generation (Fig. 5), the acoustic wave is generated in a coupling medium adjacent to the sample, usually due to heat leakage and sometimes also due to acoustic transmission from the sample; here, the coupling medium is typically a gas or a liquid, and the sample is a solid or a liquid.

Extensive reviews of the application of PA detection for spectroscopy and other measurements have been given in the literature, e.g., Rosencwaig (1980), Patel and Tam (1981), and Tam (1986).

C. Refractive - Index Gradients

PT heating of a sample can produce a refractive-index gradient (RIG) in the sample (direct effect) or in an adjacent "coupling fluid" (indirect effect). Also, there are two types of RIG produced by the PT heating of the sample, namely, a "thermal RIG" and an "acoustic RIG." The thermal RIG is produced by the decreased density of the medium (sample or coupling fluid) caused by the local temperature rise, decays in time following the diffusional decay of the temperature profile, and remains near the initial optically excited region. The acoustic RIG is associated with the density fluctuation of the medium caused by the propagation of the PA wave, decays in propagation distance following the attenuation of the PA wave, and travels at the acoustic velocity away from the initial optically excited region. Thermal RIG and acoustic RIG are related and can be used to measure similar parameters, like optical absorption coefficient, temperature, or flow velocity of the sample. However, thermal diffusivity can only be measured by the time evolution of the thermal RIG, and acoustic velocity and attenuation can only be measured by the spatial dependence of the acoustic RIG. Also, at distances farther than several thermal diffusion lengths from the excitation region, only acoustic RIG can be detected. In general, thermal RIG provides a larger signal compared to acoustic RIG, which, however, can have a narrower temporal profile and be detectable far away from the excitation region.

1. Thermal RIG

The thermal refractive-index gradient generated by the excitation beam affects the propagation of an optical beam in its vicinity, including its own propagation, resulting in the well-known effect of "self-defocusing" or "thermal blooming." Self-defocusing generally occurs instead of self-focusing because the derivative of the refractive index with respect to temperature is usually negative, so that the temperature gradient results in a negative lens. The thermal RIG also affects the propagation of another weak "probe" beam in the vicinity of the excitation beam. Thus, thermal RIG can be monitored either by self-defocusing or by PBR. Leite, Moore, and Whinnery (1964) have first shown that self-defocusing of the excitation beam provides a sensitive spectroscopic tool, and Solimini (1966) later gave a quantitative theory for this application. Swofford, Long, and Albrecht (1976) have shown that the PBR method with an additional collinear probe beam provides higher sensitivity than the single-beam self-defocusing method; this is called "thermal lensing" spectroscopy using a probe beam. Kliger (1980) and Bialkowski (1985) have reviewed the spectroscopic applications and analyzed the aberration effects of the thermal lens.

The PBR technique for probing the refractive-index gradient need not employ collinear beams as in the thermal lens experiment. A probe beam that is parallel to, but displaced from, the excitation beam can also be used (see Fig. 6a). Boccara *et al.* (1980) and Jackson *et al.* (1981a) have pointed out that the PBR method, using beams with appropriate displacements, can have higher sensitivity than thermal lens spectroscopy if the probe is positioned at the maximum refractive-index gradient, which is situated at approximately one beam radius from the axis of the excitation beam; their work indicated that an absorption fraction of 1 part in 10^8 can be detected.

In general, the probe beam need not even be parallel to the excitation beam. Although a parallel and partially overlapping pump–probe configuration gives the largest probe deflection (see e.g., discussions after Eq. (4)), this configuration may not always be possible. For example, high spatial resolution measurements necessitate the use of "crossed" beams (Fig. 6b). Also, for opaque samples, orthogonal PBR detection is generally needed because no transmission through the sample is possible (Fig. 6c).

Since the detection of thermal RIG has been extensively used in PT spectroscopy and materials probing, we shall give a more detailed description of the theoretical derivations of typical cases of parallel PBR and perpendicular PBR, and their applications. To make the mathematical forms simple and analytical in order to provide physical insights, certain conditions on the sample and optical arrangements are imposed. More

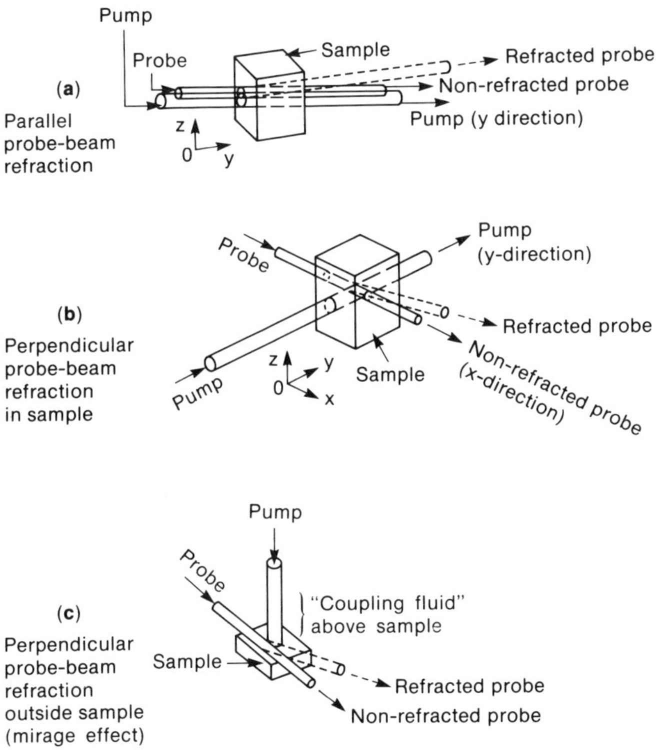

FIG. 6. Examples of various PT probe-beam refraction (PBR): (a) direct PBR in a sample for parallel pump and probe (the special case of concentric pump and probe is called "thermal lensing"); (b) direct PBR in a sample for perpendicular pump and probe; (c) indirect PBR in a coupling fluid adjacent to a sample for perpendicular pump and probe.

general cases of PBR for monitoring the thermal RIG can be done by numerical computations (Rose, Vyas, and Gupta, 1986) or by ray-tracing (Sell, 1987). While PBR measurements discussed here are all "linear" with respect to the excitation intensity, nonlinear effects can sometimes be significant (Wetsel and Spicer, 1986; Bialkowski, 1985; Bialkowski and Long, 1987; Long and Bialkowski, 1985).

a. *Parallel PBR*

We derive here the theoretical formula for the case of parallel PBR deflection angle φ (Fig. 6a) with short pulsed excitation and fast thermal de-excitation, and a narrow probe beam at a radial distance r from the axis of the excitation beam with a Gaussian profile. The energy density $E(r)$ at

r is given by

$$E(r) = \frac{2E_0}{\pi a^2} \exp(-2r^2/a^2), \tag{1}$$

where E_0 is the incident pulse energy of the excitation beam, a is the excitation beam $(1/e^2)$-radius. The corresponding temperature gradient at r immediately after the short pulsed excitation is

$$\frac{dT}{dr} = \frac{-8E_0\alpha}{\rho C_p \pi a^2} \frac{r\exp(-2r^2/a^2)}{a^2}, \tag{2}$$

where ρ = density, C_p = specific heat, and α = absorption coefficient (assumed small). For small probe deflection φ, geometrical optics gives

$$\frac{\varphi n_0}{l} = \frac{dn}{dr} = \frac{dn}{dT}\frac{dT}{dr}, \tag{3}$$

where l is the optical path length, and n is the sample refractive index of ambient value n_0. Combining Eqs. (2) and (3), we have

$$\varphi(r) = \frac{l}{n_0}\frac{dn}{dT}\frac{(-8E_0\alpha)}{\rho C_p \pi a^2}\frac{r\exp(-2r^2/a^2)}{a^2}. \tag{4}$$

Equation (4) is basically the same as given by Boccara *et al.* (1980) or by Rose, Vyas, and Gupta (1986). We see from Eq. (4) that the probe-beam deflection $\varphi(r)$ depends linearly on the temperature derivative of the refractive index, and linearly on α (for weak absorption), and is maximum at $r = a/2$.

Note that the parallel PBR angle $\varphi(r)$ given in Eq. (4) for the probe located at displacement r from the pump beam axis is valid only if the following conditions are satisfied: (a) weak absorption, (b) short excitation duration, (c) narrow probe beam, (d) sample thermal de-excitation time τ is small, (e) no thermal diffusion (i.e., observation shortly after excitation), and (f) no flow of the sample. In the case where condition (e) is not satisfied, i.e., observation is made at time t after the short excitation pulse, Eq. (4) is still valid if the following substitution for the excitation radius a is made:

$$a^2 \to (a^2 + 8Dt), \tag{5}$$

where D is the thermal diffusivity. Furthermore, if a flow velocity v_z exists, where z is the direction from the pump beam to the probe beam, Eq. (4) is still valid with the following substitution for the beam-separation parameter r:

$$r \to (z - v_z t). \tag{6}$$

Thus, in the more general case of parallel PBR observed at a delay time t in presence of a flow velocity v_z, the deflection angle $\varphi(z, t)$ is given by

$$\varphi(z, t) = \frac{l}{n_0} \frac{dn}{dT} \frac{(-8E_0\alpha)(z - v_z t)\exp\left[-2(z - v_z t)^2/(a^2 + 8Dt)\right]}{\rho C_p \pi (a^2 + 8Dt)^2}, \tag{7}$$

which agrees essentially with the "travelling thermal lens" deflection angle derived by Sontag and Tam (1986b), in which a factor of 2 was inadvertently missing. Note that Eq. (7) still requires conditions (a) to (d) stated after Eq. (4) to be satisfied.

b. *Perpendicular PBR*

Consider the geometry of Fig. 6b, where the pump beam is along the y-direction, the probe beam is along the x-direction, and the probe and pump axes are separated by a distance z. We again assume that conditions (a) to (c) listed after Eq. (4) are valid. However, we generalize here in order to examine the effects of finite thermal de-excitation time τ and finite observation time t after the short-duration excitation. The flow velocity is assumed to be zero; however, any flow v_z in the z-direction can again be accounted for by the following substitution

$$z \to (z - v_z t). \tag{8}$$

For cylindrical symmetry, the temperature rise $T(r, t)$ will obey the following differential equation

$$\frac{\partial^2 T}{\partial r^2} + \frac{1}{r}\frac{\partial T}{\partial r} - \frac{1}{D}\frac{\partial T}{\partial t} = -\frac{N(r, t)h\nu}{D\rho C_p \tau}\exp(-t/\tau), \tag{9}$$

where r is the radial coordinate, D is the thermal diffusivity, $N(r,t)$ is the concentration of excited molecules, $h\nu$ is the photon energy, ρ is the density, C_p is the specific heat, and τ is the thermal relaxation time. We have assumed that the heat generation process is characterized by the single time constant τ. The density of excited-state molecules after excitation by a short-duration Gaussian-shaped pump beam with beam parameter a is given by

$$N(r,t) = \frac{2\alpha E_0}{\pi h\nu(a^2 + 8D_m t)} \exp\left(\frac{-2r^2}{a^2 + 8D_m t}\right), \qquad (10)$$

where α is the absorption coefficient, E_0 is the pump pulse energy, $h\nu$ is the photon energy, and D_m is the molecular diffusivity. Here, we include the consideration of the diffusion of excited molecules out of the production zone; we expect the molecular diffusivity D_m and the thermal diffusivity D to be comparable. The right side of Eq. (9) contains the heat source term $Q(r, t)$ which has been expressed as

$$Q(r,t) = \frac{N(r,t)h\nu \exp(-t/\tau)}{\tau}. \qquad (11)$$

For cylindrical symmetry, the solution to Eq. (9) for the temperature rise T caused by the pulsed collimated heat source Q has been given by Carslaw and Jaeger (1959) (p. 259) as

$$T(r,t) = \int_0^t dt' \int_0^\infty 2\pi r'\, dr'\, Q(r',t')G(r',r,t-t'), \qquad (12)$$

where the Green's function $G(r', r, t - t')$ for a cylindrical shell of radius r' is given by

$$G(r',r,t-t') = \frac{1}{4\pi K(t-t')} I_0\left(\frac{rr'}{2D(t-t')}\right) \exp\left(-\frac{r^2 + r'^2}{4D(t-t')}\right). \qquad (13)$$

Here, K is the thermal conductivity, D is the thermal diffusivity of the medium, and I_0 is a modified Bessel function. By using Eqs. (10), (11), and (13), the integral in Eq. (12) can be evaluated to obtain the time-dependent temperature distribution T as

$$T(r,t) = \frac{2\alpha E_0}{\pi C_p \rho \tau} \int_0^t dt' \frac{\exp(-t'/\tau)}{A(t,t')} \exp\left(\frac{-2r^2}{A(t,t')}\right), \qquad (14)$$

where $A(t, t')$ is given by

$$A(t,t') = a^2 + 8D_m t' + 8D(t-t'). \qquad (15)$$

From Eqs. (14) and (15), it is clear that for fast thermal relaxation so that $\tau \ll a^2/8D_m$ and $\tau \ll a^2/8D$, *or* for $D_m \approx D$, the effect of excited molecular diffusion becomes negligible; in this case, we can put $A(t, t') = a^2 + 8Dt$, and perform the time integral in Eq. (14) to obtain the simple solution for the temperature distribution

$$T(r, \tau) = \frac{2\alpha E_0}{\pi\rho C_p} \frac{[1 - \exp(-t/\tau)]}{(a^2 + 8Dt)} \exp[-2r^2/(a^2 + 8Dt)]. \quad (16)$$

The perpendicular PBR angle $\psi(z, t)$ for the geometry of Fig. 6b is given by

$$\psi(z, t) = \frac{1}{n_0} \frac{\partial n}{\partial T} \int_{-\infty}^{\infty} \frac{\partial T}{\partial z} dx. \quad (17)$$

By substituting Eq. (14) into Eq. (17), we obtain the time-dependent perpendicular PBR angle

$$\psi(z, t) = \frac{1}{n_0} \frac{\partial n}{\partial T} \frac{(-8\alpha E_0 z)}{\sqrt{2\pi}\,\tau\rho C_p} \int_0^t dt' \frac{\exp(-t'/\tau)\exp[-2z^2/A(t, t')]}{A(t, t')^{3/2}}. \quad (18)$$

Again, for short τ *or* for $D_m \approx D$, this simplifies as for the case of $T(r, t)$ given by Eq. (16) to be:

$$\psi(z, t) = \frac{1}{n_0} \frac{\partial n}{\partial T} \frac{(-8\alpha E_0 z)}{\sqrt{2\pi}\,\rho C_p} \frac{(1 - e^{-t/\tau})}{(a^2 + 8Dt)^{3/2}} \exp\left(-\frac{2z^2}{a^2 + 8Dt}\right) \quad (19)$$

The calculated deflection signals, ψ for a few pump–probe separations, z, for fast thermal relaxation (i.e., $\tau \to 0$), are shown in Fig. 7a. The signal shape expected for a "slow" component (i.e., finite τ) is shown in Fig. 7b, which is taken from the work of Sontag, Tam, and Hess (1987).

As noted earlier, Eqs. (18) and (19) can be applied to a flowing fluid sample in the z-direction (i.e., in the perpendicular "travelling thermal lens" case) by the substitution of Eq. (8) in Eqs. (18) or (19).

In general, the pulsed pump beam may not be cylindrically symmetric with a Gaussian beam shape. We can estimate the importance of this effect by examining the case where the pump beam has a Gaussian intensity distribution with beam parameter a in the z-direction, but has a rectangular intensity distribution of width a_1 in the x-direction (see coordinate system in Fig. 6b). In addition, we assume that $a_1 \gg a$, such that heat flow in the

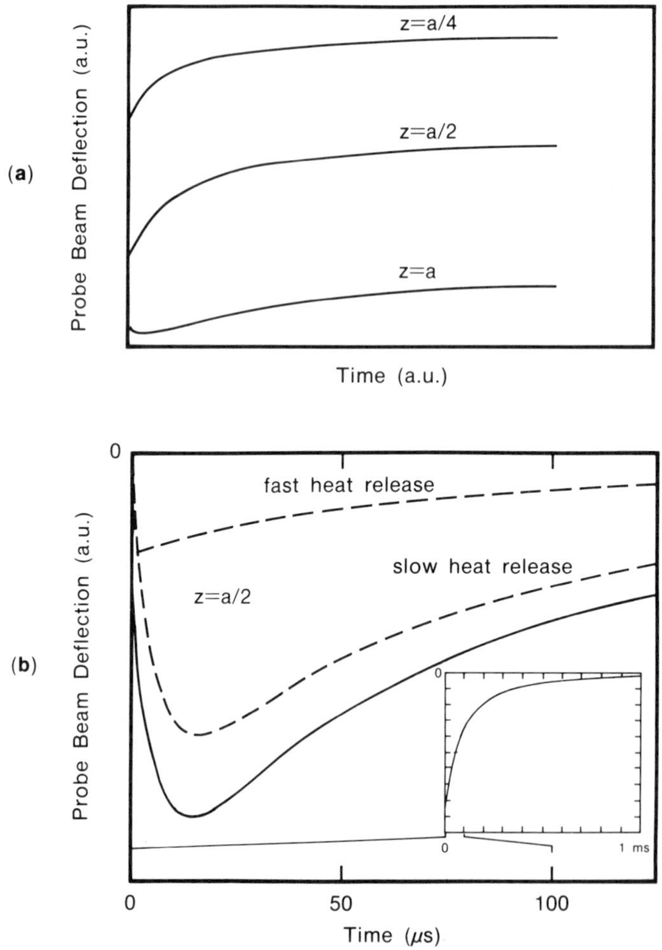

FIG. 7. (a) Calculated perpendicular PBR signals for various separations z between the pump beam and the probe beam in a stationary sample with $\tau \approx 0$. The signal is largest for $z = a/2$, where the signal decay shape is also simple and monotonic. (b) Calculated perpendicular PBR signal for $z = a/2$ in a stationary sample when the PT de-excitation has two time constants: a slow heat release component with a finite τ, and a fast heat release component with $\tau' \approx 0$. The resultant PBR signal (solid line) has a delayed peak. The inset shows that the long-time decay shape is governed by thermal diffusion.

x-direction can be neglected. In this limit, we can treat the heat flow in one dimension only. We shall show that for this case, the same probe deflection as for the cylindrical symmetry case is obtained, indicating that the intensity distribution of the pump beam in the x-direction is unimportant for PBR.

By using the result for a plane source, we can set up the general solution as (Carslaw and Jaeger, 1959, p. 259):

$$T(z,t) = \int_0^t dt' \int_{-\infty}^{\infty} dz' G(z', z, t-t') Q(z', t'), \tag{20}$$

where

$$G(z', z, t-t') = \frac{\exp\left[-(z-z')^2/(4D(t-t'))\right]}{2\sqrt{\pi D(t-t')}}, \tag{21}$$

and

$$Q(z', t') = \frac{\sqrt{2}\,\alpha E_0 \exp\left[-2z'^2/(a^2 + 8D_{\text{m}}t')\right]}{\rho C_{\text{p}} \sqrt{\pi}\,\tau a_1 \sqrt{a^2 + 8D_{\text{m}}t'}} \exp\left(-\frac{t'}{\tau}\right). \tag{22}$$

Equations (20)–(22) can be integrated to yield

$$T(z,t) = \int_0^t dt' \frac{\sqrt{2}\,\alpha E_0}{\sqrt{\pi}\,\tau a_1 \rho C_{\text{p}}} \frac{\exp(-t'/\tau)}{\sqrt{A(t,t')}} \exp\left(-\frac{2z^2}{A(t,t')}\right). \tag{23}$$

The perpendicular PBR angle ψ is again derived by inserting Eq. (23) into Eq. (17), from which the same result as in Eq. (18) is obtained. This can intuitively be understood, since in the present perpendicular beams arrangement, the interaction length of the probe beam with the thermal lens increases with a larger excitation beam waist in the x-direction, and this effect counteracts the smaller RIG. Therefore, we conclude that the intensity distribution of the pump beam in the probe direction is of minor importance in perpendicular PBR experiments if the intensity distribution of the pump beam is Gaussian along the z-direction; Eq. (18) or (19) should adequately describe the perpendicular PBR angle ψ.

c. *Application of PBR Due to Thermal RIG*

Numerous applications of PBR detection of the thermal RIG have been recently reported in the literature; such applications include spectroscopy, analytical application and trace detection, imaging and defect detection, and flow or combustion diagnostics. This great current interest arises because of the high sensitivity and spatial and temporal resolution achievable with this technique together with the desirable noncontact nature of the detection method. More detailed considerations have been given by Rose, Vyas, and Gupta (1986), Murphy and Wetsel (1986), Bialkowski (1986), and Morris and Peck (1986).

Spectroscopy. Murphy and Aamodt (1980) and Jackson *et al.* (1981a) have calculated the probe-beam deflection for cases of continuous-wave (CW) or pulsed excitation with either a perpendicular or parallel probe. Jackson *et al.* (1981a) have experimentally verified some of the predictions of their theory. For example, they have used PBR spectroscopy to measure the optical absorption of benzene at 607 nm due to the sixth harmonic of the C–H stretch. From their experimental results for a 0.1% solution of benzene in CCl_4, they concluded that the PBR method can detect an absorption coefficient of $2 \times 10^{-7}\,cm^{-1}$ for a 1-mJ pulsed-excitation laser with the detection limit being mainly due to the pointing instability of the He–Ne probe laser. Jackson *et al.* (1981b) have also reported the measurement of absorption spectra of crystalline and amorphous Si by photothermal PBR spectroscopy. Orthogonal PT PBR (or "mirage" probing) has been used for novel spectroscopic applications by Fournier, Boccara, and Badoz (1981), Murphy and Aamodt (1981), and Low *et al.* (1982). In addition, orthogonal PBR (with the probe intersecting the pump beam inside the sample) has been used by Dovichi, Nolan, and Weimer (1984) for PT microscopy applications. Nickolaisen and Bialkowski (1985) have used a pulsed infrared laser for thermal lens spectroscopy of a flowing gas sample. There are more recent reports on spectroscopic measurements on thin films (Dadarlat, Chirtoc, and Candea, 1986; Hata *et al.*, 1986) and on adsorbed species (Morterra, Low, and Severdia, 1985). The use of "hexane soot" as the reference absorber for PT spectroscopy has been investigated (Low, 1985).

Fourier transform (FT) techniques have been extensively applied to PBR spectroscopy for efficient spectroscopic measurements and signal-to-noise improvement. Earlier work was reported by Fournier, Boccara, and Badoz (1981) who showed PBR FT spectroscopy for absorption and dichroism measurements and obtained three orders of magnitude improvement in sensitivity over conventional PA spectroscopy. Low, Morterra, and Severdia (1984) have combined the infrared Fourier-multiplexed excitation technique with photothermal PBR detection for the spectroscopy of "difficult" samples like polymers, fabrics, paper, and bones. Other PBR FT spectroscopy applications include studies of thin films (Roger *et al.*, 1985), electrodes (Crumbliss *et al.*, 1985), solid–liquid interface (Varlashkin and Low,

1986a, b), and infrared spectral depth profiling (Varlashkin and Low, 1986c). Much consideration on instrumentation is given by Mandelis and co-workers (Mandelis, 1986; Mandelis, Borm, and Tiessinga, 1986a, b). Besides FT, other multiplexed techniques like the Hadamard transform technique have been reported (Fotiou and Morris, 1987).

Analytical applications. This is a particular case of spectroscopic measurement to determine compositions (perhaps in small amounts). Numerous reports have been published recently. Long and Bialkowski (1985) have shown the use of PBR PT spectroscopy for trace gas detection at the sub-part-per-billion level, and Fung and Lin (1986) and Tran (1986) have done trace gas analysis via intracavity PBR spectroscopy. Analytical applications for solutions (Uejima *et al.*, 1985), powder layer (Tamor and Hetrick, 1985), surfaces (Field *et al.*, 1985), and black inks on paper (Varlashkin and Low, 1986a) have also been reported. Comparisons of detection limits by PT spectroscopy with other types of analytical techniques have also been given by Winefordner and Rutledge (1985).

Imaging. Murphy and Aamodt (1980, 1981) have demonstrated PT imaging using perpendicular PBR (called the *mirage effect*). Kasai *et al.* (1986) have used PBR for nondestructive materials evaluation, Mundy *et al.* (1985) and Abate *et al.* (1985) have used PBR for coating evaluation, and Baumann and Tilgner (1985) have discussed the theory for the measurement of the thickness of a buried layer. McDonald, Wetsel, and Jamies (1986) have used PBR for imaging "vertical interfaces" in solids, and Rajakarunanayake and Wickramasinghe (1986) have demonstrated novel nonlinear PT imaging.

Flow diagnostics. Sontag and Tam (1985b, 1986a), Sell (1985), and Weimer and Dovichi (1985a, b) have demonstrated "traveling thermal lens" spectroscopy for monitoring flow velocity, temperature, and composition in a flowing fluid. Rose, Vyas, and Gupta (1986) have made a very detailed analysis of the PBR signal shape in a flowing fluid system for applications in combustion diagnostics. Sell and Cattolica (1986) have shown velocity imaging in a flowing gas. Optimization of velocity measurements by such techniques are given by Dasch and Sell (1986) and by Weimer and Dovichi (1985a, b). To show the feasibility of single-pulse measurement of local thermal diffusivity or temperature in a combustion system, Loulergue and Tam (1985) have used a pulsed CO_2 excitation laser beam and a continuous He–Ne probe beam that is parallel to, but displaced from, the excitation beam for thermal diffusivity measurements in an unconfined hot gas.

Other applications. The use of time-resolved photothermal PBR techniques to detect delayed heat release, for example, due to energy transfer or photochemical reactions, is also possible. This has been demonstrated by Tam, Sontag, and Hess (1985) for the case of photochemical particulate production in a CS_2 vapor induced by a pulsed N_2 laser, where the delayed heat release is attributed to the nucleation and growth of particulates. A

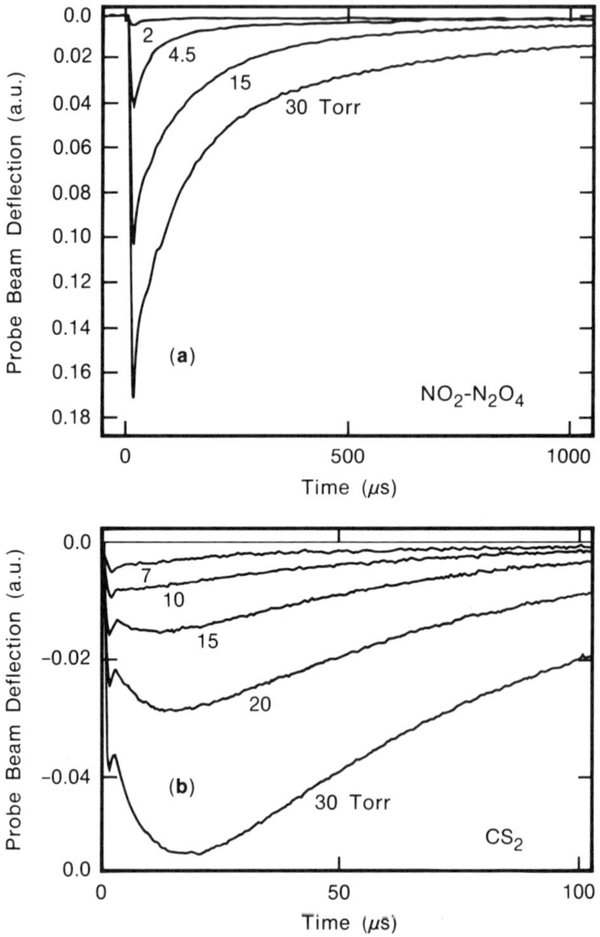

FIG. 8. (a) Observed perpendicular PBR signals in a nitrogen dioxide gas of various pressure excited at 337 nm, with $z = a/2$, corresponding to the theoretical curve shown in Fig. 7a, middle trace. (b) Observed perpendicular PBR signals in a carbon disulphide gas of various pressures excited at 337 nm, with $z = a/2$; the higher pressure signals show the delayed peak indicated in Fig. 7b.

delayed peak can be observed in the probe-beam deflection signal when delayed heat release occurs, as shown in Fig. 8; more details on this use of time-resolved PBR for thermal relaxation measurements in chemically active systems are reported by Sontag, Tam, and Hess (1987). We also note that PBR can be used to detect thermal release due to radio-frequency or microwave-frequency excitations, e.g., in ferromagnetic resonance (Netzelmann *et al.*, 1986).

2. Acoustic RIG

Similar to the optical probing of the thermal RIG, we can also detect the RIG associated with the PA pulse. The detection of refractive-index variations caused by acoustic waves is not new; for example, this has been reported by Lucas and Biquard (1932) and Davidson and Emmony (1980). The present discussion is focused on its use for PA detection, which is similar to the above PT detection scheme. This method can be called optical probing of the acoustic refractive-index gradient (OPARIG); it is applicable for "direct detection" inside a material that is transparent to the probe beam, or to "indirect detection" in a transparent coupling medium adjacent to an opaque sample. Here, the transient deflection of the probe due to the traversal of the PA pulse is detected. By using two or more probe beams at different displacements at different times, acoustic velocity and attenuation of the sample or of the coupling medium can be detected. Quantification of the OPARIG signal $S(t)$ has been given by Sullivan and Tam (1984), using a photodiode with a small active area as a fast deflection sensor (see Fig. 9). When the PA pulse produced by the excitation beam crosses the probe beam, the transient angular deflection φ (Klein, 1970) of the probe is true only when probe diameter is much smaller than PA pulse width:

$$\varphi(r, t) \approx \frac{l}{n_0} \frac{\partial n(r, t)}{\partial r} \propto \frac{\partial P(r, t)}{\partial t}, \tag{24}$$

where $n(r, t)$ is the refractive index of normal value n_0, $P(r, t)$ is the PA pressure, and l is the interaction length of the PA pulse with the probe laser. The small transient probe deflection φ causes the probe beam to move across the detection photodiode. The observed probe-beam deflection signal $S(t)$ from the photodiode is given by

$$S(t) = GI_p'(r_d)L\varphi, \tag{25}$$

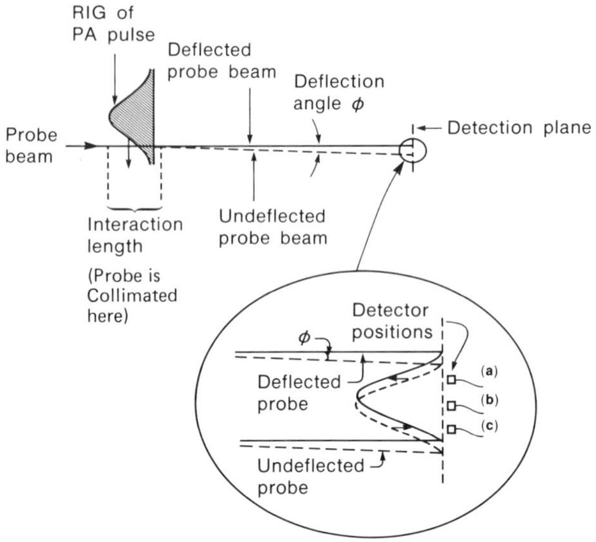

FIG. 9. A probe-beam refraction (also called probe-beam deflection) method for noncontact transmission detection of PA pulses; transient deflection of a *narrow collimated* probe beam is caused by the PA refractive-index gradient (RIG). Three possible locations of a small detector, (a), (b), and (c) are indicated. The centered (b) position gives nearly zero signal for a *collimated* probe undergoing small deflections, while (a) and (c) give the maximum signals (with opposite phase) for a *collimated* probe if the detectors are situated at the positions of maximum-intensity slopes.

where G is a constant depending on the photodiode sensitivity and electronic gain of the detection system, $I_p'(r_d)$ is the lateral spatial derivative of the probe-beam intensity distribution at the photodiode position r_d (the photodiode active area is assumed to be small), and L is the "lever arm" of the probe beam (i.e., distance from the interaction region in the cell to the photodiode). Combining Eqs. (24) and (25), we have

$$S(t) \propto \frac{\partial P(r, t)}{\partial t}, \tag{26}$$

which means that our experimental probe-beam deflection signal is a measure of the *time derivative* of the PA pulse at the probe-beam position. This present discussion assumes that the detection is made sufficiently far away from the excitation region so that thermal RIG effects are negligible. Sullivan and Tam (1984) have used OPARIG for a reliable determination of PA pulse profiles generated by laser beams of various durations in the μsec or nsec regimes to verify the theoretical PA pulse profiles calculated by Lai and Young (1982) and by Heritier (1983).

A novel application of OPARIG for acoustic absorption spectroscopy in gases, to be called "PA spectroscopy of the second kind" (to distinguish it from the well-known first kind concerned with optical absorption spectroscopy) has been reported by Tam and Leung (1984). Here, the acoustic absorption spectrum of a gas sample is obtained by measuring the Fourier spectrum of $S(t)$ at different distances from the PA source; the decay of the Fourier spectrum with propagation distance gives the acoustic absorption spectrum of the gas. Thus, optical monitoring of PA pulse profiles should open up new noncontact ultrasonic velocity, relaxation, and dispersion measurements. For example, chemical reactions, nucleations, precipitations, etc., in a system frequently cause changes in the ultrasonic absorption or dispersion spectra, detectable by monitoring the profiles of the probe-beam deflection signals at several propagation distances.

Zapka and Tam (1982b, c) have demonstrated a new application of the laser-induced acoustic source for measurements in a flowing fluid. They show that the flow velocity of a pure particle-free gas (as well as of a liquid) can be measured to an accuracy of 5 cm/sec, and simultaneously the fluid temperature can be obtained to an accuracy of 0.1°C. Such noncontact measurements were not possible previously by other known laser scattering methods (like laser Doppler velocimetry, coherent anti-Stokes–Raman scattering, or stimulated Raman-gain spectroscopy). In their experimental arrangement, an acoustic pulse in a flowing air stream is produced at position 0 by a pulsed-excitation laser (Nd: YAG laser, ~10-mJ energy and 10-nsec duration). Three probe He–Ne laser beams, 1, 2, and 3, at distances l_1, l_2, and l_3 from 0, are used to monitor the acoustic pulse arrival times t_1, t_2, and t_3, respectively. The acoustic pulse arrival at each beam is measured by the transient deflections of the beam and can be detected by using a small photodiode with suitable risetime. Zapka and Tam (1982c) have shown that the deflection signals provide enough data to give both the flow velocity and the fluid temperature simultaneously, and at the same time minimizing possible errors due to a blast wave or a shock wave produced at the origin 0.

Note that both the acoustic RIG detection and the thermal RIG detection can be used for noncontact flow, temperature, and composition measurements in fluids. In fact, both the PA and PT deflection (i.e., refraction) can be generally observed with the same experimental setup, with the PA deflection occurring earlier for displaced excitation and probe beams. For flow diagnostics, PT deflection is more suitable for slow flow velocities, and PA deflection is more suitable for fast flows.

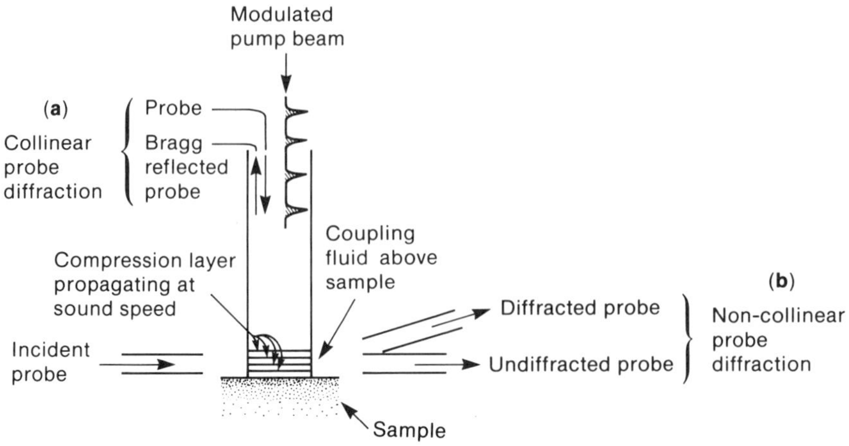

FIG. 10. Examples of probe-beam diffraction for the case of a pump beam consisting of a train of short pulses incident on a sample, creating a layered refractive-index structure in the adjacent coupling fluid. The probe beam may be collinear or noncollinear with the pump beam, as indicated is (a) and (b), respectively.

3. Other Detection Schemes for RIG by PT Generation

So far, we have described only PBR techniques for detecting PT-generated refractive-index variations (thermal or acoustic); however, such variations can also be probed by other optical techniques, for example, the phase-fluctuation heterodyne interferometry technique (Davis, 1980) and the Moire deflectometry technique (Glatt, Karny, and Kafri, 1984). Also, Nelson and Fayer (1980) have used two intersecting coherent pump beams for PT generation of a thermal refractive-index grating in a sample, and a probe-beam diffraction scheme to measure the decay of this grating. Furthermore, Williams (1984) has used a high-frequency (~ 1 GHz) train of short laser pulses to irradiate an opaque sample adjacent to a gas (Fig. 10), and observed Bragg scattering of a probe beam due to the modulated refractive index in the gas. This method has the interesting application of measuring ultrahigh-frequency sound propagation in gases or liquids.

D. Surface Deformation

As for the case of the RIG, the deformation of a sample surface due to PT excitation can be of two kinds, thermal deformation and acoustic deformation. Although these two deformations are related, the thermal one decays in time as thermal diffusion, whereas the acoustic one propagates into the sample bulk and also along the sample surface as bulk and surface acoustic waves, respectively.

1. Thermal Surface Deformation

The PT heating of a surface causes distortions due to thermal expansion. The distortions can be very small, e.g., 10^{-3} Å. However, Amer and coworkers (Amer, 1983; Olmstead *et al.*, 1983; Olmstead and Amer, 1984) have observed that even such small surface deformations can be detected, providing new sensitive spectroscopic applications. This has been called PT displacement spectroscopy.

The technique of photothermal displacement spectroscopy can be understood from Fig. 11. Figure 11a indicates the case when a laser beam of power P and area A chopped at a frequency f is incident on a solid. Let us assume the simple case of weak absorption by a thin coating (of thickness l and absorption coefficient α) on a transparent substrate. The modulated heating is "spread" through a thermal diffusion length μ given by

$$\mu = \left[\frac{D}{\pi f}\right]^{1/2}, \tag{27}$$

where D is the thermal diffusivity of the solid. The average temperature rise ΔT in this thermal diffusion volume can be estimated as

$$\Delta T = \frac{\alpha l P}{2 f \rho C \mu A}, \tag{28}$$

where $P/2f$ is approximately the incident energy in one cycle, αl is the fraction of light absorbed, ρ is the density, and C is the specific heat of the solid. The magnitude of the maximum surface displacement h_0 is estimated as

$$h_0 = \beta\mu\,\Delta T = \frac{\beta\alpha l P}{2 f \rho C A}, \tag{29}$$

where β is the linear thermal expansion coefficient of the solid. We see that the surface displacement is proportional to the coating absorption coefficient α, and hence absorption spectroscopy of the coating can be performed by monitoring the distortion as a function of the excitation laser wavelength. Equation (29) is derived semiquantitatively and shows the correct functional dependence of h_0 on the various given parameters for thermally

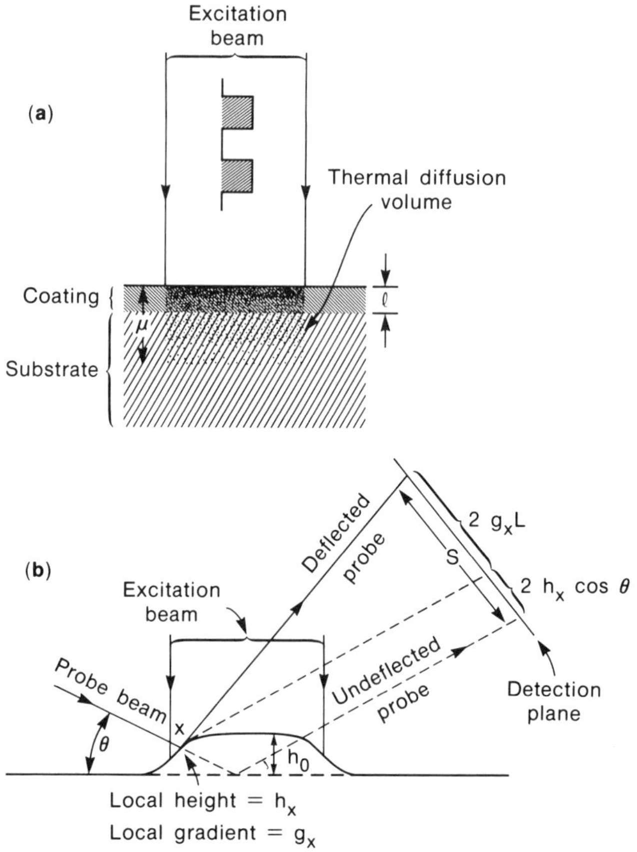

FIG. 11. Photothermal displacement spectroscopy exemplified for the case of an absorbing layer of thickness l on a transparent substrate. (a) Schematic. The local temperature rise is indicated by the density of the "dots." (b) The probe beam incident at position X of the displaced surface (exaggerated in the figure) is deflected, and the displacement at the detection plane is $2h_x \cos\theta + 2g_x L$.

thick samples; more detailed analyses have been given by Miranda (1983).

Amer and coworkers show that the surface displacement can be conveniently monitored by a probe-beam deflection method indicated in Fig. 11b. The continuous probe is deflected by the deformation, and the displacement S at a detector plane (perpendicular to the probe beam) is given by (see Fig. 11b)

$$S = 2h_x \cos\theta + 2g_x L. \tag{30}$$

Here, the subscript x indicates the location where the probe beam meets the solid surface, h_x is the height of the distortion at x, g_x is the local gradient due to the distortion at x, L is the distance from x to the detector, and θ is the grazing angle of the probe beam. Both h_x and g_x are proportional to h_0 given by Eq. (29) for small distortion. The gradient deflection term $(2g_xL)$ can be made much larger than the height deflection term $(2h_x \cos\theta)$ by making the optical lever arm L sufficiently long.

The advantages of PT displacement spectroscopy for absorption measurements are that it is noncontact and compatible with a vacuum environment. The disadvantages are that it is applicable only to reflective and smooth surfaces, and careful alignment of the probe beam is required. Olmstead and Amer (1984) have used photothermal displacement spectroscopy for measuring absorption at silicon surfaces and have achieved a new understanding of the surface atomic structure. Karner, Mandel, and Traeger (1985) have used pulsed-laser PT displacement spectroscopy for surface analysis.

It is not always possible to totally separate surface deformation effects from the refractive-index gradient effects described in Section C. For example, Rosencwaig, Opsal, and Willenborg (1983) and Opsal, Rosencwaig, and Willenborg (1983) have examined the case of a coated sample (SiO_2 on Si) in which the probe beam can be significantly affected by several PT effects, including deformations of the SiO_2 and of the Si surfaces, refractive-index gradients in the gas and in the SiO_2, optical interference effects in the SiO_2 film, and PT-induced reflectivity variations.

2. Acoustic Surface Deformation

Noncontact optical probing of surface movement due to acoustic waves has been studied by numerous workers because of its important materials-testing applications. Earlier reviews were given by Whitman and Korpel (1969) and Stegeman (1976), and a general summary of noncontact ultrasonic transducers for nondestructive testing has been given by Hutchins (1983). The optical detection of PA surface deformation is similar to the earlier work on optical probing of surface acoustic waves; the more commonly used methods are probe-beam deflection, interferometry, and Doppler velocimetry measurements. Review of the optical detection of PA surface waves or deformations are given by Monchalin (1986) and by Sontag and Tam (1986b).

The probe-beam deflection measurement of surface acoustic deformation is similar to the method of optical detection of PT surface distortion developed by Amer (1983) described above. It is based on the deflection of a probe beam reflected from the surface. The distinctions between PT distortion and PA distortion may be described as follows: PT distortion is due to the thermal expansion associated with the local temperature rise, follows the temperature decay via diffusion, and remains close to the excitation region; PA distortion, on the other hand, propagates at a sound speed away from the excitation region. Only PA monitoring can provide values of sound speed and attenuation. At distances more than several thermal diffusion lengths away from the excitation regime, only PA distortion can be detected. Sontag and Tam (1985a) have extended the probe-beam deflection technique of Amer (1983) for detecting multiple reflections of a PA pulse in a silicon wafer excited by a pulsed N_2 laser. Such noncontact techniques (extending the usefulness of contact PA testing in Tam, 1984, and Tam and Ayers, 1986) are useful for fast ultrasonic testing and imaging and for remote sensing of samples that are "inaccessible" (e.g., in a vacuum chamber, in hostile environments, or whenever contamination must be minimized).

Another method of optical sensing of PA surface distortion relies on interferometry. In this case, a probe beam is split into two parts, one being reflected from the sample surface and another from a reference surface. The reflected beams are recombined, and the resultant intensity is monitored. Kino and his associates (Jungerman *et al.*, 1982, Jungerman, Khuri-Yakub, and Kino, 1983; Bowers, 1982) have developed a coherent fiber-optic interferometry technique for measuring acoustic waves on a polished or even on a rough surface. Also, Cielo (1981) has described an optical detection method of acoustic waves for the characterization of samples with unpolished surfaces. Palmer, Claus, and Fick (1977) have analyzed a Michelson interferometry system for measuring displacement amplitudes in acoustic emission; their analysis was subsequently criticized by Kim and Park (1984). Bondarenko *et al.* (1976), Calder and Wilcox (1980), and Hutchins and Nadeau (1983) have used Michelson interferometry for detecting the PA pulse shape excited by a powerful pulsed laser (e.g., ruby, Nd:glass, or Nd:YAG laser) in metal plates, while Suemune, Yamamoto, and Yamanishi (1985) have extended this work to detect photoacoustic vibration in a GaAs plate produced by a focused diode laser beam modulated at ~ 100 Hz.

Other novel methods for optical sensing of PA surface deformation have been reported, notably the technique of "piezo-reflectance" measurement of Eesley, Clemens, and Paddock (1987); they showed that ultrashort PA pulses of picosecond duration produce observable surface reflectivity changes, thus opening up the possibility of optical ultrasonic measurements in the picosecond domain.

E. Photothermal Radiometry

Photothermal radiometry (PTR) relies on the detection of variations in the infrared thermal radiation emitted from a sample that is excited by an electromagnetic "pump" beam (typically from a laser or from an arc lamp) of varying intensity or wavelength. A simple theory of PTR is given by Nordal and Kanstad (1979). The total radiant energy W emitted from a grey body of emissivity ε and absolute temperature T is given by the Stefan–Boltzmann law

$$W = \varepsilon\sigma T^4, \tag{31}$$

where σ is the Stefan–Boltzmann constant. Suppose the body is irradiated by an optical pulse of energy E at wavelength λ that is absorbed by the body with an absorption coefficient $\alpha(\lambda)$, resulting in a small temperature rise $\delta T(E, \alpha)$. By Eq. (31), the total radiant energy is increased by

$$\delta W(E, \alpha) = 4\varepsilon\sigma T^3 \delta T(E, \alpha). \tag{32}$$

If $\delta T(E, \alpha)$ varies linearly with αE, spectroscopic measurement is possible by defining the "normalized" PTR signal S as

$$S(\alpha) = \delta W(E, \alpha)/E. \tag{33}$$

An excitation spectrum called a PTR spectrum can be obtained by monitoring S for various excitation wavelengths λ.

In a typical PTR measurement, the excitation beam (of photons, or more generally, of some form of energy) is either continuously modulated, with about 50% duty cycle, or pulse modulated, with low duty cycle and high peak power. The observation spot can, in principle, be anywhere on the sample; however, the infrared (IR) emission is usually detected at the excitation spot in a backward direction (called "backscattering PTR"), or from a spot that is "end-on" through the sample thickness from the excitation spot (called "transmission PTR"). Thus, there are four common varieties of PTR in the literature, classified according to the excitation mode (continuously modulated or pulsed) and to the detection mode (transmission or backscattered). These are summarized in Fig. 12, and some applications published in the literature are reviewed by Tam (1985) and Kanstad and Nordal (1986).

A common problem with PTR and all other PT monitoring techniques is that the signal magnitude normalized to the energy of the excitation pulse is not necessarily proportional to the absorption coefficient of the sample at

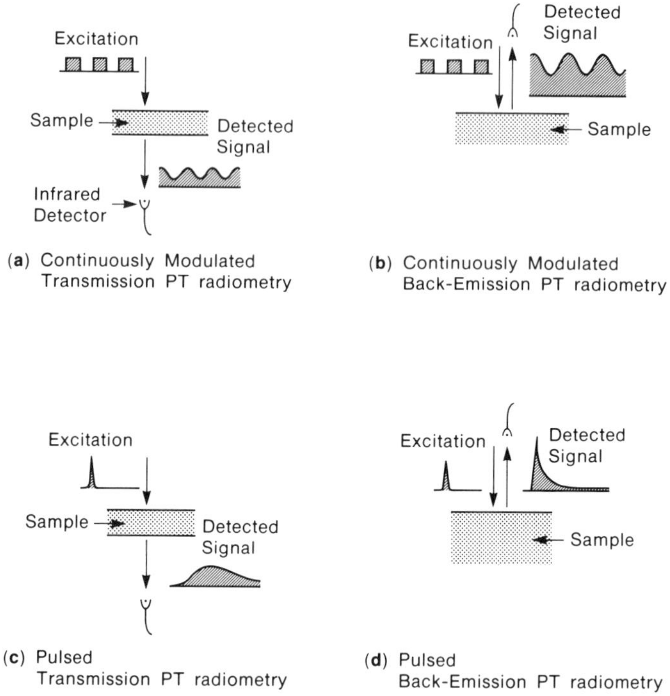

FIG. 12. Variations of the photothermal radiometry techniques.

the excitation wavelength. In other words, the excitation spectrum obtained by measuring the normalized PTR signal as a function of excitation wavelength may not be proportional to the true absorption spectrum. For a semi-infinite sample, this proportionality is valid if the sample absorption coefficients at the detected infrared wavelengths are very large compared to those at the excitation wavelengths (see Tam, 1985).

1. APPLICATIONS

Early work in PTR was done by Deem and Wood (1962), who used pulsed ruby lasers for excitation and transmission radiometric monitoring to perform noncontact thermal diffusivity measurements of nuclear fuel material. Nordal and Kanstad (1979, 1981) and Kanstad and Nordal (1986) have extensively investigated the many applications of PTR, resulting in its modern popularity; they have demonstrated very sensitive spectroscopic absorption measurements, e.g., due to less than a monolayer of molecules on a surface. Moreover, such spectroscopic measurements can be performed on "difficult" samples like powders or materials at high temperature. By using continuously modulated transmission PTR, Busse (1980), Busse and Eyerer (1983), and Busse and Renk (1983) have detected voids inside opaque solids.

By using pulsed transmission PTR, Deem and Wood (1962) have measured thermal diffusivity of "dangerous" materials like nuclear fuels. With pulsed backscattering PTR, Tam and Sullivan (1983) and Leung and Tam (1984a, b) have demonstrated several remote-sensing applications, including the measurement of absolute absorption coefficients, monitoring of layered structure and film thickness, and detection of the degree of aggregation in powdered materials. By using intense pulsed lasers for excitation and single-ended monitoring, such measurements should be possible for samples that are ~ 1 km away.

There is rapidly expanding interest in PTR mainly because of its remote-sensing nature, depth-profiling capability, absolute spectroscopic measurement possibilities, and adaptability to measure the spectra of aerosols, powders, and films. Examples of such work include the quantitative measurement of opaque layered structures and subsurface air gaps or contact thermal resistance (Tam and Sontag, 1986; Leung and Tam, 1987, 1988; Cielo, Rousset, and Bertrand, 1986; Egee *et al.*, 1986; Beaudoin *et al.*, 1986), aerosol and flow measurements (Lin and Campillo, 1985; Sontag and Tam, 1986a), semiconductor defect detection and imaging (Nakamura, Tsubouchi, and Mikoshiba, 1985) and small absorption measurements (Lopatkin, Sidoryuk, and Skvortsov, 1985).

F. Other PT Changes

Besides the more common PT effects described above that have been extensively used for PT spectroscopy and material-testing applications, many other PT effects are possible, especially in special circumstances. For example, large "bimorph"-type mechanical vibrations can be produced for a suitably supported thin plate that is PT excited at one side only (Rousset, Charbonnier, and Lepoutre, 1983), and these vibrations, especially large at mechanical resonances, can be detected by probe-beam deflection, interferometry, or Doppler velocity methods. Also, modulated PT heating of many types of metal or semiconductor surfaces causes modulated reflectivity changes (Rosencwaig *et al.*, 1985) or transmission and scattering changes (Rosencwaig *et al.*, 1986) that can be due to the density change or the photocarrier generation at the surface; this technique of "transient thermal reflectance" has been extended by Paddock and Eesley (1986) to the picosecond time-resolution regime for monitoring thermal properties of thin films (~ 100 Å) and of its interface conditions. Also, PT heating can cause changes in absorptivity in a sample; Zapka and Tam (1982a) have used a probe-beam absorption measurement to detect the change in the Boltzmann molecular population distribution due to the PT heating of a gaseous sample.

III. Conclusion

We have given here an overview of the different versions of PT spectroscopy and material characterizations that can be classified as follows:

(1) PT excitation spectroscopy: In this class, the PT signal amplitude is measured for a range of optical excitation wavelengths, producing an excitation spectrum. For quantitative interpretation, the PT generation efficiency must be known or at least kept fixed while the PT spectrum is obtained. Otherwise, the PT excitation spectrum at best shows the absorption peak positions qualitatively.

(2) PT monitoring of de-excitation processes: Here, the thermal decay branch is monitored to provide information on a competing decay branch. After optical excitation, four decay branches are generally possible: luminescence, photochemistry, photoelectricity, and heat that may be generated directly or through energy-transfer processes. For example, if luminescence and heat are the only two competing branches, and if the branching ratio can be controllably varied, PT monitoring of the heat branch can provide the quantum efficiency of luminescence.

(3) PT probing of thermoelastic and other physical properties of materials: Various information can be obtained conveniently with the help of the optical generation of thermal waves or acoustic waves. Such information includes sound velocity, elasticity, temperature, flow velocity, specific heat, thermal diffusivity, thickness of a thin film, subsurface defects, layered structures, delamination or air-gap thickness, thermal contact resistance, and so on.

(4) PT generation of mechanical motions: This is a small area of application now. PT effects can produce motions like liquid droplet ejection (Tam and Gill, 1982) or structural vibrations. Such effects can be enhanced by taking advantage of PT melting or boiling, light-induced chain-reaction effects, or mechanical resonances.

In general, the choice of a PT source for heating rather than a Bunsen burner is justified by some of the following reasons: (a) PT heating can provide convenient, noncontact, and sensitive methods for detecting optical absorptions in matter, and is applicable for traditionally difficult samples (e.g., highly opaque, transparent, or scattering); (b) information concerning de-excitation mechanisms and quantum yields can be obtained; (c) very localized or very rapid photothermal heating can be achieved to provide novel measurements or produce new effects with high spatial or temporal resolution, as well as providing subsurface and depth-dependent information.

This article has given an overview on PT probes with emphasis on spectroscopic applications and probe-beam refraction techniques to detect PT refractive-index gradients because of their quantitative, noncontact, and highly sensitive nature. More detailed reviews of other PT probes have been given elsewhere, e.g., reviews on photoacoustics (Rosencwaig, 1980; Patel and Tam, 1981; Tam, 1986) and photothermal radiometry (Kanstad and Nordal, 1986; Tam, 1985).

ACKNOWLEDGMENTS

This work is supported in part by the U.S. Office of Naval Research. The author sincerely thanks Dr. Heinz Sontag for his many valuable contributions to this work.

REFERENCES

Abate, J. A., Schmid, A. W., Guardalben, M. J., Smith, D. J., and Jacobs, S. D. (1985). *NBS Spec. Publ. (U.S.)* **688**, 385.
Amer, N. M. (1983). *J. Phys. (Paris) Colloq.* **C6**, 185.
Bass, M., and Liou, L. (1984). *J. Appl. Phys.* **56**, 184.
Bass, M., Van Stryland, E. W., and Stewart, A. F. (1979). *Appl. Phys. Lett.* **34**, 142.
Baumann, J., and Tilgner, R. (1985). *J. Appl. Phys.* **58**, 1982.
Baumann, T., Dacol, F., and Melcher, R. L. (1983). *Appl. Phys. Lett.* **43**, 71.
Beaudoin, J. L., Merienne, E., Raphael, D., and Egee, M. (1986). *Proc. SPIE Int. Soc. Opt. Eng.* **590**, 285.
Bialkowski, S. E. (1985). *Appl. Opt.* **24**, 2792.
Bialkowski, S. E. (1986). *Spectroscopy* **1**, 26.
Bialkowski, S. E., and Long, G. R. (1987). *Anal. Chem.* **59**, 873.

Boccara, A. C., Fournier, D., Jackson, W., and Amer, N. M. (1980). *Opt. Lett.* **5**, 377.
Bondarenko, A. N., Yu B. Drobot, and Kruglov, S. V. (1976). *Sov. J. Nondestr. Test.* **12**, 655.
Bowers, J. E., (1982). *Appl. Phys. Lett.* **41**, 231.
Brilmyer, G. H., Fujishima, A., Santhanam, K. S. V. and Bard, A. J. (1977). *Anal. Chem.* **49**, 2057.
Busse, G. (1980). *Infrared Phys.* **20**, 419.
Busse, G., and Eyerer, P. (1983). *Appl. Phys. Lett.* **43**, 355.
Busse, G., and Renk, K. F. (1983). *Appl. Phys. Lett.* **42**, 366.
Calder, C. A., and Wilcox, W. W. (1980). *Mater. Eval.* **38**, 86.
Carslaw, H. S., and Jaeger, J. C. (1959). *Conduction of Heat in Solids*. 2nd ed., Clarendon Press, London.
Cielo, P. (1981). *J. Acoust. Soc. Am.* **25**, Suppl. 1 **70**, 546.
Cielo, P., Rousset, G., and Bertrand, L. (1986). *Appl. Opt.* **25**, 1327.
Coufal, H. (1984). *Appl. Phys. Lett.* **44**, 59.
Coufal, H., and Hefferle, P. (1985). *Appl. Phys.* **A38**, 213.
Crumbliss, A. L., Lugg, P. S., Childers, J. W., and Palmer, R. A. (1985). *J. Phys. Chem.* **89**, 482.
Dadarlat, D., Chirtoc, M., and Candea, R. M. (1986). *Phys. Status Solidi A* **98**, 279.
Dasch, C. J., and Sell, J. A. (1986). *Opt. Lett.* **11**, 603.
Davidson, G. P., and Emmony, D. C. (1980). *J. Phys. E* **13**, 92.
Davis, C. C. (1980). *Appl. Phys. Lett.* **36**, 515.
Deem, H. W., and Wood, W. D. (1962). *Rev. Sci. Instrum.* **33**, 1107.
Dovichi, N. J., Nolan, T. G., and Weimer, W. A. (1984). *Anal. Chem.* **56**, 1700.
Eesley, G. L., Clemens, B. M., and Paddock, C. A. (1987). *Appl. Phys. Lett.* **50**, 717.
Egee, M., Dartois, R., Marx, J., and Bissieux, C. (1986). *Can. J. Phys.* **64**, 1297.
Field, R. S., Leyden, D. E., Masujima, T., and Eyring, E. M. (1985). *Appl. Spectrosc.* **39**, 753.
Fotiou, F. K., and Morris, M. D. (1987). *Anal. Chem.* **59**, 185.
Fournier, D., Boccara, A. C., and Badoz, J. (1981). *Appl. Opt.* **21**, 74.
Fung, K. H., and Lin, H. B. (1986). *Appl. Opt.* **25**, 749.
Glatt, I., Karny, Z., and Kafri, O. (1984). *Appl. Opt.* **23**, 274.
Hata, T., Hatsuda, T., Miyabo, T., and Hasegawa, S. (1986). *J. Appl. Phys. Japan* **25**, Suppl. 25-1, 226.
Heritier, J. M. (1983). *Opt. Commun.* **44**, 267.
Hutchins, D. A. (1983). *Nondestr. Test. Commun.* **1**, 37.
Hutchins, D. A., and Nadeau, F. (1983). *IEEE Ultrasonic Symposium Proceedings*, IEEE. Piscataway, New Jersey, 1175.
Hutchins, D. A. and Tam, A. C. (1986). *IEEE Trans. UFFC*-**33**, 429.
Jackson, W. B., Amer, N. M., Boccara, A. C., and Fournier, D. (1981a). *Appl. Opt.* **20**, 1333.
Jackson, W. B., Amer, N. M., Fournier, D., and Boccara, A. C. (1981b). In *Technical Digest, Second International Conference on Photoacoustic Spectroscopy*. Berkeley, Optical Society of America, Washington, D.C., Paper WA3.
Jungerman, R. L., Bowers, J. E., Green, J. B., and Kino, G. S. (1982). *Appl. Phys. Lett.* **40**, 313.
Jungerman, R. L., Khuri-Yakub, B. T., and Kino, G. S. (1983). *J. Acoust. Soc. Am.* **73**, 1838.
Kanstad, S. O., and Nordal, P. E. (1986). *Can. J. Phys.* **64**, 1155.
Karner, C., Mandel, A., and Traeger, F. (1985). *Appl. Phys. A* **38**, 19.
Kasai, M., Sawada, T., Gohshi, Y., Watanabe, T., and Furuya, K. (1986). *J. Appl. Phys. Japan* **25**, Suppl. **25**-1, 229.
Kim, H. C. and Park, H. K. (1984). *J. Phys. D* **17**, 673.
Klein, M. V. (1970). *Optics*. Wiley, New York.

Kliger, D. S. (1980). *Accs. Chem. Res.* **13**, 129.
Lai, H. M., and Young, K. (1982). *J. Acoust. Soc. Am.* **72**, 2000.
Leite, R. C. C., Moore, R. S., and Whinnery, J. R. (1964). *Appl. Phys. Lett.* **5**, 141.
Leung, W. P., and Tam, A. C. (1984a). *Opt. Lett.* **9**, 93.
Leung, W. P., and Tam, A. C. (1984b). *J. Appl. Phys.* **56**, 153.
Leung, W. P., and Tam, A. C. (1987). *Appl. Phys. Lett.* **51**, 2085.
Leung, W. P. and Tam, A. C. (1988). *J. Appl. Phys.* **63**, 4505.
Lin, H. B., and Campillo, A. J. (1985). *Appl. Opt.* **24**, 422.
Long, G. R., and Bialkowski, S. E. (1985). *Anal. Chem.* **57**, 1079.
Lopatkin, V. N., Sidoryuk, O. E., and Skvortsov, L. A. (1985). *Kvantovaya Elektron.* (*Moscow*) **12**, 339.
Loulergue, J.-C., and Tam, A. C. (1985). *Appl. Phys. Lett.* **46**, 457.
Low, M. J. D. (1985). *Spectrosc. Lett.* **18**, 619.
Low, M. J. D., Morterra, C., Severdia, A. G., and Lacroix, M. (1982). *Appl. Surf. Sci.* **13**, 429.
Low, M. J. D., Morterra, C., and Severdia, A. G. (1984). *Mat. Chem. and Phys.* **10**, 519.
Low, M. J. D., Morterra, C., and Khosrofian, J. M. (1986). *IEEE Trans. UFFC*-**33**, 573.
Lucas, R., and Biquard, P. (1932). *J. Phys. Radium* **3**, 464.
McDonald, F. A., Wetsel, G. C. Jr., and Jamies, G. E. (1986). *Can. J. Phys.* **64**, 1265.
Mandelis, A. (1986). *Rev. Sci. Instrum.* **57**, 617.
Mandelis, A., Borm, L. M. L., and Tiessinga, J. (1986a). *Rev. Sci. Instrum.* **57**, 622.
Mandelis, A., Borm, L. M. L., and Tiessinga, J. (1986b). *Rev. Sci. Instrum.* **57**, 630.
Miranda, L. C. M. (1983). *Appl. Opt.* **22**, 2882.
Monchalin, J. P. (1986). *IEEE Trans. UFFC*-**33**, 485.
Morris, M. D., and Peck, K. (1986). *Anal. Chem.* **58**, 811A.
Morterra, C., Low, M. J. D., and Severdia, A. G. (1985). *Appl. Surf. Sci.* **20**, 317.
Mundy, W. C., Ermshar, J. E. L., Hanson, P. D., and Hughes, R. S. (1985). *NBS Spec. Publ.* (*U.S.*) **688**, 360.
Murphy, J. C., and Aamodt, L. C. (1980). *J. Appl. Phys.* **51**, 4580.
Murphy, J. C., and Aamodt, L. C. (1981). *Appl. Phys. Lett.* **38**, 196.
Murphy, J. C., and Wetsel, G. C. (1986). *Mater. Eval.* **44**, 1224.
Nakamura, H., Tsubouchi, K., and Mikoshiba, N. (1985). *J. Appl. Phys. Japan* **24**, Suppl. **24-1**, 222.
Nelson, K. A., and Fayer, M. D. (1980). *J. Chem. Phys.* **72**, 5202.
Netzelmann, U., Pelzl, J., Fournier, D., and Boccara, A. C. (1986). *Can. J. Phys.* **64**, 1307.
Nickolaisen, S. L., and Bialkowski, S. E. (1985). *Anal. Chem.* **57**, 758.
Nordal, P. E., and Kanstad, S. O. (1979). *Phys. Scr.* **20**, 659.
Nordal, P. E., and Kanstad, S. O. (1981). *Appl. Phys. Lett.* **38**, 486.
Olmstead, M. A., and Amer, N. M. (1984). *Phys. Rev. Lett.* **52**, 1148.
Olmstead, M. A., Amer, N. M., Kohn, S. E., Fournier, D., and Boccara, A. C. (1983). *Appl. Phys. A* **32**, 141.
Opsal, J., Rosencwaig, A., and Willenborg, D. L. (1983). *Appl. Opt.* **22**, 3169.
Paddock, C. A., and Eesley, G. L. (1986). *J. Appl. Phys.* **60**, 285.
Palmer, C. H., Claus, R. O., and Fick, S. E. (1977). *Appl. Opt.* **16**, 1849.
Patel, C. K. N., and Tam, A. C. (1981). *Rev. Mod. Phys.* **53**, 517.
Rajakarunanayake, Y. N., and Wickramasinghe, H. K. (1986). *Appl. Phys. Lett.* **48**, 218.
Roger, J. P., Fournier, D., Boccara, A. C., Noufi, R., and Cahen, D. (1985). *Thin Solid Films* **128**, 11.
Rose, A., Vyas, R., and Gupta, R. (1986). *Appl. Opt.* **24**, 4626.
Rosencwaig, A. (1980). *Photoacoustics and Photoacoustic Spectroscopy*. Wiley, New York.
Rosencwaig, A., Opsal, J., and Willenborg, D. L. (1983). *Appl. Phys. Lett.* **43**, 166.

Rosencwaig, A., Opsal, J., Smith, W. L., and Willenborg, D. L. (1985). *Appl. Phys. Lett.* **46**, 1013.
Rosencwaig, A., Opsal, J., Smith, W. L., and Willenborg, D. L. (1986). *J. Appl. Phys.* **59**, 1392.
Rousset, G., Charbonnier, F., and Lepoutre, F. (1983). *J. Phys. (Paris) Colloque* **C6**, 39.
Sell, J. A. (1985). *Appl. Opt.* **24**, 3725.
Sell, J. A. (1987). *Appl. Opt.* **26**, 366.
Sell, J. A., and Cattolica, R. J. (1986). *Appl. Opt.* **25**, 1420.
Solimini, D. (1966). *Appl. Opt.* **5**, 1931.
Sontag, H., and Tam, A. C. (1985a). *Appl. Phys. Lett.* **46**, 725.
Sontag, H., and Tam, A. C. (1985b). *Opt. Lett.* **10**, 436.
Sontag, H., and Tam, A. C. (1986a). *Can. J. Phys.* **64**, 1121.
Sontag, H., and Tam, A. C. (1986b). *IEEE Trans. UFFC*-**33**, 500.
Sontag, H., Tam, A. C., and Hess, P. (1987). *J. Chem. Phys.* **86**, 3950.
Stegeman, G. I. (1976). *IEEE Trans. Sonics Ultrason. SU*-**23**, 33.
Suemune, I., Yamamoto, H., and Yamanishi, M. (1985). *J. Appl. Phys.* **58**, 615.
Sullivan, B., and Tam, A. C. (1984). *J. Acoust. Soc. Am.* **75**, 437.
Swofford, R. L., Long, M. E., and Albrecht, A. C. (1976). *J. Chem. Phys.* **65**, 179.
Tam, A. C. (1984). *Appl. Phys. Lett.* **45**, 510. V. S. Teslenko, (1977). Sov. J. Quantum Electron. **7**, 981
Tam, A. C. (1985). *Infrared Phys.* **25**, 305.
Tam, A. C. (1986). *Rev. Mod. Phys.* **58**, 381.
Tam, A. C., and Ayers, G. (1986). *Appl. Phys. Lett.* **49**, 1420.
Tam, A. C., and Gill, W. (1982). *Appl. Opt.* **21**, 1891.
Tam, A. C., and Leung, W. P. (1984). *Phys. Rev. Lett.* **53**, 560.
Tam, A. C., and Sontag, H. (1986). *Appl. Phys. Lett.* **49**, 1761.
Tam, A. C., and Sullivan, B. (1983). *Appl. Phys. Lett.* **43**, 333.
Tam, A. C., Sontag, H., and Hess, P. (1985). *Chem. Phys. Lett.* **120**, 280.
Tamor, M. A., and Hetrick, R. E. (1985). *Appl. Phys. Lett.* **46**, 460.
Teslenko, V. S. (1977). *Sov. J. Quantum Electron.* **7**, 981.
Tran, C. D. (1986). *Appl. Spectrosc.* **40**, 1108.
Uejima, A., Habiro, M., Itoga, F., Sugitani, Y., and Kato, K. (1986). *Anal. Sci.* **2**, 389.
Varlashkin, P. G., and Low, M. J. D. (1986a). *Appl. Spectrosc.* **40**, 507.
Varlashkin, P. G., and Low, M. J. D. (1986b). *Appl. Spectrosc.* **40**, 1170.
Varlashkin, P. G., and Low, M. J. D. (1986c). *Infrared Phys.* **26**, 171.
Weimer, W. A., and Dovichi, N. J. (1985a). *Appl. Spectrosc.* **39**, 1009.
Weimer, W. A., and Dovichi, N. J. (1985b). *Appl. Opt.* **24**, 2981.
Wetsel, G. C., Jr., and Spicer, J. B. (1986). *Can. J. Phys.* **64**, 1269.
Whitman, R. L., and Korpel, A. (1969). *Appl. Opt.* **8**, 1567.
Williams, C. C. (1984). *Appl. Phys. Lett.* **44**, 1115.
Winefordner, J. D., and Rutledge, M. (1985). *Appl. Spectrosc.* **39**, 377.
Zapka, W., and Tam, A. C. (1982a). *Opt. Lett.* **7**, 86.
Zapka, W., and Tam, A. C. (1982b). *Appl. Phys. Lett.* **40**, 310.
Zapka, W., and Tam, A. C. (1982c). *Appl. Phys. Lett.* **40**, 1015.

CHAPTER 2

PHOTOTHERMAL INVESTIGATION OF SOLIDS: BASIC PHYSICAL PRINCIPLES

D. FOURNIER AND A. C. BOCCARA
Laboratoire d'Optique Physique,
Centre Nationale de la Recherche Scientifique, E.S.P.C.I
Paris, France

Introduction 35
I. Physical Parameters Involved in a Photothermal Experiment 36
II. Detection of the Photothermal Signal 37
A. Photothermal Radiometry (PTR) 38
B. Photorefractive Methods 39
III. Photothermal Signal Generation: Theoretical Approach 43
A. 1-D Calculation of the Periodic Temperature Field in the Case of Thick or Coated Samples 44
B. 1-D Calculation of the Periodic Temperature Field in the Case of Multilayered Samples 52
C. 3-D Calculation of the Periodic Temperature Rise 53
D. Application of 3-D Theory to the Calculation of Mirage Detection 61
IV. Specific Problems 63
A. Influence of Energy Migration or Carrier Diffusion 63
B. Heat Diffusion, Rough Surfaces, and Heterogeneous Media: Fractal Approach 69
V. Spectral and Spatial Multiplexing of the Photothermal Signals 76
References 78

Introduction

The aim of this chapter is to provide the reader with a link between the physics relevant to the photothermal experiments and the methodology involved in such investigations.

Photothermal measurements have been proved to be very useful in various fields such as spectroscopy, thermal characterization, and nondestructive evaluation. Nevertheless, they always need to be associated with a theoretical model leading to a realistic view of the heat diffusion processes.

In this chapter, we shall first review the main parameters that have to be accounted for in a photothermal experiment. Then we shall briefly describe

ISBN 0-12-636345-5

a few experimental schemes allowing the detection of the photothermal signal. Heat diffusion equations will be considered in 1-D and 3-D geometry in Section III. Specific problems of energy migration and of heat diffusion in heterogeneous samples (fractals) will be analyzed in Section IV. Finally, we shall conclude with the various possibilities to improve the sensitivity and the speed of the experiments by signal multiplexing.

I. Physical Parameters Involved in a Photothermal Experiment

We shall first analyze the main parameters that may affect the photothermal signal and that can be measured by means of a photothermal experiment.

When a light beam irradiates the surface of an absorbing sample, the heating will be a function of numerous physical parameters, both of the sample itself and of the surrounding medium; this latter dependence distinguishes photothermal from other spectroscopic experiments.

The amount of energy deposited in the system depends first on the absorption coefficient ($\alpha\,\mathrm{cm}^{-1}$) of the media that are crossed by the light beam and of the specular (R_s) or diffuse (R_D) reflection coefficients of the various interfaces and on the efficiency η of the conversion of the absorbed energy into heat; η could be correlated to different relaxation schemes: radiative, photochemical reaction, photovoltaic, etc. The resulting temperature rise, which constitutes the object of the measurement, is obviously related to the specific heat C and to the specific gravity ρ. Moreover, the parameters that account for heat diffusion within the bulk of a sample, or at the interfaces, are the heat diffusivity $D = k/\rho C$ and the effusivity $e = \sqrt{k\rho C}$, respectively (with k being the thermal conductivity). The spatial and temporal heat distribution $T(r, t)$ is also strongly dependent on the geometrical parameters (dimensions of the beam and of the various parts of the sample and its surroundings) as well as on the heterogeneity of the material and of the properties of the surfaces and interfaces (roughness) that may be described by statistical parameters or by fractal dimensions.

Finally, in a photothermal experiment, $T(r, t)$ will also be a function of the parameters that account for the diffusion and delocalization of heat-generating centers within the bulk of a solid sample (carrier diffusion, energy migration, fluorescence trapping, etc.).

The parameters we have mentioned are shown in Table I. Although the parameters that come into play are rather numerous, we shall show that

TABLE I

PHYSICAL PARAMETERS INVOLVED IN A PHOTOTHERMAL EXPERIMENT

α	Optical absorption coefficient	cm^{-1}
C	Specific heat	J/g degree
ρ	Specific gravity	g/cm^3
D	Heat diffusivity	cm^2/s
D_i	Heat diffusivity in medium i	i = s, f, b
e	Effusivity	$J/\sqrt{s}\ cm^2$ degree
k	Thermal conductivity	W/cm degree
η	Light–heat conversion efficiency	
T	Temperature	degree
d	Euclidean dimensionality	
$\bar{d}$	Fractal dimensionality	
$\bar{\bar{d}}$	Spectral dimensionality	
D_c	Carrier diffusivity	cm^2/s
s	Surface recombination	cm/s
τ	Carrier lifetime	s
N	Carrier density	cm^{-3}

photothermal experiments can lead to quantitative results for various applications such as:

- Spectroscopy (absorption coefficients);
- Thermal measurements (diffusivity, effusivity, etc.);
- Transport properties (carrier diffusivity, spectral dimension, etc.);
- Geometrical parameters (thickness, fractal dimension, etc.).

II. Detection of the Photothermal Signal

When a sample is irradiated by a periodic or a pulsed light source, and a temperature rise is induced within the sample, we can measure it by probing the variation of some physical parameters either of the sample itself or of the surrounding medium. Because most of the applications developed throughout this book deal with noncontact methods using optical probes, we shall restrict ourselves to such techniques concerning in particular the beam deflection or "mirage" detection that we proposed a few years ago (Boccara, Fournier, and Badoz, 1979).

A. Photothermal Radiometry (PTR)

In this method, the time-varying part of the sample surface temperature, induced by the absorption of an intensity-modulated or a pulsed light beam, is measured by the modulated or pulsed variations in radiant emittance (Kanstad and Nordal, 1986).

If the total radiant emittance from a gray body is $W = \varepsilon\sigma T^4$ and its fluctuation $\Delta W = 4\varepsilon\sigma T^3 \Delta T$, by collecting the infrared (IR) modulated energy, one can reach the surface temperature fluctuations ($\sigma = 5.67\ 10^{-12}$ W cm^{-2} K^{-1}, ε thermal emissivity). Experiments either with laser sources or incoherent sources have been performed; Figure 1 shows an experimental arrangement for PTR.

The great advantage of this noncontact probing of the sample is that no cell is needed and that vacuum operation is possible. Among the few limitations, one finds the necessity for the sample to be a "gray-like" one, opaque at least in the probed infrared region. If it is not the case (e.g., for semiconductors), the infrared emission may be due not only to surface but to volume emittance, and one expects a complicated behaviour and a "saturation" of the photothermal signal when the optical absorption length for the heating radiation is smaller than the optical absorption length for the infrared probe radiation that defines the emitting volume. The sensitivity of this method at room temperature is large enough to allow experiments with light sources in the mW range.

The illumination area of the sample must be set as small as possible to match the small areas of low-noise infrared detectors. Minimum temperature fluctuations of a few 10^{-5} degrees have been measured (1-Hz bandwidth).

Let us point out two interesting characteristics of PTR:

The sensitivity increases as the third power of the temperature; the method is well adapted to high-temperature studies.

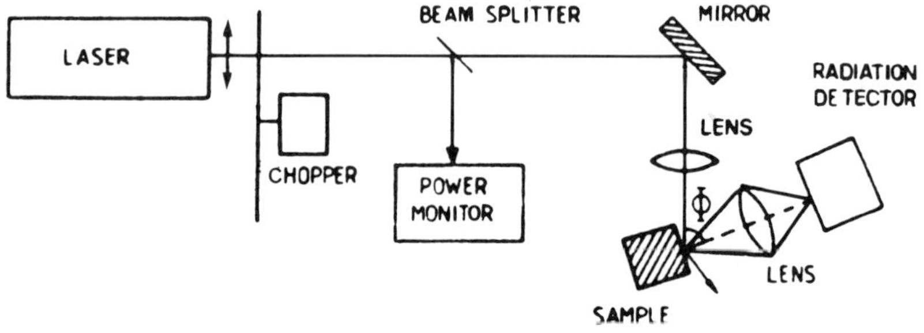

Fig. 1. Experimental arrangement for PTR. (From Kanstad *et al.*, 1986.)

- The risetime response may be very short, because it is unnecessary to wait for the heat transfer from the sample to a surrounding fluid (e.g., mirage detection scheme).

Also, the use of infrared cameras and arrays of detectors coupled to signal-processing units increases the interest in photothermal radiometry.

B. Photorefractive Methods

The basic idea is to take advantage of the effects of the temperature distribution on the probe-beam propagation. With the index of refraction being, in general, much more affected by the thermal effects than by acoustic ones (Jackson *et al.*, 1981), we shall neglect the pressure contribution.

The index of refraction associated with the temperature field can be developed as a Taylor series of the form

$$n(\mathbf{r}_0 - \mathbf{r}, t) = n(\mathbf{r}_0, t) + \mathbf{r}\left(\frac{\partial n}{\partial r}\right)_{r_0} + \frac{\mathbf{r}^2}{2}\left(\frac{\partial^2 n}{\partial r^2}\right)_{r_0} + \cdots, \tag{1}$$

where

$$n(\mathbf{r}, t) = \frac{\partial n}{\partial T} T(\mathbf{r}, t), \tag{2}$$

with $T(\mathbf{r}, t)$ being the light-induced temperature rise that will be calculated in Section III. This photo-induced perturbation leads to the above development, where each of the first three terms is associated with three different techniques: interferometry, beam deflection, and thermal lensing, respectively.

1. Interferometry

The optical path difference is

$$\delta(t) = \int_{\text{path}} n(\mathbf{r}, t)\, ds = \frac{\partial n}{\partial T} \int_{\text{path}} T(\mathbf{r}, t)\, ds, \tag{3}$$

where $n(\mathbf{r}, t)$ represents the variation of the index of refraction induced by the photothermal effect. The optical path difference can be monitored easily by using an interferometric technique (Stone, 1972; Lepoutre *et al.*, 1983). If one considers the recent progress, e.g., in heterodyne detection (Wickramasinghe *et al.*, 1983), $\delta \sim 10^{-4}$ Å$/\sqrt{\text{Hz}}$ has been already achieved with a simple and compact setup (Royer, Dieulesaint, and Martin, 1986).

The main limitations of such a setup come from the low-frequency acoustical noise that must be carefully damped. The minimum detectability of a few 10^{-5} Å/ $\sqrt{\text{Hz}}$ when using a ~ 1-mW probe beam is due to the shot noise and can be reached at frequencies higher than ~ 1 KHz. When using a solid for which $\partial n/\partial T \sim 10^{-5}\,^{\circ}\text{C}^{-1}$ and when probing over a 1-cm path length, one gets:

$$T_{\min} \sim 10^{-7}\,\text{degree}/\sqrt{\text{Hz}}\,.$$

Nevertheless, interferometry is still relatively difficult to align and needs to be used in a medium of good optical quality (distortion of the wavefront $< \lambda$), otherwise the sensitivity goes rapidly to zero.

Except in the case of frequency shifting, (e.g., heterodyne detection), the noise induced by the laser probe intensity fluctuation is difficult to eliminate.

2. Beam Deflection: Mirage Effect

We proposed and demonstrated a few years ago (Boccara, Fournier, and Badoz, 1979) that beam deflection can be advantageously used for monitoring the temperature-gradient field close to a sample surface or within the bulk of a sample.

This technique, which was found to be sensitive and very simple to set up, has had many theoretical and experimental developments (Jackson *et al.*, 1981; Aamodt and Murphy, 1981; Grice *et al.*, 1983). It is based on the measurement of the time-dependent beam deflection, which can be computed for various kinds of probe beams. For a Gaussian beam propagating in inhomogeneous media, most of the beam parameters can be deduced from the analysis of Casperson (Casperson, 1973).

The propagation of the beam through the spatially varying index of refraction is given by (Born and Wolf, 1970)

$$\frac{d}{ds}\left(n_0 \frac{d\mathbf{r}_0}{ds}\right) = \nabla_{\perp} n(\mathbf{r}, t), \tag{4}$$

where $\mathbf{r}_0$ is the perpendicular displacement of the beam from its original direction, n_0 is the uniform index of refraction, and $\nabla_{\perp} n(\mathbf{r}, t)$ is the gradient of the index of refraction perpendicular to $\mathbf{S}$ (the ray path). This relation can be integrated over the ray path $\mathbf{S}$:

$$\frac{d\mathbf{r}_0}{ds} = \frac{1}{n_0}\int_{\text{path}} \nabla_{\perp} n(\mathbf{r}, t)\, ds. \tag{5}$$

Since the deviation is small, one can get the expression of the deflection $\theta(t)$:

$$\frac{d\mathbf{r}_0}{ds} - \theta(t) - \frac{1}{n_0}\frac{\partial n}{\partial T}\int_{\text{path}} \nabla_{\perp} T(\mathbf{r}, t)\, ds. \tag{6}$$

This deflection is usually decomposed into two components θ_n and θ_t, where θ_n and θ_t, respectively, are the deflections normal and parallel to the sample surface.

Figure 2 shows the experimental scheme used for the mirage (or photothermal deflection) detection outside the sample under study. The sample is irradiated by a modulated (or pulsed) laser beam focused on the sample surface by a spherical or a cylindrical lens. In the fluid surrounding the sample, a modulated temperature gradient is induced by the diffusion process. This temperature gradient depends both upon the spatial distribution of the sample illumination and upon the sample and fluid thermal properties. It is detected by measuring the deflection θ of a probe beam (focused He–Ne laser) that propagates close and parallel to the sample surface.

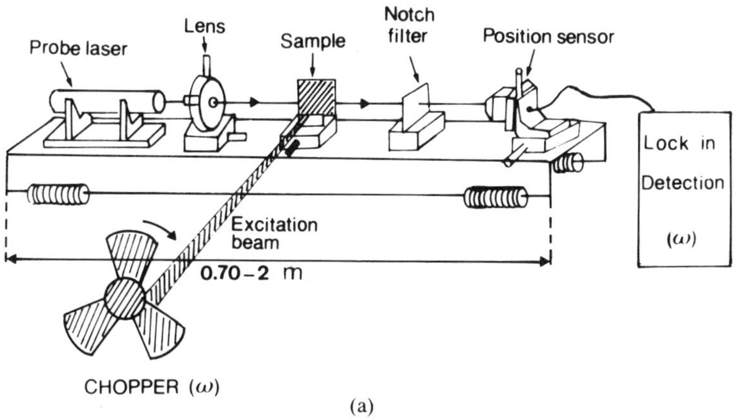

(a)

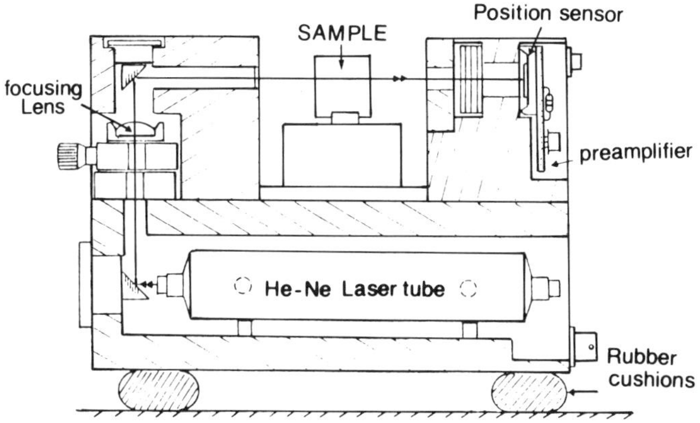

(b)

FIG. 2. (a) Experimental arrangement for mirage detection. (b) Compact mirage detection setup. (From Charbonnier *et al.*, 1986.)

The one-dimensional case corresponds to the well-known formula:

$$\theta(t) = \frac{l}{n}\frac{\partial n}{\partial T}\frac{\partial T}{\partial z}(z, t), \tag{7}$$

where $\partial T/\partial z$ is the temperature gradient perpendicular to the probe-beam propagation, and l is the path length.

We can now get an order of magnitude of the sensitivity of the mirage detection. The temperature gradient, either induced by heat diffusion or by the inhomogeneous power distribution in the cross-section of the heating beam (e.g., Gaussian distribution), is:

$$\frac{T}{\mu} \quad \text{or} \quad \frac{T}{w_0},$$

where μ is the thermal diffusion length, w_0 the beam waist of the heating beam, and T the surface temperature or the temperature at the center of the beam.

By using a compact setup (Charbonnier and Fournier 1986), such as the one described in Fig. 2b, we have obtained a shot-noise-limited detection level at frequencies higher than ~ 30 Hz.

The minimum $\theta_{\min}$ that can be reached is a function of many parameters, such as the mechanical quality of the setup, and the pointing beam stability of the laser; thus:

$$\theta_{\min} < 10^{-10} rd/\sqrt{\text{Hz}}\,.$$

With a beam waist of ~ 60 μ and a confocal length of ~ 1 cm,

$$T_{\min} \sim 10^{-7}\ \text{deg}/\sqrt{\text{Hz}}\ \left(1\ \text{cm path}, \frac{\partial n}{\partial T} \sim 10^{-5}\right),$$

which is equivalent to the best results obtained by interferometry, but with a much simpler setup. Let us point out that the use of a differential position sensor (i.e., a quadrant cell) allows elimination of the probe-laser intensity fluctuations. It is obvious that in order to get the best from such detection, the beam size w_1 must be adapted to the area to be probed, otherwise (e.g., if $w_1 > w_0$ or $w_1 > \mu$) one has to take into account the convolution of the heat distribution with the probe-beam profile (Aamodt and Murphy, 1981; Roger *et al.*, 1987).

3. Thermal Lensing

After taking advantage of the gradient of the distribution, one can also use the curvature of the heat distribution (the second derivative with respect to the spatial variables). The effect of this curvature is to change the divergence of the probe beam. More precisely, following Casperson paper (Casperson, 1973), the change in the complex beam parameter, q, is given by:

$$\frac{d}{ds}\left(\frac{1}{q_{S_{i\perp}}}\right) = -\left(\frac{1}{q_{S_{i\perp}}}\right)^2 - \frac{\partial^2 n}{n_0 \partial S_{i\perp}^2} \qquad i = 1,2, \tag{8}$$

where

$$\frac{1}{q_{S_{i\perp}}} = \frac{1}{R_{S_{i\perp}}} - \frac{i\lambda}{\left(n_0 \pi w_0^2\right)}. \tag{9}$$

The index i designates the two principal directions of curvature for an elliptical gaussian beam; $R_{S_{i\perp}}$ is the radius of curvature of the phase fronts, w_0 is the $1/e^2$–spot size, and λ is the vacuum wavelength of the probe beam. Integrating Eq. (8) over the ray path S gives:

$$\left(\frac{1}{q_{S_{i\perp}}}\right)_{\substack{\text{end of}\\ \text{interaction}}} - \left(\frac{1}{q_{S_{i\perp}}}\right)_{\substack{\text{beginning of}\\ \text{interaction}}} = \int_{\text{path}} ds \left(-\frac{1}{q_{S_{i\perp}}^2} - \frac{\partial^2 n}{n_0 \partial S_{i\perp}^2}\right)$$

$$i = 1,2. \tag{10}$$

From this equation, we see that the effect of the curvature of the index of refraction is equivalent to an astigmatic lens of focal length F_i, in the S_i direction, where F_i is given by:

$$1/F_i = -\frac{1}{n_0}\int_{\text{path}} \frac{\partial^2 n}{\partial S_{i\perp}^2} ds = -\frac{1}{n_0}\frac{\partial n}{\partial T}\int_{\text{path}} \frac{\partial^2 T}{\partial S_{i\perp}^2} ds \qquad i = 1,2, \tag{11}$$

when the heating and the probe beams are collinear,

$$1/F_1 = 1/F_2 = 1/F. \tag{12}$$

The thermal lensing detection has been widely used for investigation of low-absorption liquids (Fang and Swofford, 1983).

For a situation in which the probe and pump beams are well matched, the sensitivity is equivalent to interferometry or mirage detection; nevertheless, for such an experiment that needs a careful optical alignment of the beam, the signal is affected by the noise intensity of the probe beam.

III. Photothermal Signal Generation: Theoretical Approach

We have demonstrated in the preceding sections that the temperature rise measurement of the sample (and of its surface) irradiated by a light flux allows the determination of numerous parameters such as optical parameters, thermal parameters, or geometrical ones. This determination is not possible without the previous setting of a theoretical model in which the heat diffusion is carefully analyzed, within the sample and its surroundings. In this section, we shall determine the expressions of the temperature field in the sample and in the surrounding fluid, where the measurement is usually achieved in the cases of a time-varying irradiation.

Indeed, it would be very difficult to get absolute and very precise measurements in the case of a time-independent excitation, whereas it is easy to record modulated or pulsed temperature variations in the range of 10^{-4}–10^{-5} degrees, and even in special cases, in the range of 10^{-6}–10^{-7} degrees. A time-varying excitation allows us either to take account of the time-varying heat diffusion and to get optical information from opaque samples, for instance, and/or to use the usual experimental technique of signal processing (lock-in amplification in the case of periodic excitation, or averaging in the case of pulsed excitation). We shall illustrate the calculations with photothermal spectroscopic and nondestructive evaluation examples.

A. 1-D Calculation of the Periodic Temperature Field in the Case of Thick or Coated Samples

In this model, we shall suppose that the system extends infinitely in the xy-directions, and we shall only consider the heat diffusion along the z-direction.

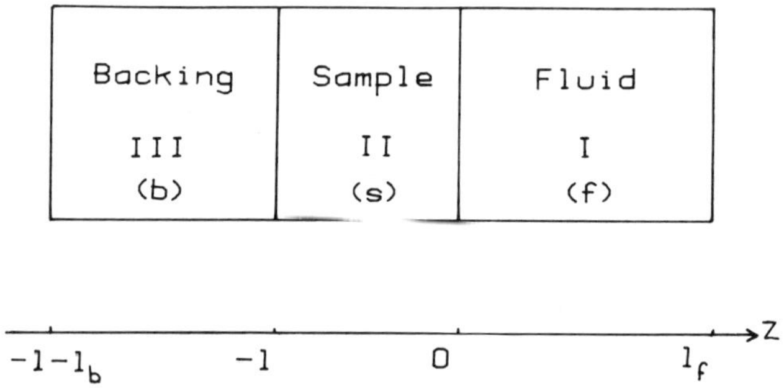

FIG. 3. Geometry of the 1-D Rosencwaig and Gersho (1976) model (see text).

Figure 3 describes the geometry of the system. The absorbing sample (II) is uniformly irradiated by a modulated plane wave whose intensity is:

$$I = \frac{I_0}{2}(1 + \cos \omega t). \tag{13}$$

(I_0 in W/cm^2).

The backing (III) and the fluid (I) are supposed to be optically transparent.

The heat diffusion equations in the three regions are:

$$\frac{\partial^2 T_f}{\partial z^2} = \frac{1}{D_f}\frac{\partial T_f}{\partial t} \qquad \text{Region I} \quad 0 \le z \le l_f,$$

$$\frac{\partial^2 T_s}{\partial z^2} = \frac{1}{D_s}\frac{\partial T_s}{\partial T} - A\exp(\alpha z)(1 + \exp(j\omega t)) \tag{14}$$

$$\text{Region II} \quad -l \le z \le 0,$$

$$\frac{\partial^2 T_b}{\partial z^2} = \frac{1}{D_b}\frac{\partial T_b}{\partial t} \qquad \text{Region III} \quad -l - l_B \le z \le -l$$

with the following boundary conditions:

$$T_f(z = 0) = T_s(z = 0) \qquad k_f\frac{dT_f}{dz}(z = 0) = k_s\frac{dT_s}{dz}(z = 0),$$

$$T_s(z = -l) = T_b(z = -l) \qquad k_s\frac{dT_s}{dz}(z = -l) = k_b\frac{dT_b}{dz}(z = -l), \tag{15}$$

$$T_f(z = l_f) = T_b(z = -(l - l_B)) = 0$$

with l_f and l_b being very much larger than the lengths on which the periodic heat diffuses.

D_i, k_i have been defined in Section I, and $T(z, t)$ is the temperature field. The term $A\exp(\alpha, z)(1 + \exp j\omega t)$ represents the modulated heat source due to the light absorbed by the sample whose optical coefficient is α,

$$A = \frac{\alpha I_0 \eta}{2k_s},$$

where η is the light–heat conversion efficiency by nonradiative de-excitation processes.

These calculations were made by Rosencwaig and Gersho in 1976 for the purpose of photoacoustic detection (Rosencwaig and Gersho, 1976) and have been extended by several authors to the cases of other detections (Jackson *et al.*, 1981; Aamodt and Murphy, 1981). Let us outline the actual temperature field given by the real part of the solution $T(z, t)$.

The resolution of these equations will lead to a temperature field that will contain both static and periodic steady solution and transient solution. We shall restrict ourselves to the determination of the steady periodic solution that alone is important for the photothermal signal.

In order to solve Eqs. 14 and their boundary conditions, we can use the following procedure:

a) Replace the modulated heat source $I = I_0/2\ (1 + \cos \omega t)$ by a unit source $(I_0/2)\ \delta(t)$ in Eq. 14.
b) Use the Laplace transformation in order to reduce the partial differential equations to ordinary differential equations.

$$T(z, t) \to \mathcal{F}(z, p) \qquad \text{with} \qquad \begin{array}{l} T_i(z, 0) = 0, \\ i = \mathrm{f, s, b}, \end{array}$$

$$\begin{aligned} \frac{\partial^2 \mathcal{F}_f}{\partial z^2} &= \frac{p}{D_f} \mathcal{F}_p(z, p), \\ \frac{\partial^2 \mathcal{F}_s(z, p)}{\partial z^2} &= \frac{p}{D_s} \mathcal{F}_s(z, p) - A \exp(\alpha, z), \\ \frac{\partial^2 \mathcal{F}_b(z, p)}{\partial z^2} &= \frac{p}{D_b} \mathcal{F}_b(z, p), \end{aligned} \tag{16}$$

with the boundary conditions:

$$\begin{aligned} &\mathcal{F}_f(0, p) = \mathcal{F}_s(0, p) \qquad && k_f \frac{\partial \mathcal{F}_f}{\partial z}(0, p) = k_s \frac{\partial \mathcal{F}_s}{\partial_z}(0, p), \\ &\mathcal{F}_s(-l, p) = \mathcal{F}_b(-l, p) \qquad && k_s \frac{\partial \mathcal{F}_s}{\partial z}(-l, p) = k_b \frac{\partial \mathcal{F}_b}{\partial z}(-l, p), \\ &\mathcal{F}_f(l_f, p) = \mathcal{F}_b(-l - l_b, p) = 0. && \end{aligned} \tag{17}$$

c) Solve the differential equations and use the boundary conditions to determine the integration constants T_s, U, V, E, W, B, and C defined below:

$$\begin{aligned}
\mathscr{F}_f(z, p) &= T_s(p)e^{-\sqrt{p/D_f}z} + B(p)e^{+\sqrt{p/D_f}z}, \\
\mathscr{F}_s(z, p) &= U(p)\exp\left(\sqrt{p/D_s}\,z\right) + V(p)\exp\left(-\sqrt{p/D_s}\,z\right) \\
&\quad - E(p)\exp(\alpha z), \qquad (18) \\
\mathscr{F}_b(z, p) &= W(p)\exp\left(\sqrt{p/D_b}\,(z + l)\right) \\
&\quad + C(p)\exp\left(-\sqrt{p/D_b}\,(z + l)\right),
\end{aligned}$$

where B and C are zero because the fluid and the backing are supposed to be very thick.

d) It is then straightforward to determine the steady periodic state when a periodic source appears in the diffusion equations.

$$T(z, t) = e^{+j\omega t}\mathscr{F}(z, p)\big|_{p=j\omega}. \qquad (19)$$

Notice that this method also allows the impulse response of the system by taking the inverse transformation of $\mathscr{F}(z, p)$.

The periodic steady-state solutions are then given by:

$$\begin{aligned}
T_f(z, t) &= T_s\exp(-\sigma_f z + j\omega t), \\
T_s(z, t) &= \left[U\exp(\sigma_s z) + V\exp(-\sigma_s z) - E\exp(\alpha z)\right]\exp(j\omega t), \qquad (20) \\
T_b(z, t) &= W\exp\left[\sigma_b(z + l) + j\omega t\right],
\end{aligned}$$

where T_s, U, V, E, and W are complex.

Now σ_i is given by:

$$\sigma_i = \sqrt{\frac{j\omega}{D_i}} = \frac{1 + j}{\mu_i}, \qquad (21)$$

where μ_i is called the thermal diffusion length and characterizes the length upon which the heat diffuses during a period ($T = 2\pi/\omega = 1/f$).

$$\mu_i = \sqrt{\frac{D_i}{\pi f}} = \sqrt{\frac{2D_i}{\omega}} = \sqrt{\frac{2k_i}{\rho_i C_i \omega}}. \qquad (22)$$

Notice that some authors have called this heat diffusion phenomenon "thermal wave." Indeed, one can say that thermal waves are "propagating" in the system with a wavelength $2\pi\mu_i$ and a speed $\sqrt{2\omega D_i}$. These waves can reflect, diffract, and interfere like optical waves, but their amplitude is strongly decreasing due to the $e^{-x/\mu}$-term.

The complex amplitude of the modulated temperature T in the three regions are thus given by:

$$\begin{aligned} T(z) &= T_s \exp(-\sigma_f z), \\ T_s(z) &= U \exp(\sigma_s z) + V \exp(-\sigma_s z) - E \exp(\alpha z), \\ T_b(z) &= W \exp(\sigma_b (z + l)), \end{aligned} \tag{23}$$

with

$$T_s = E \frac{(r-1)(1+b)\exp(\sigma_s l) - (r+1)(b-1)\exp(-\sigma_s l)}{D} + \frac{2(b-r)\exp(-\alpha l)}{D},$$

$$W = E \frac{2(r+g) - \exp(-\alpha l)[(r+1)(1+g)\exp(\sigma_s l) + (r-1)(1-g)\exp(-\sigma_s l)]}{D},$$

$$U = E \frac{(r+g)(1+b)\exp(\sigma_s l) - (r-b)(1-g)\exp(-\alpha l)}{D},$$

$$V = E \frac{(r+g)(1-b)\exp(-\sigma_s l) - (r-b)(1+g)\exp(-\alpha l)}{D},$$

with

$$D = \exp(\sigma_s l)(1+g)(1+b) - \exp(-\sigma_s l)(1-g)(1-b),$$

$$E = +\frac{\alpha I_0}{2k_s(\alpha^2 - \sigma_s^2)},$$

$$b = \frac{k_b \sigma_b}{k_s \sigma_s}, \qquad g = \frac{k_f \sigma_f}{k_s \sigma_s}, \qquad r = \frac{\alpha}{\sigma_s}.$$

Most of the time, measurements are achieved by using a probe beam propagating in the transparent fluid in contact with the sample surface. Thus the temperature field along the probe-beam path is both a function of the surface temperatures $T_s(0)$ and of the sample surface probe-beam distance ($e^{-\sigma_f z}$). This last term, which is frequency dependent, has to be taken into account in an absolute measurement.

The complex amplitude of the surface temperature is given by:

$$T_s(0) = \frac{\alpha I_0}{2k_s(\alpha^2 - \sigma_s^2)} \times \frac{(r-1)(b+1)\exp(\sigma_s l) - (r+1)(b-1)\exp(-\sigma_s l) + 2(b-r)\exp(-\alpha l)}{(\exp(\sigma_s l)(1+g)(1+b) - \exp(-\sigma_s l)(1-g)(1-b))}. \quad (24)$$

This formula is complicated, and we shall restrict ourselves to the examination of two cases of practical interest: bulk and coated samples.

1. Thick Samples

In this case, the thermal diffusion length μ_s is much smaller than the sample thickness l; and in these conditions, the surface temperature $T_s(0, t)$ becomes ($\exp(-\sigma_s l) \to 0, \exp(\sigma_s l) \to \infty$)

$$T_s(0, t) = \frac{\alpha I_0 (r-1)\exp(j\omega t)}{2k_s(\alpha^2 - \sigma_s^2)(g+1)}. \quad (25)$$

It is interesting to write the amplitude and the phase of $T_s(0, t)$ as an explicit function of ($\alpha\mu_s$),

$$|T_s(0, t)| = \frac{I_0}{2k_s} \frac{1}{(1+g)} \alpha\mu_s^2 \left[\frac{(\alpha\mu_s/2 - 1)^2 + \alpha^2\mu_s^2/4}{\alpha^4\mu_s^4 + 4} \right]^{1/2}. \quad (26)$$

$$\mathrm{Arg}(T_s(0, t)) = \mathrm{Atn}(2/\alpha^2\mu_s^2) + \mathrm{Atn}(\alpha\mu_s/(2 - \alpha\mu_s)).$$

The phase variation between very small absorption coefficients and large ones is $\pi/4$. Careful analysis of the signal phase evolution versus frequency will allow an absolute determination of α.

Figure 4 is an illustration for an absolute optical coefficient determination of a semiconductor (GaAs sample) by "mirage" detection. The sample is immersed in cedar oil and illuminated by a modulated plane source (Yacoubi, 1986).

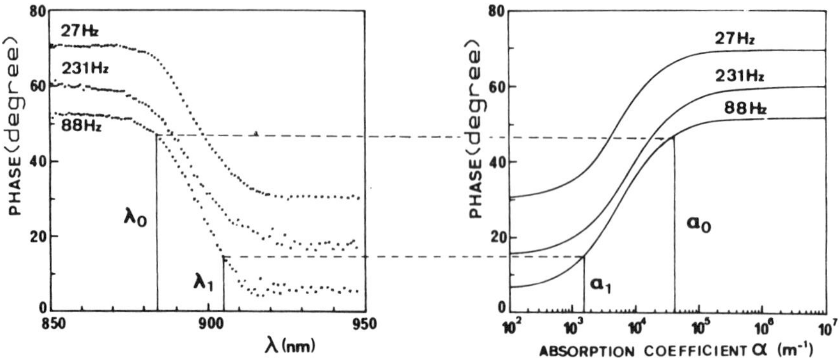

FIG. 4. Experimental (a) and theoretical (b) phase shift of the photothermal (mirage) signal for a GaAs single crystal (very low to very high values of the absorption coefficient).

2. Absorbing Layer Deposited on a Transparent Substrate

This is the case of a thermally thin sample; thus

$$\sigma_s l < 1 \rightarrow \exp(\pm\sigma_s l) \sim 1 \pm \sigma_s l.$$

The surface temperature can be written as:

$$T_s(0, t) = \frac{\alpha I_0\left[(r - b)(1 - \exp(-\alpha l)) + \sigma_s l(rb - 1)\right]}{2k_s\left(\alpha^2 - \sigma_s^2\right)(b + g)} \exp(j\omega t). \quad (27)$$

For very thin samples $\sigma_s l \ll 1$, such as micronic or submicronic semiconductor films deposited on a transparent substrate, the phase of the above formula reduces to $\pi/4$ independently of α, and thus, we can only record the amplitude of the surface temperature. In order to get absolute information, we have to calibrate in the strong absorption region with a classical spectrometer or with the signal itself. Figure 5 shows the photothermal deflection signal amplitude versus wavelength for an amorphous silicon 0.7-μ-thick sample immersed in CCl_4 (Fournier *et al.*, 1982).

For thin samples, the thermal mismatch between the absorbing film (10–1000 μ) and the substrate is the main parameter that determines the signal phase variation when going from a region of transparency to a very absorbing spectral region (Fig. 6) (Yacoubi, 1986).

In this case, great care must be taken when determining absolute optical coefficients, and a thermal analysis is essential.

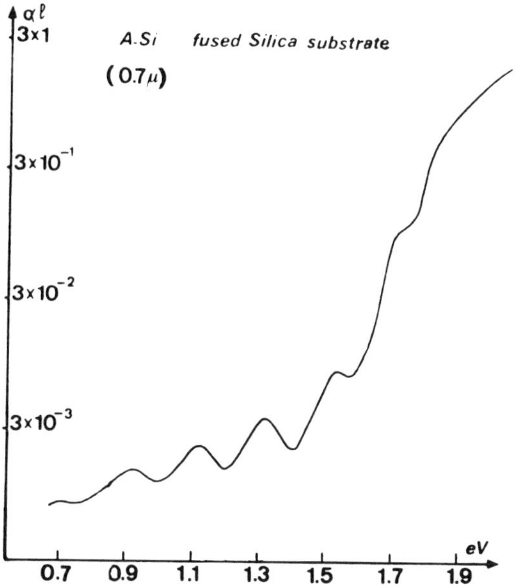

FIG. 5. Photothermal spectrum of amorphous silicon (corrected from saturation).

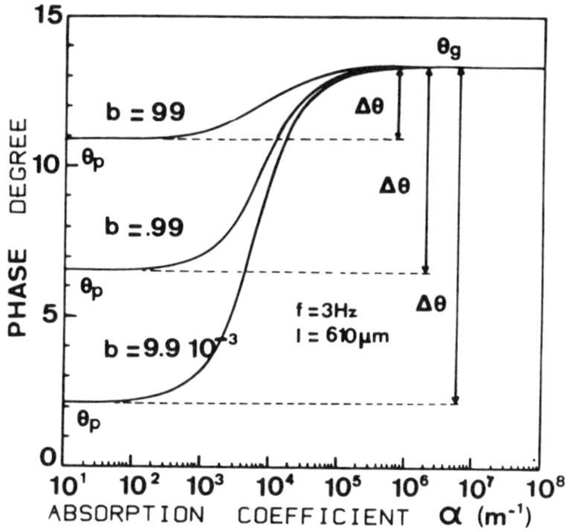

FIG. 6. Phase shift of the photothermal signal for a GaAs 610-μ-thick sample on a transparent backing for various values of $k_b\sigma_b/k_s\sigma_s$.

B. 1-D Calculation of the Periodic Temperature Field in the Case of Multilayered Samples

Fernelius (1980) extended the Rosencwaig and Gersho (1976) model to the case of an absorbing layer (thickness h, absorption coefficient α_c) deposited on an absorbing substrate (thickness l, absorption coefficient α_s) (Fig. 7).

The sample surface temperature $T_s(0, t)$ can be obtained by solving heat equations in the four media with suitable boundary conditions. The expression is rather complicated:

$$
\begin{aligned}
T_s(0,t)\Bigg\{\bigg[(1-b)\exp(-\sigma_s l)\bigg[(1-c)\left(1+\frac{g}{c}\right)\exp(\sigma_c h) \\
+(1-c)\left(1-\frac{g}{c}\right)\exp(-\sigma_c h)\bigg] \\
-(1+b)\exp(\sigma_s l)\bigg[(1+c)\left(1+\frac{g}{c}\right)\exp(\sigma_c h) \\
+(1-c)\left(1-\frac{g}{c}\right)\exp(-\sigma_c h)\bigg]\Bigg\} \\
= 2E\Bigg\{[2(r_s-b)\exp(-\alpha_s l)+(1+b)(1-r_s)\exp(\sigma_s l) \\
-(1-b)(1+r_s)\exp(-\sigma_s l)] \\
+Z\bigg[2(1-b)(1+r_c)\exp(-\sigma_s l-\alpha_c h) \\
-2(1+b)(1-r_c)\exp(\sigma_s l-\alpha_c h) \\
-(1-b)(1-c)\left(1-\frac{r_c}{c}\right)\exp(-\sigma_s l+\sigma_c h) \\
-(1-b)(1+c)\left(1+\frac{r_c}{c}\right)\exp(-\sigma_s l-\sigma_c h) \\
+(1+b)(1+c)\left(1-\frac{r_c}{c}\right)\exp(\sigma_s l+\sigma_c h) \\
+(1+b)(1-c)\left(1+\frac{r_c}{c}\right)\exp(\sigma_s l-\sigma_c h)\bigg]\Bigg\}e^{j\omega t},
\end{aligned}
$$

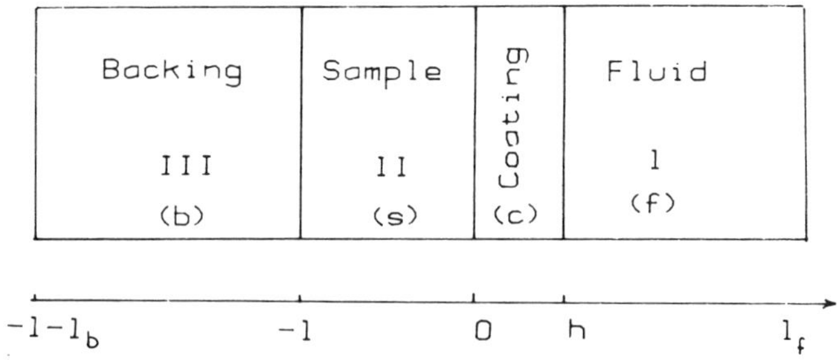

FIG. 7. Geometry of the system under study from the Fernelius (1980) model.

where

$$c = \frac{k_c\sigma_c}{k_s\sigma_s}, \qquad b = \frac{k_b\sigma_b}{k_s\sigma_s},$$

$$g = \frac{k_f\sigma_f}{k_s\sigma_s}, \qquad r_s = (1-j)\frac{\alpha_s\mu_s}{2},$$

$$r_c = (1-j)\frac{\alpha_c k_c \mu_s}{2k_s},$$

$$E = \frac{\alpha_s I_0 \exp(-\alpha_c h)}{2k_s(\alpha_s^2 - \sigma_s^2)}, \qquad Z = \frac{\alpha_c I_0}{2k_c(\alpha_c^2 - \sigma_c^2)}$$

and with

$$\sigma_i = \frac{1+j}{\mu_i} \qquad (i = \mathrm{f, c, s, b}).$$

This formula is very useful when a multiabsorbing layer sample is under spectroscopic investigation.

Yacoubi has used this approach to determine the absolute optical coefficients of a two-layer semiconductor sample (Yacoubi 1986). Figure 8 shows the results obtained for a GaAlAs layer deposited on GaAs.

C. 3-D CALCULATION OF THE PERIODIC SAMPLE TEMPERATURE RISE

Most of the time, the sample is illuminated by a cylindrical source such as a laser beam or Fourier-transform infrared (FTIR) interferometer beam.

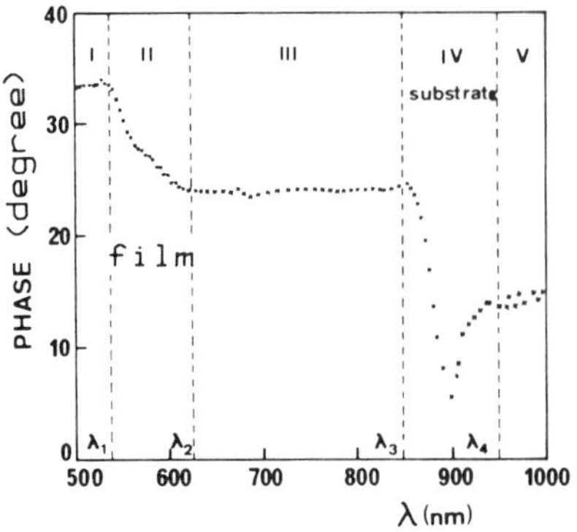

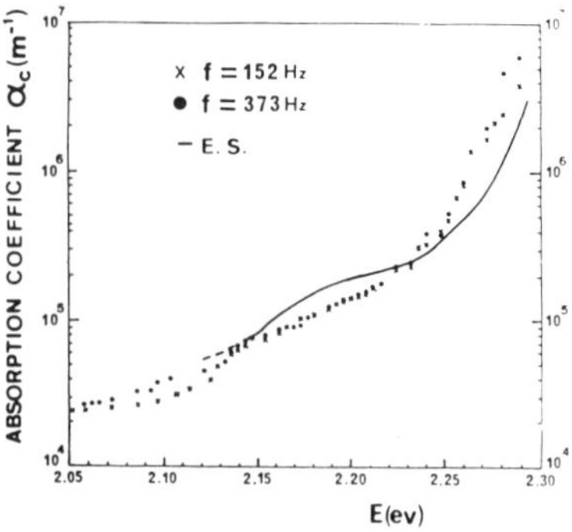

FIG. 8. Wavelength dependence of the phase shift of a mirage signal for a GaAlAs layer on a GaAs substrate (a), and the absorption coefficient of the layer deduced from the Fernelius (1980) model (b) compared with ellipsometric (E. S.) data.

When the beam is focused, the above 1-D calculations are no longer valid. Thus we have to resolve the heat diffusion equation in cylindrical geometry:

$$\frac{\partial^2 T}{\partial r^2} + \frac{1}{r}\frac{\partial T}{\partial r} + \frac{\partial^2 T}{\partial z^2} = \frac{1}{D}\frac{\partial T}{\partial t}. \tag{29}$$

We still assume that the homogeneous sample is the absorbing medium, whereas the fluid and the backing are transparent. For simplicity, we assume that all three regions extend infinitely in the radial direction, with the irradiated area usually being limited and small compared to the radial size of the sample.

We can write the same equations as in the 1-D case (Fig. 3):

$$\begin{aligned}
&\frac{\partial^2 T_f}{\partial r^2} + \frac{1}{r}\frac{\partial T_f}{\partial r} + \frac{\partial^2 T_f}{\partial z^2} = \frac{1}{D_f}\frac{\partial T_f}{\partial t}, \qquad 0 \le z \le l_f, \\
&\frac{\partial^2 T_s}{\partial r^2} + \frac{1}{r}\frac{\partial T_s}{\partial r} + \frac{\partial^2 T_s}{\partial z^2} = \frac{1}{D_s}\frac{\partial T_s}{\partial t} - A(r,t)\exp(\alpha z)(1 + \exp(j\omega t)), \\
&\qquad\qquad -l \le z \le 0, \\
&\frac{\partial^2 T_b}{\partial r^2} + \frac{1}{r}\frac{\partial T_b}{\partial r} + \frac{\partial^2 T_b}{\partial z^2} = \frac{1}{D_b}\frac{\partial T_b}{\partial t}, \qquad -(l + l_b) \le z \le -l,
\end{aligned} \tag{30}$$

with the boundary conditions:

$$k_s \frac{\partial T_s}{\partial z}(z=0) = k_f \frac{\partial T_f}{\partial z} \quad (z=0),$$
$$k_s \frac{\partial T_s}{\partial z}(z=-1) = k_b \frac{\partial T_b}{\partial z} \quad (z=-l), \tag{31}$$
$$T_s(z=-l,t) = T_b(z=-l,t), \qquad T_s(z=0,t) = T_f(z=0,t),$$
$$T_f(\infty,t) = T_b(-\infty,t) = 0 \quad \text{with} \quad l_f \sim \infty \quad l_b \sim -\infty.$$

The following equation,

$$A(r,t) = A(r)(1+\cos\omega t) = \frac{\alpha P \eta}{k_s \pi a^2} \exp\left(-\frac{2r^2}{a^2}\right)(1+\cos\omega t), \tag{32}$$

represents the energy intensity deposited in the absorbing sample, where P is the optically exciting beam power (W), and a is the $1/e^2$-radius of the Gaussian heating beam.

We shall assume as in the preceding section that l_f and l_b are very large compared to the heated area and shall neglect the backwards heat propagation in these two regions. In cylindrical geometry, it is more difficult to find an analytical expression of the temperature field than in the 1-D case.

We can use integral transformations, such as Hankel and/or Laplace, to get ordinary differential equations. Let us recall that the Hankel transform of a function $T(r)$ is defined as:

$$T_0(\lambda) = \int_0^\infty T(r) J_0(\lambda r) r\, dr, \tag{33}$$

and its inversion formula is:

$$T(r) = \int_0^\infty T_0(\lambda) J_0(\lambda r) \lambda\, d\lambda. \tag{34}$$

By Hankel transformation, the diffusion equation

$$\frac{\partial^2 T(r,z,t)}{\partial r^2} + \frac{1}{r}\frac{\partial T(r,z,t)}{\partial r} + \frac{\partial^2 T(r,z,t)}{\partial z^2} = \frac{1}{D}\frac{\partial T(r,z,t)}{\partial t} \tag{35}$$

becomes

$$-\lambda^2 T_0(\lambda,z,t) + \frac{\partial^2 T_0}{\partial z^2}(\lambda,z,t) = \frac{1}{D}\frac{\partial T_0}{\partial t}(\lambda,z,t). \tag{36}$$

We suggest the following steps to get the 3-D periodic steady solutions:

a) Replace the modulated source by the unit source $A(r)\delta(t)$.
b) Use the Hankel transformation

$$T(r, z, t) \to T_0(\lambda, z, t)$$

in order to reduce the partial differential equation to a simpler partial differential equation:

$$-\lambda^2 T_0(\lambda, z, t) + \frac{\partial^2 T_0}{\partial z^2}(\lambda, z, t) = \frac{1}{D}\frac{\partial T_0}{\partial t}(\lambda, z, t), \tag{37}$$

or with a source term:

$$-\lambda^2 T_0(\lambda, z, t) + \frac{\partial^2 T_0}{\partial z^2}(\lambda, z, t)$$
$$= \frac{1}{D}\frac{\partial T_0}{\partial t}(\lambda, z, t) - A_0(\lambda)\exp(\alpha z)\delta(t), \tag{38}$$

where $A_0(\lambda)$ is the Hankel transform of $A(r)$.
c) Use the Laplace transformation

$$T_0(\lambda, z, t) \to \mathscr{F}_0(\lambda, z, p)$$

to reduce the partial differential equations to ordinary differential equations

$$-\lambda^2 \mathscr{F}_0(\lambda, z, p) + \frac{\partial^2 \mathscr{F}_0}{\partial z^2}(\lambda, z, p) = \frac{p}{D}\mathscr{F}_0(\lambda, z, p), \tag{39}$$

$$-\lambda^2 \mathscr{F}_0(\lambda, z, p) + \frac{\partial^2 \mathscr{F}_0}{\partial z^2}(\lambda, z, p) = \frac{p}{D}\mathscr{F}_0(\lambda, z, p) - A_0(\lambda)\exp(\alpha z) \tag{40}$$

$$(T_0(\lambda, z, t) = 0 \qquad \text{for} \quad t \leq 0).$$

d) Resolve the ordinary differential equations in z:

$$\mathscr{F}_0(\lambda, z, p) = A(\lambda, p)\exp\left(-\sqrt{\lambda^2 + \frac{p}{D}}\, z\right)$$
$$+ B(\lambda, p)\exp\left(\sqrt{\lambda^2 + \frac{p}{D}}\, z\right), \tag{41}$$

or

$$\mathscr{F}_0(\lambda, z, p) = A'(\lambda, p)\exp\left(-\sqrt{\lambda^2 + \frac{p}{D}}\,z\right) + B'(\lambda, p)\exp\left(\sqrt{\lambda^2 + \frac{p}{D}}\,z\right) + C'(\lambda, p)A_0(\lambda)\exp(\alpha z), \tag{42}$$

with $C'(\lambda, p) = \dfrac{1}{\lambda^2 + (p/D) - \alpha^2}$.

e) Use the Hankel inversion to obtain:

$$\mathscr{F}_0(r, z, p) = \int_0^\infty \left[A(\lambda, p)\exp\left(-\sqrt{\lambda^2 + \frac{p}{D}}\,z\right) + B(\lambda, p)\exp\left(\sqrt{\lambda^2 + \frac{p}{D}}\,z\right)\right] \times J_0(\lambda r)\lambda\, d\lambda, \tag{43}$$

and

$$\mathscr{F}_0(r, z, p) = \int_0^\infty \left[A'(\lambda, p)\exp\left(-\sqrt{\lambda^2 + \frac{p}{D}}\,z\right) + B'(\lambda, p)\exp\left(\sqrt{\lambda^2 + \frac{p}{D}}\,z\right) + \frac{A_0(\lambda)\exp(\alpha z)}{\lambda^2 + \frac{p}{D} - \alpha^2}\right] J_0(\lambda r)\lambda\, d\lambda. \tag{44}$$

f) In order to get the periodic steady state, use the same procedure as for the 1-D resolution

$$T(r, z, t) = \mathscr{F}_0(r, z, p)\big|_{p=j\omega}\exp(j\omega t) \tag{45}$$

with the constants being determined by the boundary conditions.

We then obtain the expressions of the modulated temperature field in the three regions.

$$\begin{aligned} T_f(r, z, t) &= \int_0^\infty T_s(\lambda)\exp(-\beta_f z)\exp(j\omega t)J_0(\lambda r)\lambda\, d\lambda, \\ T_b(r, z, t) &= \int_0^\infty W(\lambda)\exp(\beta_b(z + l) + j\omega t)J_0(\lambda r)\lambda\, d\lambda, \\ T_s(r, z, t) &= \int_0^\infty [U(\lambda)\exp(\beta_s z) + V(\lambda)\exp(-\beta_s z) - E(\lambda)\exp(\alpha z)] \\ &\quad \times \exp(j\omega t)J_0(\lambda r)\lambda\, d\lambda, \end{aligned} \tag{46}$$

with

$$E(\lambda) = +\frac{P\eta}{\pi k_s} \frac{\exp\left(-\frac{\lambda^2 a^2}{8}\right)}{\left(-\lambda^2 - j\frac{\omega}{D_s} + \alpha^2\right)}, \tag{47}$$

and

$$a^2 \exp\left(-\frac{\lambda^2 a^2}{8}\right) = \int_0^\infty \exp\left(-\frac{2r^2}{a^2}\right) J_0(r\lambda) r\, dr,$$

and

$$\beta_i^2 = \lambda^2 + \frac{j\omega}{D_i}.$$

The final temperature distribution is obtained by substituting the following expressions in the above equations:

$$\begin{aligned} T_s(\lambda) &= -E(\lambda) + U(\lambda) + V(\lambda), \\ W(\lambda) &= -E(\lambda)\exp(-\alpha l) + U(\lambda)\exp(-\beta_s l) + V(\lambda)\exp(+\beta_s l), \\ U(\lambda) &= \left[(1-g)(b-r)\exp(-\alpha l) + (g+r)(1+b)\exp(+\beta_s l)\right] \\ &\quad \times \frac{E(\lambda)}{H(\lambda)}, \\ V(\lambda) &= \left[(1+g)(b-r)\exp(-\alpha l) + (g+r)(1-b)\exp(-\beta_s l)\right] \\ &\quad \times \frac{E(\lambda)}{H(\lambda)}, \end{aligned} \tag{48}$$

with

$$g = \frac{k_f \beta_f}{k_s \beta_s} \qquad b = \frac{k_b \beta_b}{k_s \beta_s} \qquad r = \frac{\alpha}{\beta_s},$$

and

$$H(\lambda) = (1+g)(1+b)\exp(+\beta_s l) - (1-g)(1-b)\exp(-\beta_s l).$$

Let us point out that in order to obtain the 1-D solution of Section I, we take $2\pi \int_0^\infty r\, dr$ of Eq. (46). Since $\int_0^\infty \lambda\, d\lambda \int_0^\infty r J_0(\lambda, r) T(\lambda, r)\, dr = T(0)$, we obtain:

$$T_s(z,t) = \left[-E(0)\exp(\alpha z) + U(0)\exp(\sigma_s z) + V(0)(-\sigma_s z)\right]\exp(j\omega t), \tag{49}$$

with $\sigma_s = \beta_s(\lambda)|_{\lambda=0} = j\omega/D_s$ and $g = k_f\sigma_f/k_s\sigma_s$, $b = k_b\sigma_b/k_s\sigma_s$ and $r = \alpha/\sigma_s$, and with $T_s(z, t) = 2\pi\int_0^\infty T(r, z, t) r\, dr$, which is the same formula as in Section I.

Let us remark that the surface temperature can be written as

$$T_s(0, t) = \int_0^\infty E(\lambda) \left[\frac{\begin{array}{l} -(1 + b)(1 - r)\exp(\beta_s l) + (1 - b) \\ \times(1 + r)\exp(-\beta_s l) \\ -2(r - b)\exp(-\alpha l) \end{array}}{\begin{array}{c} (1 + g)(1 + b)\exp(\beta_s l) \\ -(1 - g)(1 - b)\exp(-\beta_s l) \end{array}} \right]$$

$$\times J_0(\lambda r)\lambda\, d\lambda \exp(j\omega t)$$

$$= \int_0^\infty \overline{T_s(\lambda)}\, J_0(\lambda r)\lambda\, d\lambda \exp(j\omega t). \tag{50}$$

In this expression, one can distinguish between two terms in the Hankel spectrum of the surface temperature $(\overline{T_s(\lambda)})$:

- $E(\lambda)$, which is related to the excitation spatial spectrum (a narrow excitation beam corresponds to a spatial pulse, for instance).
- the second term describing the thermal response of the three media to a unit spatial pulse (see the analogy with the 1-D calculation when letting $\lambda = 0$ in the above formula).

We can then easily find the surface temperature of thick samples by simply replacing this second term by:

$$\frac{(r - 1)}{g + 1} \qquad \text{(see formula 25)},$$

or of thermally thin samples:

$$\frac{(r - b)(1 - \exp(-\alpha l)) + \sigma_s l(rb - 1)}{(b + g)} \qquad \text{(see formula 27)}.$$

This simple procedure is limited to the cases for which cylindrical geometry holds. In the presence of a defect that breaks the symmetry, the Green's function formalism is more suitable.

Moreover, let us point out that the physical interpretation of such a formula,

$$T_f(r, z, t) = \int_0^\infty T_s(\lambda)\exp(-\beta_f z) J_0(\lambda r)\lambda\, d\lambda \exp(j\omega t), \tag{51}$$

is that any temperature distribution can be decomposed into distributions of the form $J_0(\lambda r)\exp(-\beta_f z)$. These distributions have to be superposed to create the temperature field and are characterized by an effective thermal length $1/Re(\beta_f)$. When λ equals zero, one gets a radially uniform distribution corresponding to the 1-D case.

Some authors have used a series instead of the integral form described above, such as:

$$T_s(r, z, t) = \sum_m N_m J_0(\gamma_m r)\left[C_{im}\exp(-\sigma_{im} z) + D_{im}\exp(\sigma_{im} z) + F_{im}\exp(\alpha z)\right]. \quad (52)$$

This kind of formula obtains the same results, because the numerical integration converts the integral into the same series (McDonald, 1982).

The knowledge of the exact 3-D temperature field induced by a focused laser beam can be very important in the case of a nondestructive evaluation

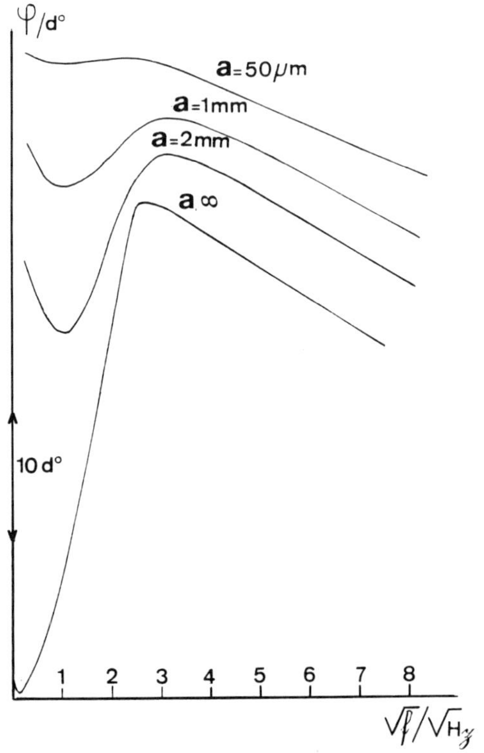

FIG. 9. Theoretical phase shift versus $\sqrt{f}$ for different diameters a of a beam illuminating on sample with a plane defect.

(NDE) experiment. Let us assume that we have to detect a plane defect (slice of air, for instance, located below the surface and parallel to the surface). Figure 9 shows that it is not useful and even harmful to focus the excitation beam over a certain size: The more the focusing is increased, the less the defect appears (Lepoutre, Fournier, and Boccara, 1987).

A large excitation beam contains few high spatial frequency modes and can propagate further compared to a focused beam for which the high spatial frequency modes are more important and exhibit a shorter effective thermal length.

In conclusion, for usual spectroscopic experiments, 1-D calculations are sufficient to account for heat diffusion in the different regions and to allow absolute measurements of optical absorption coefficients. However, for NDE experiments, even in a rather simple case, i.e., defects, it is essential to make 3-D calculations in order to keep the cylindrical geometry (slice parallel to the surface).

D. Application of 3-D Theory to the Calculation of Mirage Detection

We shall describe the two cases of practical interest in which the beam deflection detection will be performed either outside the heated sample and parallel to its surface or within the bulk with a probe-beam propagation collinear (or almost collinear) to the excitation beam.

1. Transverse Photothermal Deflection (Fig. 10).

If the probe beam is propagating in the fluid f along the x direction (unit vector i), the probe-beam deflection is given in the two directions z and y by

$$\theta_n = \frac{1}{n_0}\frac{dn}{dT}\int_{-\infty}^{+\infty}\frac{\partial T_f}{\partial z}\,dx, \qquad \text{(Normal deflection, parallel to } z\text{)}, \tag{53}$$

and

$$\theta_t = \frac{1}{n_0}\frac{dn}{dT}\int_{-\infty}^{+\infty}\sin\alpha\frac{\partial T_f}{\partial r}\,dx. \qquad \text{(transverse deflection, parallel to } y\text{)}. \tag{54}$$

Following Aamodt and Murphy (1981), we can express θ_n and θ_t by using

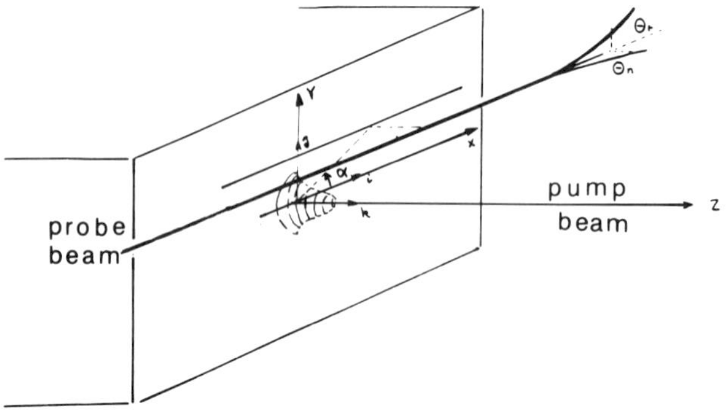

FIG. 10. 3-D experimental scheme for transverse θ_t and normal θ_n photothermal deflection.

the Hankel spectrum of the surface temperature (Eq. 50).

$$\theta_n = \frac{2}{n_0}\frac{dn}{dT}\exp(j\omega t)\int_0^{\infty}\overline{T_s}(\lambda)\beta_f\exp(-\beta_f z)\cos(\lambda y)\,d\lambda, \quad (55)$$

$$\theta_t = \frac{2}{n_0}\frac{dn}{dT}\exp(j\omega t)\int_0^{\infty}\overline{T_s}(\lambda)\lambda\exp(-\beta_F z)\sin(\lambda y)\,d\lambda. \quad (56)$$

These two formulas derive directly from the transform theory

$$\int_{-\infty}^{+\infty} T_s(x)\,dx = 2\int_0^{\infty}\overline{T_s}(\lambda)\cos\lambda y\,d\lambda. \quad (57)$$

Let us point out that θ_n is related to the heat diffusion processes perpendicular to the surface, whereas θ_t describes the heat diffusion processes parallel to the sample surface. These two measurements are particularly useful for NDE investigation (Lepoutre *et al.*, 1985) and for thermal diffusivity measurements (Kuo *et al.*, 1986).

A detailed analysis of the probe-beam size effects can be found in the paper by Lasalle, Lepoutre, and Roger (1988). Let us note that in the framework of the preceding theory, the finite-size beam effect could be introduced by a multiplicative correction factor (Aamodt and Murphy, 1981).

2. Collinear Photothermal Deflection

For this experimental scheme, the pump- and the probe-beam laser are nearly parallel, and we integrate the radial deflection over the beam path.

This experimental scheme has been found to be useful for coatings deposited on transparent substrates (Boccara *et al.*, 1980) or for weakly-absorbing media. Indeed, by using laser excitation, we take advantage of the tight focusing to increase the thermal gradient, which is probed within the condensed medium that usually exhibits a dn/dT factor larger than the one of air. In the first case (a transparent sample), the energy density can be considered as constant along the pump-beam path. So the radial probe-beam deflection is given by

$$\theta_c = \frac{1}{n_0}\frac{dn}{dt}\int_{-\infty}^{+\infty}\frac{\partial T}{\partial r}\,dz \tag{58}$$

when the probe and the pump beam can be considered as parallel.

When the two beams are tilted by an angle β and are propagating in a unique very-low absorbing medium, the probe-beam deflection in the x-direction is given by:

$$\theta_c = \frac{1}{n_0}\frac{dn}{dT}\frac{1}{\sin\beta}\int_{-\infty}^{+\infty}\frac{\partial T}{\partial r}\frac{x_0}{r}\,dy, \tag{59}$$

where x_0 is the transverse offset of the two beams; θ_c can be written in the same formalism (Jackson *et al.*, 1981):

$$\theta_c = \frac{1}{n_0}\frac{dn}{dT}\frac{1}{\sin\beta}\frac{\alpha}{\pi k_s}\exp(j\omega t)\int_0^{\infty}\exp\left(-\frac{\lambda^2 a^2}{8}\right)\frac{\lambda\sin(\lambda x_0)}{\lambda^2 + j\dfrac{\omega}{D}}\,d\lambda. \tag{60}$$

This formula can be integrated. For instance, in the high-frequency case:

$$\theta_c = \frac{1}{n_0}\frac{dn}{dT}\frac{1}{\sin\beta}\frac{\alpha x_0}{\sqrt{\pi}\, jC_s\rho_s\omega a^3}\exp\left(-\frac{2x_0^2}{a^2}\right)\exp(j\omega t). \tag{61}$$

IV. Specific Problems

A. Influence of Energy Migration or Carrier Diffusion

In the preceding sections, we supposed that the relaxation of the absorbing system is localized at the place where the light is absorbed. We know that in gases and sometimes in liquids, mass diffusion can take place before the relaxation processes of the excited species. Although such delocalizations of heat sources are uncommon at room temperature, in solids they have been observed with semiconductors for which carriers diffuse (over a length that competes with the thermal diffusion length) before they recombine nonradiatively. One can also imagine that energy migration, or fluorescence trapping, may lead to the same kind of behaviour.

We shall focus the model on semiconductors, which have been studied both theoretically and experimentally.

The main processes that have to be considered when a low-intensity light beam is absorbed at room temperature by a doped semiconductor sample (which will be useful to set our model) are represented in Fig. 11.

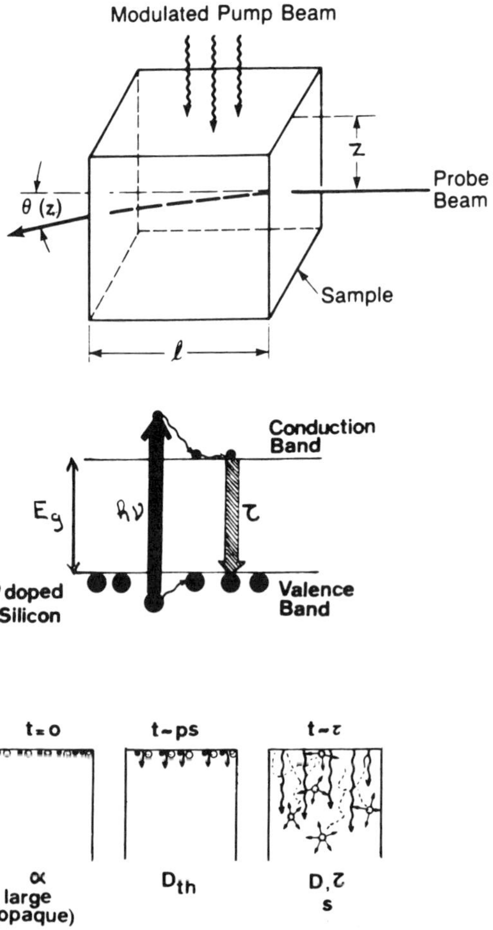

FIG. 11. Experimental scheme and main parameters obtained when probing plasma wave and thermal wave in a silicon sample.

The monochromatic light flux impinging on the sample is composed of photons whose energy $h\nu$ is larger than the band gap E_g. The absorbed photon gives rise to an electron–hole pair. We shall operate with a doped sample (e.g., "p"-doped in Fig. 11) and a low light-level intensity that maintains the photo-induced carrier population at a level lower than the majority-carrier population (Smith, 1978).

The excess of energy $(h\nu - E_g)$ is converted by the fast (~ 1 ps) nonradiative processes into thermal energy. Thus, the sample exhibits at first a heat source that reflects the distribution of the absorbed energy. For example, if the sample is opaque, a thermal wave will be generated from the sample surface. Then, during their lifetime τ, the photo-induced carriers diffuse through the crystal. During this random walk, they can hit the sample surface and thus have a probability to recombine there (surface recombination velocity s). At last, after a time delay of τ, the recombination processes take place, leading to a second expanded heat source whose distribution reflects the carrier diffusion processes that will interfere with the first one.

1. Modulated Excitation: 1-D Model (Fig. 11)

The sample is supposed to extend infinitely in the positive z-direction, and we shall ignore the thermal coupling between sample and gas. In order to calculate the carrier ($N(z, t)$) density and temperature ($T(z, t)$) distributions through the opaque sample whose surface is illuminated by a periodic uniform light flux, we have to couple two diffusion equations (Fournier *et al.*, 1986).

a. *Free-Carrier Diffusion*

The carrier population density distribution $N(z, t)$ is given by the nonhomogeneous diffusion equation

$$D_c \frac{d^2 N}{dz^2} = \frac{dN}{dt} + \frac{N}{\tau}, \tag{62}$$

with the boundary conditions

$$\begin{aligned} N(z,0) &= 0, \\ D_c \frac{dN}{dz}(0, t) &= -\frac{\phi_0}{h\nu} e^{j\omega t} + sN(0, t), \end{aligned} \tag{63}$$

where D_c is the minority-carrier diffusivity (cm^2/s). For simplification, we have assumed here that surface recombination is only localized on the sample surface, and we have neglected the influence of a space charge region.

For the minority-carrier distribution, one obtains:

$$N(z, t) = \mathrm{Re}\left(\frac{\phi_0 e^{-z/\lambda_{el}}}{h\nu D_c(s/D_c + 1/\lambda_{el})} e^{j\omega t} \right), \tag{64}$$

where $1/\lambda_{el}^2 = (1 + j\omega\tau)/D_c\tau$.

b. *Thermal Diffusion*

The equation describing the thermal diffusion follows from the standard heat diffusion with two source terms, one due to carrier recombination (65) and another due to a surface source (66):

$$D_s \frac{d^2T}{dz^2} = \frac{dT}{dt} - \frac{E_g D_s}{k_s} \frac{N(z, t)}{\tau}, \tag{65}$$

with the boundary conditions

$$\begin{gathered} T(z, 0) = 0, \\ -k_s \frac{dT}{dz}(0, t) = sN(0, t)E_g + \frac{h\nu - E_g}{h\nu} \phi_0 e^{j\omega t}, \end{gathered} \tag{66}$$

where D_s is the thermal diffusivity (cm^2/s), k_s is the thermal conductivity (W/cm K), of the semiconductor, E_g is the semiconductor gap energy (J), $h\nu$ the photon energy (J), ϕ_0 is the light flux (W/cm^2), s is the surface recombination velocity (cm/s), and τ is the minority-carrier lifetime (s).

The steady-state solution for the temperature distribution is given by:

$$\begin{aligned} T(z, t) = \mathrm{Re}\Bigg[& \frac{(\phi_0 E_g/\lambda_{el}) \left(\dfrac{e^{-z/\lambda_{th}}}{1/\lambda_{th}} - \dfrac{e^{-z/\lambda_{el}}}{1/\lambda_{el}} \right) e^{j\omega t}}{h\nu\tau D_c k_s (s/D_c + 1/\lambda_{el}) [1/\tau D_c + j(\omega/D_c) - j(\omega/D_s)]} \\ & + \left(\frac{\phi_0 E_g s}{h\nu D_c k_s (s/D_c + 1/\lambda_{el})} + \frac{(h\nu - E_g)\phi_0}{h\nu k_s} \right) \\ & \times \lambda_{th} e^{-z/\lambda_{th}} e^{j\omega t} \Bigg], \end{aligned} \tag{67}$$

where:

$$1/\lambda_{\text{th}}^2 = j\omega/D_s \quad \text{and} \quad 1/\lambda_{\text{el}}^2 = (1 + j\omega\tau)/D_c\tau.$$

The minority-carrier distribution exhibits a purely electronic term characterized by an exponential dependence on the depth away from the illuminated surface. This decay is characterized by the complex electronic diffusion λ_{el}. The temperature distribution is a linear combination of two kinds of terms:

- the thermal wave terms with the exponential dependence characterized by the complex thermal diffusion length λ_{th};
- the electronic wave (sometimes called "plasma wave") with the complex diffusion length λ_{el} (Opsal *et al.*, 1987).

This simple 1-D calculation with opaque samples can be extended to the case of 1-D and 3-D excitation of samples whose absorption coefficient α is finite.

This analysis has been performed by quite a few authors for photoacoustic detection (Mikoshiba, Nakamura, and Tsubouchi, 1982; Miranda, 1982), with the theoretical work of Sablikov and Sandomirskii being the most complete physical analysis (Sablikov and Sandomirskii, 1983).

2. Pulsed Excitation

It is relatively easy to solve free-carrier population equations with a pulsed excitation by using the Laplace transform. In the case of an opaque sample uniformly irradiated by a short light pulse, the time and spatial dependence of the carriers is given by:

$$N(z,t) = \frac{Q_0 E_g}{D_c} e^{-t/\tau} \left[\sqrt{\frac{D_c}{\pi t}}\, e^{-z^2/4D_c t} - s e^{(sz/D_c + s^2 t/D_c)} \times \operatorname{erfc}\left(z/2\sqrt{D_c t} + s\sqrt{t/D_c} \right) \right] \tag{68}$$

(Q_0 pulse energy density, photon energy $h\nu = E_g$).

The temperature distribution is more difficult to get in an analytical form for practical use. We have thus simply computed the Fourier transform of expression (68) with suitable parameters to get the pulsed response to an excitation (Pelzl, Fournier, and Boccara, 1986, 1987).

3. "Mirage" Probing in the Bulk of Semiconductor Samples

The time-dependent deflection of a narrow probe beam propagating through an inhomogeneous medium at a given depth z is given by:

$$\theta(z,t) = \frac{l}{n}\frac{\partial n(z,t)}{\partial z},$$

where θ is the angular deflection, l is the interaction length ($l \gg z$, and also much larger than both the thermal μ, and the carrier $\sqrt{\tau D_c}$ diffusion lengths), n is the local index of refraction, and $\partial n(z,t)/\partial z$ is the photo-induced gradient in the index of refraction. In the case of semiconductors, $\partial n(z,t)/\partial z$ has two contributions, a thermal and a free-carrier term, and is given by:

$$\frac{\partial n(z,t)}{\partial z} = \frac{\partial n}{\partial T}\frac{\partial T(z,t)}{\partial z} + \frac{\partial n}{\partial N}\frac{\partial N(z,t)}{\partial z}, \tag{69}$$

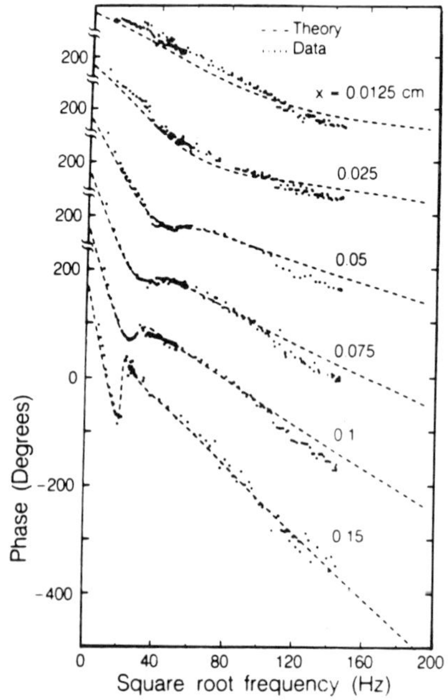

FIG. 12. Experimental and theoretical phase shift of photothermal signal for various distances from the illuminated surface for p-type silicon sample: ($s = 450$ cm/s, $D_c = 30$ cm^2/s, $\tau = 4.5\ 10^{-5}$ s, $D_s = 0.9$ cm^2/s).

where $T(z, t)$ and $N(z, t)$ are the time-dependent temperature and the minority-carrier density distribution, respectively. From their values calculated in Eqs. (67) and (68), one obtains the thermal gradient as well as the minority-carrier gradient.

The experimental setup represents an extension of the "mirage" setup. The excitation light source, i.e., pump beam, was the output from an Ar^+-ion laser (5145 Å) uniformly illuminating the front face of the sample. The probe beam used to detect the index-of-refraction gradient within the bulk of the sample was the 1.15- or 3.39-μm line of a He–Ne laser ($\sim$ 1–3 mW) focused to $\sim$ 30 μm (Fig. 11).

The solids examined were p-type Si crystals. The Si sample had one face mechanically polished and treated with methanol to minimize the surface recombination. The power density on the Si sample varied from 30 to 800 mW/cm^2. The latter value corresponds to a minority-carrier density roughly one order of magnitude smaller than that of the majority carriers ($\rho \sim 6\Omega - \text{cm}$, $N_1 \sim 3 \times 10^{15}$ carriers per cm^3).

In Fig. 12, one can see the good accordance between experimental data and the theoretical model developed above.

B. Heat Diffusion, Rough Surfaces, and Heterogeneous Media: Fractal Approach

Heat sources and samples with usual geometrical shapes such as planes, points, and cylinders, have previously been considered (Carslaw and Jaeger, 1959). When highly-complex systems have to be studied, it is difficult to account for the individual properties of each component to get the global properties of the system. For instance, when one deals with heat diffusion, one can account for a perturbed sample surface by introducing an "equivalent layer" whose thermal properties are different from the bulk ones (Bein, Krieger, and Pelzl, 1986). Or one can try to model the complex sample by an assembly of elements of simple geometrical shapes (e.g., spheres (Lavecek, 1980)).

Recently, macroscopic self-similar objects have been considered in the framework of fractal theory developed by Mandelbrot (1983). We would like, at this point, to suggest how this geometrical approach can be of great help in understanding the thermal behaviour of rough surfaces and heterogeneous samples. These problems are of great interest both from a fundamental point of view (physics of disordered systems), and for practical applications to characterize divided materials of high technological interest (ceramics, sintered materials).

FIG. 13. Self-similar geometrical construction called *Sierpinskii gasket.*

Let us recall that a structure is self-similar if we cannot tell the difference in the structure as we change the scale. Among the fractal objects, one can distinguish two kinds of self-similar structures, the geometrical ones, which we shall use as an example to define the fractal dimension $\bar{d}$, and the random self-similar structures, which are of actual importance to describe physical systems. It is obvious that in real physical systems, the scale range at which this self-similarity occurs is limited both in the small dimension limit (e.g., grains or atoms) and in the large dimension one where the objects behave like Euclidean entities.

Figure 13 shows an example of self-similar geometrical construction called the Sierpinski gasket. We can see that if we change the unit length by $1/2$, we have three equal pieces. Let us recall that for a straight line interval, a square or a cube, if we change the scale by $1/2$, we have $(1/2)^{-d}$ equal pieces of the initial object, with d being equal to 1, 2, or 3, respectively. Here d is the usual Euclidean dimension. We can use this

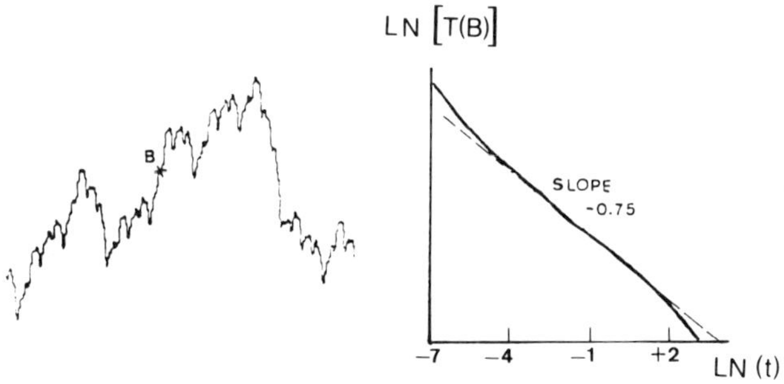

FIG. 14. Brownian scalar motion curve ($d = 1.5$) and the temperature at B as a function of time after a pulsed excitation.

definition to get the dimension of the Sierpinski gasket.

$$(1/2)^{-\bar{d}} = 3 \qquad \text{thus} \qquad \bar{d} = 1.58,$$

$\bar{d}$ is called the fractal dimension.

Now for application to heat diffusion, it is worth focusing our attention on random self-similar structures. For such structures, the above properties must be considered as an average. As an example, Fig. 14 shows a so-called Brownian scalar motion curve whose fractal dimension is 1.5.

1. Euclidean and Fractal Heat Sources

In order to get insight into the behaviour of fractals, let us consider the following simple situation: When a point is heated at $t = 0$ by a fast heat pulse, the temperature in the surrounding three-dimensional medium whose diffusivity is D is given by:

$$T(r, t) = \frac{e^{-r^2/4Dt}}{(Dt)^{3/2}} \qquad \text{(geometrical dimension of the source: } d = 0 \text{ unit source excitation).} \tag{70}$$

Integrating this contribution over a line source will lead to:

$$T(t) \sim \frac{1}{Dt} \qquad d = 1 \tag{71}$$

for a point located on the line source, and over a plane source to:

$$T(t) \sim \frac{1}{(Dt)^{1/2}} \qquad d = 2, \tag{72}$$

for a point located on the plane source.

More generally, for a source of dimension $\bar{d}$ (fractal or Euclidean if $\bar{d} = d$), one finds the temperature of a point located on the source to be:

$$T(t) \sim \frac{1}{(Dt)^{3/2 - \bar{d}/2}}. \tag{73}$$

In order to illustrate this result, we have used the curve of Fig. 14 as a unit source at time $t = 0$, and we have followed the temperature of a central point of this source (in fact, we have averaged the results over 20 curves) with the source being limited in space, we expect a crossover for long times ($Ln(t) > 2$), and because of quantification (1000 points to build the curve), crossover for short times. In between, the slope is close to the expected value $-(3/2 - \bar{d}/2) = -0.75$.

2. Diffusion within Random Fractal Structures

When one is concerned with diffusion processes within random fractal structures, one needs three dimensionalities:

d the Euclidean dimensionality;
$\bar{d}$ the fractal dimensionality;
$\bar{\bar{d}}$ the spectral (or "fracton") dimensionality.

The dimension, $\bar{\bar{d}}$, sometimes called the spectral dimension, is introduced in order to account for the peculiar diffusion processes (Orbach, 1986).

Indeed, looking at a diffusion process requires one to consider the mean square displacement $\langle r^2 \rangle$ of the diffusers (phonons, electrons).

In a Euclidean space, $\langle r^2 \rangle \sim t$, and thus the diffusion coefficient $D = d/dt(\langle r^2 \rangle)$ is a constant. (See, for instance, the diffusion equations in Section III.)

In a fractal space, $\langle r^2 \rangle \sim t^{\bar{\bar{d}}/\bar{d}}$, and the diffusion coefficient D is time dependent, thus the usual heat diffusion solutions are no longer valid.

Among the random fractal structures, particular attention has been devoted to the so-called percolation network. To create a percolation network, each intersection of a d-dimensional grid is occupied, for example, at random with probability p. A critical probability p_c is found such that, as for $p > p_c$, a connected cluster will cross the grid (an infinite cluster in the case of an infinite grid). Such a structure, which can be easily generated by computer calculation, is found to exhibit a fractal structure ($\bar{d} = 1.9$ for $d = 2$, and $\bar{d} = 2.6$ for $d = 3$). Moreover, Alexander and Orbach have conjectured that $\bar{\bar{d}} = 4/3$ for such a percolating network.

Let us go back to heat diffusion and suppose that at time $t = 0$, the sample surface is (uniformly) heated by a short pulse of heat, with the heat source dimensionality being d_s (e.g., $d_s = 0, 1, 2$ for a point, a line, and a plane source, respectively). After a time t, the heated mass $M \sim \langle r^2 \rangle^{\bar{d}/2}$, with $\langle r^2 \rangle \sim t^{\bar{\bar{d}}/\bar{d}}$ assuming a compact exploration of the space (Orbach, 1986; Zallen, 1983).

Thus, with the temperature being proportional to $\langle r^2 \rangle^{d_s/2}$ (total energy deposited), its time dependence is:

$$T(t) \sim \langle r^2 \rangle^{-\bar{d}/2 + d_s/2} = \left(t^{\bar{\bar{d}}/\bar{d}}\right)^{-(\bar{d} - d_s/2)} \tag{74}$$

$$T(t) \sim t^{-\bar{\bar{d}}/2 + d_s \bar{\bar{d}}/2\bar{d}}. \tag{75}$$

One can verify that for a Euclidean sample of dimension 3 ($d = \bar{d} = \bar{\bar{d}} = 3$) excited by a point, a line, or a plane ($d_s = 0$, 1, or 2), one finds as expected:

$$T \sim t^{-3/2}, \qquad t^{-1}, \quad \text{and} \quad t^{-1/2}, \text{ respectively.}$$

3. Experimental Results

In order to check the time behaviour of the surface temperature of various optically opaque samples (both in the visible and in the IR), we have used the experimental setup described in Fig. 15. The sample surface is heated by a short (~ 10 ns) light pulse (0.53 nm), and its average temperature is monitored by a fast IR detector after being collected by an elliptic mirror. The signal is recorded and averaged with a Lecroy 7600 digital oscilloscope.

For an Euclidean sample and a fractal source such as a rough surface $3 > d_s > 2$, formula (75) leads to:

$$T(r) \sim t^{-3/2 + d_s/2} \sim t^{-\alpha} \qquad \text{with} \quad 0 < \alpha < 1/2. \tag{76}$$

Thus the slope in log log scale is smaller than the usual $1/2$ for a plane excitation.

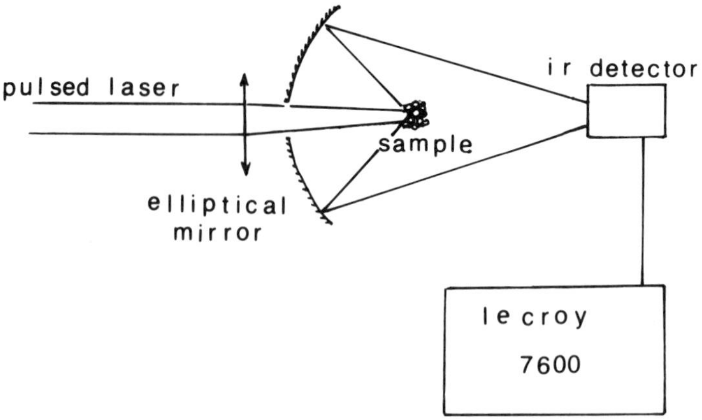

FIG. 15. Experimental scheme for recording the surface temperature after a pulsed excitation.

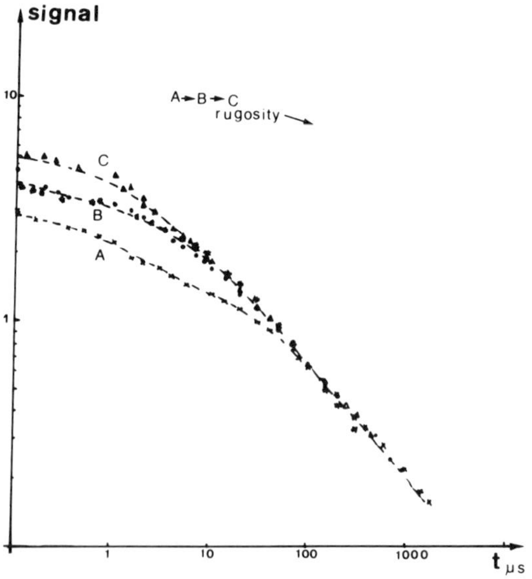

FIG. 16. Surface temperature versus time for different roughnesses of a carbon sample.

In a real physical situation, we expect a crossover corresponding to a diffusion over a distance equal to the deepest structures of the surface. Indeed, such behaviour has been observed by our group on opaque rough surfaces of carbon samples (Fig. 16). The crossover between the two lines varies towards the short time scales as the polishing is improved (smaller grain size of the polishing paper).

For a fractal sample, compact enough to assimilate its surface to a plane ($d_s = 2$), one gets:

$$T \sim t^{-(\bar{\bar{d}}/2)+(\bar{\bar{d}}/\bar{d})}. \tag{77}$$

To give an order of magnitude of $\bar{\bar{d}}/2 - \bar{\bar{d}}/\bar{d}$, let us consider the case of the 3-D percolating network ($\bar{d} = 2.6$, $\bar{\bar{d}} = 4/3$), which has been used many times (Zallen, 1983) as a model for disordered systems. For such a system:

$$\frac{\bar{\bar{d}}}{2} - \frac{\bar{\bar{d}}}{\bar{d}} = 0.154.$$

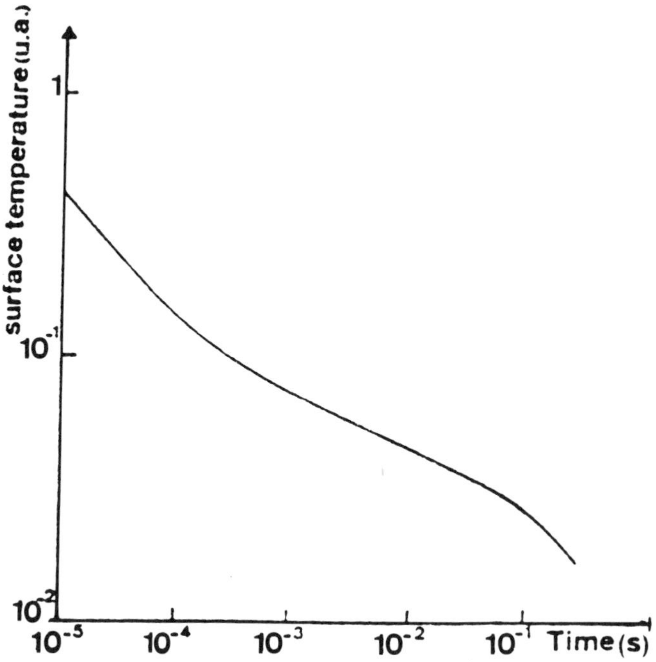

FIG. 17. Surface temperature versus time for an assembly of copper spheres (100-μ diameter).

The exponent is thus much lower than 0.5, which is normally expected in a Euclidean case. We have looked at a large variety of random disordered structures and have found in many instances a power law over a large time scale with an exponent typically in the range 0.15–0.25. As an example, Fig. 17 shows the surface temperature of an assembly of copper spheres (100-μ diameter). This experiment was done by using the setup of Fig. 15.

We do not claim that the structures we have studied are fractal systems, but now that it is well established that nature very often exhibits a fractal geometry (Mandelbrot, 1983), it should be interesting to examine heat diffusion in the framework of such a theory.

The important point that has to be emphasized is that, when diffusion processes take place in random media (random structures, random bonds), the average square length covered by the diffuser during its random walk, $\langle r^2 \rangle$, is a power function of the time whose exponent accounts for the dimension of the walk (Pynn and Skjeltorp, 1985; Stanley and Ostrowsky, 1985).

V. Spectral and Spatial Multiplexing of the Photothermal Signals

Photothermal techniques have been proved to be efficient for spectroscopic investigation, thermal characterization, and nondestructive evaluations.

Nevertheless, when they are used in the low-frequency domain (a few Hz $\leftrightarrow$ a few KHz), each measurement requires a delay that is typically of the order of 1 s. This delay is both imposed by the time of integration associated with the modulation frequency and by signal-to-noise considerations. Moreover, the signal amplitude cannot always be increased by increasing the power of the light source without changing the physical properties or even destroying the sample. Thus, an effort to improve the speed of the measurement has to be made by multiplexing the signal. Indeed one knows that when N samples have to be recorded (e.g., N points on a sample surface in NDE, or N spectral elements in a spectrum), the speed is increased by N if all N measurements are performed simultaneously (or the signal-to-noise ratio is multiplied by $\sqrt{N}$ if we keep the total measurement time constant).

The technique of multiplexing is already effective in spectroscopy. Indeed, the use of either home-made or commercially available FTIR spectrometers has opened the way to many new powerful measurements on solids (semiconductors, plastics, adsorbed species).

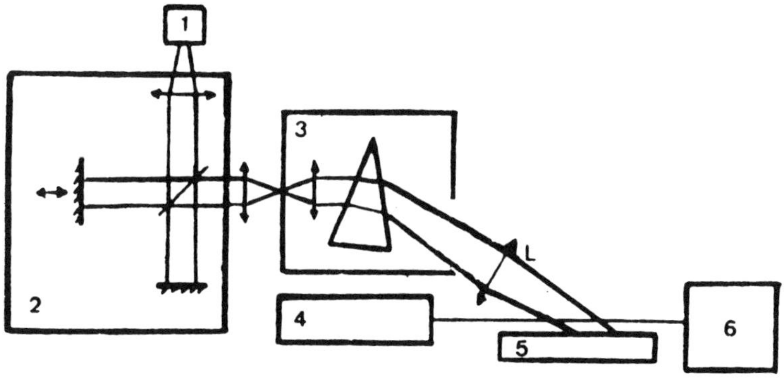

FIG. 18. Schematic experimental setup used for spatially multiplexing the excitation beam in a mirage experiment. (1) Quartz halogen source. (2) Fourier transform interferometer. (3) Dispersive spectrograph. (4) He–Ne laser. (5) Opaque sample. (6) Position sensor.

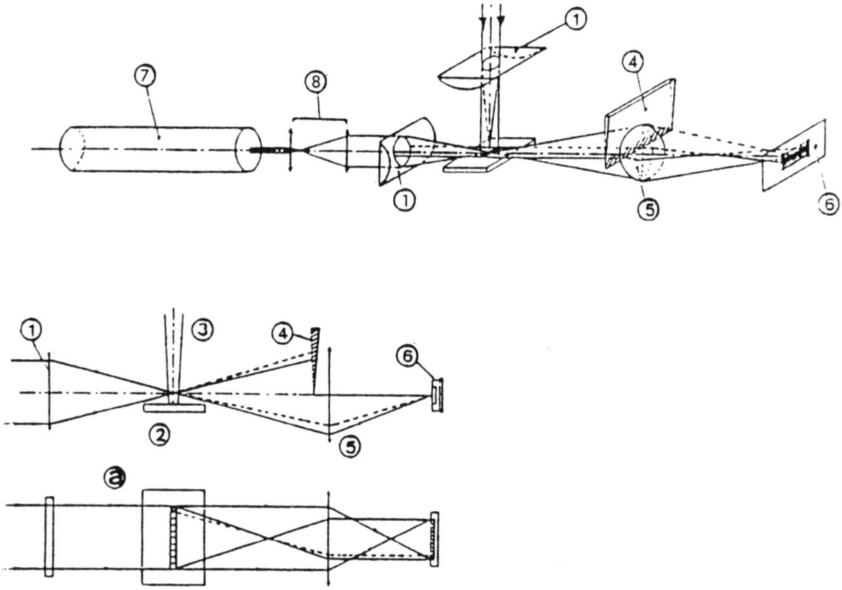

FIG. 19. Schematic experimental setup used to multiplex the mirage detection probe beam. (1) Cylindrical lens. (2) Sample. (3) Excitation beam. (4) Razor blade. (5) Spherical lens. (6) Charge Coupled Device (CCD) Photodiode array. (7) Laser. (8) Beam expander.

With respect to imaging problems for nondestructive evaluation, multi-element IR detectors are now available as linear arrays or surface arrays; however, they are still expensive.

We have recently used two different approaches for multiplexing the mirage signal. In the first, the output of a FT interferometer is spread out by a dispersive system on an opaque sample under examination, with each point of the sample being "coded" by a specific wavelength (Fig. 18). In a second approach, we have used a linear array of 256 photodiodes to get the deflection of 256 small heated areas on the sample surface simultaneously (Fig. 19).

From these preliminary experiments, we expect a large gain in speed, and/or in signal-to-noise ratio, and/or a reduction of the average level of illumination on the sample surface.

REFERENCES

Aamodt, L. C., and Murphy, J. C. (1981). "Photothermal measurements using a localized excitation source," *J. Appl. Phys.* **52**, 4903.

Bein, B. K., Krieger, S., and Pelzl, J. (1986). "Photoacoustic measurements of effective thermal properties of rough and porous limiter graphite," *Can. J. Phys.* **64**, 1208.

Boccara, A. C., Fournier, D., and Badoz, J. (1979). "Thermo-optical spectroscopy: Detection by the "mirage effect," *Appl. Phys. Lett.* **36**, 130.

Boccara, A. C., Fournier, D., Jackson, W., and Amer, N. (1980). "Sensitive photothermal deflection technique for measuring absorption in optically thin media," *Opt. Lett.* **5**, 377.

Born, M., and Wolf, E. *Principle of Optics*. Pergamon Press, Oxford. (1970).

Carslaw, H. S., and Jaeger, J. C. (1959). *Conduction of Heat in Solids*. Oxford, Clarendon.

Casperson, L. W. (1973). "Gaussian light beams in inhomogeneous media," *Appl. Opt.* **12**, 2434.

Charbonnier, F., and Fournier, D. (1986). "Compact design for photothermal deflection (mirage): Spectroscopy and imaging," *Rev. Sci. Instrum.* **57**, 1126 (1986).

Fang, H. L. and Swofford, R. L. (1983). "The thermal lens in absorption spectroscopy," in *Ultrasensitive Laser Spectroscopy* (D. S. Kliger, Ed.). Academic Press, New York.

Fernelius, N. C. (1980). "Extension of the Rosencwaig-Gersho photoacoustic spectroscopy theory to include effects of a sample coating," *J. Appl. Phys.* **51**.

Fournier, D., Boccara, A. C., and Badoz, J. (1982). "Photothermal deflection Fourier transform spectroscopy: a tool for high-sensitivity absorption and dichroism measurements," *Appl. Opt.* **21**, 74.

Fournier, D., Boccara, A. C., Skumanic, A., and Amer, N. (1986). "Photothermal investigation of transport in semiconductors: theory and experiment," *J. Appl. Phys.* **59**, 787.

Grice, K. R., Inglehart, L. J., Favro, L. O., Kuo, P. K., and Thomas, R. L. (1983). "Thermal wave imaging of closed slanted craks in opaque solids," *J. Appl. Phys.* **54**, 6245.

Jackson, W. B., Amer, N. M., Boccara, A. C., and Fournier, D. (1981) "Photothermal deflection spectroscopy and detection," *Appl. Opt.* **20**, 1333.

Kanstad, S. O., and Nordal, P. E. (1986). "Experimental aspects of photothermal radiometry," *Can. J. Phys.* **64**, 1155.

Kuo, P. K., Lin, M. J., Reyes, C. B., Favro, L. D., Thomas, R. L., Kim, D. S., Zhang, Shu-yi, Inglehart, L. J., Fournier, D., Boccara, A. C., and Yacoubi, N. (1986). "Mirage effect measurement of thermal diffusivity. Part I: experiment," *Can. J. Phys.* **64**, 1165.

Lasalle, E. L., Lepoutre, F., and Roger, J. P. (1988). "Probe beam size effects in photothermal deflection experiments," *J. Appl. Phys.* **64**, 1.

Lavacek, M. H. (1980). "Heat conduction in porous materials," *Arch. Mech.* (Varszawa) **32**, 491–504.

Lepoutre, F., Roger, J. P., Fournier, D., and Boccara, A. C. (1983). "Measurement of the temperature field in the gas of a photoacoustic cell," *J. Appl. Phys.* **54**, 4586.

Lepoutre, F., Fournier, D., and Boccara, A. C. (1985). "Nondestructive control of weldings using the Mirage detection," *J. Appl. Phys.* **57**, 1009.

Lepoutre, F., Fournier, D., and Boccara, A. C. (1987). "Influence of 3-D effects on the contrast in NDE by photothermal methods." To be published in *Non-Destructive Evaluation of Materials* (J. F. Bussiere, Ed.).

McDonald, F. A. (1982). "Three-dimensional, one-dimensional and piston models for the photoacoustic effects," *J. of Photoacoustics* **1**, 21.

Mandelbrot, B. (1983). "*The Fractal Geometry of Nature*." Freeman, New York.

Mikoshiba, N., Nakamura, H., and Tsubouchi, K. (1982). "Investigation of nonradiative processes in semiconductors by photoacoustic spectroscopy with ZnO transducer," *Proceedings of the IEEE Ultrasonic Symposium San Diego*, San Diego, CA.

Miranda, L. C. (1982). "Theory of the photoacoustic effect in semiconductors. Influence of carrier diffusion and recombination," *Appl. Opt.* **21**, 2923.

Opsal, J., Taylor, M. W., Smith, W. L., and Rosencwaig, A. (1987). "Temporal behaviour of modulated optical reflectance in silicon," *J. Appl. Phys.* **61**, 240.

Orbach, R. (1986). "Dynamics of fractal networks," *Science* **231**, 814.

Pelzl, J., Fournier, D., and Boccara, A. C. (1986). "Spatially- and time-resolved study of carrier and heat transport in Si by photothermal methods," *Verhandl. DPG* **5**, 1083.

Pelzl, J., Fournier, D., and Boccara, A. C., "Time and spatially resolved investigation of transport in semiconductors using the mirage effect." *5th International Topical Meeting on Photoacoustic and Photothermal Phenomenon* (Heidelberg 1987).

Pynn, R., and Skjeltorp, ed. (1985). "Scaling phenomena in disordered systems," in *NATO ASI Series*. Plenum.

Roger, J. P., Lepoutre, F., Fournier, D., and Boccara, A. C. (1987). "Thermal diffusivity measurement of micron-thick semiconductor films by mirage detection." *Thin Solid Films* **155**, 165.

Rosencwaig, A. and Gersho, A. (1976). "Theory of the photoacoustic effect with solids," *J. Appl. Phys.* **47**, 64.

Royer, D., Dieulesaint, E., and Martin, Y. (1986). "Improved version of polarized beam heterodyne interferometer," *IEEE Ultrason. Symp. Proc., San Francisco* **1**, 432.

Sablikov, V. A., and Sandomirskii, V. A. (1983). "The photoacoustic effect in semiconductors," *Phys. Status Solidi* **B120**, 471.

Smith, R. A. (1978). *Semiconductors*. Cambridge University Press.

Stanley, H. E., and Ostrowsky, N., eds. (1985). "On growth and forms," in *NATO ASI Series*. Nijhoff, Amsterdam.

Stone, J. (1972). "Absorption of light in low loss liquids," *J. Opt. Soc. Amer.* **62**, 327.

Wickramasinghe, H. K., Martin, Y., Spear, D. A., and Ash, E. A. (1983). "Optical heterodyne technics for photoacoustic and photothermal detection" (J. Badoz and D. Fournier Editors), *J. Phys.* **C-6**, 1191.

Yacoubi, N. (1986). "Contribution à l'étude des propriétés thermiques et optiques des semiconducteurs par la méthode photothermique," Ph. D. Thesis, Montpellier.

Zallen, R. (1983). *The Physics of Amorphous Materials*. Wiley, New York.

CHAPTER 3

THE THEORY OF PHOTOTHERMAL EFFECT IN FLUIDS

R. GUPTA

Department of Physics
University of Arkansas
Fayetteville, AR

I. Formation of the Thermal Image 82
A. A Simple Model 82
B. Thermal Diffusion in the Presence of Fluid Flow 84
C. Solution of the Differential Equation for a Pulsed Source 86
D. Solution of the Differential Equation for a CW Source 90
II. Detection of the Thermal Image 93
A. Photothermal Phase-Shift Spectroscopy 94
B. Photothermal Deflection Spectroscopy 101
C. Photothermal Lensing Spectroscopy 112
References 126

In photothermal spectroscopy, absorption of a laser beam (pump beam) results in an increase in the temperature of the irradiated region. The increase in temperature is normally accompanied by a decrease in the refractive index of the medium. The changes in the refractive index can be monitored in several different ways. For example, if the absorbing medium is placed in one beam of an interferometer (or inside a Fabry–Perot cavity), the change in the refractive index would cause a fringe shift that can conveniently be detected as an intensity change of the central fringe. The nonuniform refractive index produced by the absorption of the pump beam can also be detected by the deflection of a probe laser beam passing through this medium. The nonuniform refractive index may also produce a lensing effect, that is, a probe beam passing through the medium (as well as the pump beam itself) may change shape. This can be detected as a change in the intensity of the probe (or the pump) beam passing through a pinhole.

The theory of photothermal spectroscopy of fluids naturally breaks down into two parts: formation of the thermal image and the optical detection of the thermal image. The former subject is discussed in Section I, while the latter is discussed in Section II.

ISBN 0-12-636345-5

I. Formation of the Thermal Image

In this section, an expression for the temperature distribution $T(r, t)$ produced by the absorption of a laser beam (pump beam) and the subsequent evolution of this temperature distribution with time will be derived. A derivation based on a simple model appropriate for a short laser pulse (Gupta, 1987) will be given in Section I.A. This model has the advantage over the rigorous solution that it gives good physical insight into the problem. A rigorous solution to the problem (Rose, Vyas, and Gupta, 1986) will be given in Sections I.B and I.C. A solution for the case of a continuous-wave (CW) laser beam (Vyas *et al.*, 1988) will be derived in Section I.D.

A. A Simple Model

Let us assume that the pump-laser beam propagates through the medium in the z-direction, and the medium is flowing with velocity v_x in the x-direction. The pump beam is assumed to be centered at the origin of the coordinate system. If we assume that the medium is weakly absorbing, the heat produced per unit time per unit volume by the absorption of laser energy is given by

$$Q(x, y, t) = \alpha I(x, y, t) = \begin{cases} \dfrac{2\alpha E_0}{\pi a^2 t_0} e^{-2(x^2+y^2)/a^2} & \text{for } 0 \le t \le t_0 \\ 0 & \text{for } t > t_0, \end{cases} \tag{1}$$

where α is the absorption coefficient of the medium, and $I(x, y, t)$ is the intensity of the laser beam. Implicit in Eq. (1) is the assumption that all of the absorbed laser energy is converted to heat and only a negligible portion is radiated as fluorescence. It is further assumed that this conversion of optical energy into heat takes place on a time scale much shorter than the typical thermal diffusion and convection times in that medium. The total energy in each laser pulse is assumed to be E_0. The spatial profile of the beam is assumed to be a Gaussian with $1/e^2$-radius a. It is further assumed that the laser pulse turns on sharply at $t = 0$ and turns off sharply at $t = t_0$. The assumption of a rectangular temporal profile is a good one if the rise and the fall times of the laser pulse are very short compared to the thermal diffusion and convection times.

In this simple model, we assume the laser pulse to be so short that during the period that the laser pulse is on, no appreciable heat is transferred out of the laser-irradiated region either by thermal diffusion or by the flow of the medium. The temperature rise of the laser-irradiated region above the ambient is then given by

$$\begin{aligned} T(x, y) &= \frac{1}{\rho C_p} \int_0^{t_0} Q(x, y, t)\, dt \\ &= \frac{2\alpha E_0}{\pi a^2 \rho C_p} e^{-2(x^2+y^2)/a^2}. \end{aligned} \tag{2}$$

Here ρ is the density and C_p is the specific heat at constant pressure of the medium. Equation (2) shows that a cylindrical volume of heated medium with a Gaussian cross-section is produced by the absorption of the laser pulse. After the laser pulse is over, this "line of heat" moves downstream with the flow of the medium, and as it does so, it broadens spatially due to thermal diffusion. By ignoring the effect of the thermal diffusion for the moment, the spatial and temporal behavior of the heated region may simply be written by replacing x by $(x - v_x(t - t_0))$ as

$$T(x, y, t) = \frac{2\alpha E_0}{\pi a^2 \rho C_p} e^{-2[(x - v_x(t-t_0))^2 + y^2]/a^2}. \tag{3}$$

Now consider the effect of the thermal diffusion. At $t = t_0$, the laser-irradiated region has a Gaussian spatial profile with a $1/e^2$-radius a. At $t > t_0$, the heated region broadens, and its new radius $a'(t)$ is given by

$$a'(t) = \left[a^2 + \eta D(t - t_0)\right]^{1/2}, \tag{4}$$

where D is the thermal diffusivity of the medium (m^2/s) and η is a constant. Equation (4) has been written down simply from dimensional arguments. Rigorous calculations show that $\eta = 8$ (see Section I.C). Therefore, to take the effect of thermal diffusion into account, the radius of the heated region a in Eq. (3) must be replaced with $a'(t)$. The result is

$$T(x, y, t) = \frac{2\alpha E_0}{\pi\left[a^2 + 8D(t - t_0)\right]\rho C_p} e^{-2\{[x - v_x(t-t_0)]^2 + y^2\}/[a^2 + 8D(t-t_0)]}. \tag{5}$$

Figure 1 shows the spatial profile of the heat pulse in a flowing gas at various times after the end of the laser pulse, as calculated from Eq. (5)

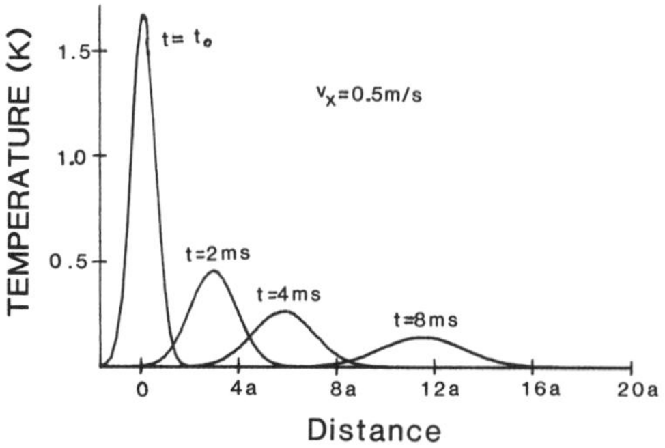

FIG. 1. Spatial profile of the heat pulse in a flowing gas at different times after the end of the excitation pulse. The medium was assumed to be N_2 at room temperature seeded with 1000 ppm of NO_2 (absorption coefficient $\alpha = 0.39\ \text{m}^{-1}$) and flowing with velocity $v_x = 0.5$ m/s. The pump laser was assumed to have an energy $E_0 = 1$ mJ in a pulse of $t_0 = 1$-μs duration. $(1/e^2)$-radius of the pump beam was 0.35 mm. (From Rose, Vyas, and Gupta, 1986.)

(Rose, Vyas, and Gupta, 1986). The distance has been plotted in units of the $1/e^2$-radius of the pump beam. As expected, the heat pulse moves downstream with the flow, and as it does, it broadens due to thermal diffusion. Various parameters used in this calculation are given in the figure caption.

B. Thermal Diffusion in the Presence of Fluid Flow

In this section, a *general* equation describing the temperature distribution $T(\mathbf{r}, t)$ in a fluid medium will be derived. The fluid is assumed to be moving with velocity $\mathbf{v}$ in an arbitrary direction, and the heat is supplied by a source $Q(\mathbf{r}, t)$. Consider a small element of volume $dV = dx\,dy\,dz$ at x, y, z with temperature at its center $T(x, y, z, t)$, as shown in Fig. 2. To begin with, assume that the medium is stationary and that there is no source of heat present. Let us consider the flow of heat in the x-direction. The temperature at the two boundaries of the volume element perpendicular to the x-direction is given by

$$T_{\pm} = T \pm \left(\frac{\partial T}{\partial x}\right)\frac{dx}{2}. \tag{6}$$

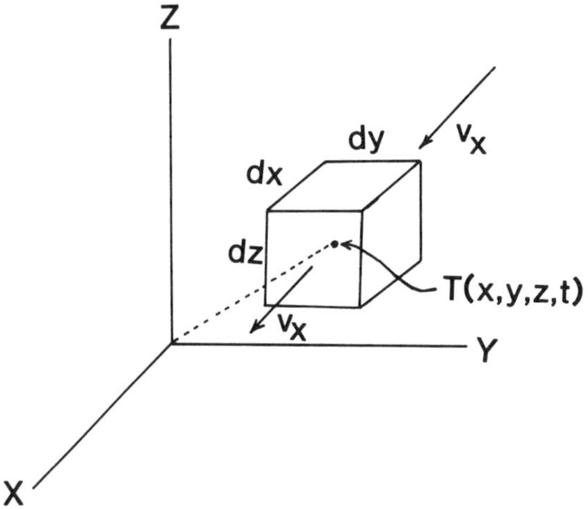

FIG. 2. Transport of heat through a volume element $dx\,dy\,dz$ due to thermal conduction and forced convection.

The rate of heat transport across the two boundaries is then given by

$$\begin{aligned}\frac{dq_\pm}{dt} &= -k\,dA\,\frac{\partial T_\pm}{\partial x}\\ &= -k\,dy\,dz\,\frac{\partial}{\partial x}\left(T \pm \frac{\partial T}{\partial x}\frac{dx}{2}\right),\end{aligned} \tag{7}$$

where k is the thermal conductivity of the medium. The net heat gain of the volume element due to thermal conduction is then the difference between dq_-/dt and dq_+/dt:

$$\frac{dq}{dt} = k\,dV\,\frac{\partial^2 T}{\partial x^2}. \tag{8}$$

Now consider the heat gain of the volume element due to the flow of the medium in the x-direction with velocity component v_x. The rate at which the heat flows in and out of the volume element is given by

$$\left(\frac{dq}{dt}\right)_{\substack{\text{in}\\ \text{out}}} = \rho C_\text{p}\,dy\,dz\,v_x\left(T \mp \frac{\partial T}{\partial x}\frac{dx}{2}\right), \tag{9}$$

where the plus sign applies to heat flow out of the volume element, and the minus sign applies to heat flow into the volume element. The net gain of heat by the volume element due to the flow is then

$$\left(\frac{dq}{dt}\right) = -\rho C_{\mathrm{p}}\, dV \frac{\partial T}{\partial x} v_x. \tag{10}$$

The total heat gain due to thermal conduction and the flow is then given by the sum of Eqs. (8) and (10):

$$\frac{dq}{dt} = k\, dV \frac{\partial^2 T}{\partial x^2} - \rho C_{\mathrm{p}}\, dV \frac{\partial T}{\partial x} v_x. \tag{11}$$

Eq. (11) can easily be generalized to three dimensions:

$$\frac{dq}{dt} = k\, dV\, \nabla^2 T - \rho C_{\mathrm{p}}\, dV\, \mathbf{v} \cdot \nabla T. \tag{12}$$

If there is a source of heat present that deposits an amount of heat $Q(\mathbf{r}, t)$ per unit volume per second, then Eq. (12) modifies to

$$\frac{dq}{dt} = k\, dV\, \nabla^2 T - \rho C_{\mathrm{p}}\, dV\, \mathbf{v} \cdot \nabla T + Q\, dV. \tag{13}$$

Now, since the heat gain of the volume element is related to the temperature rise by

$$\frac{dq}{dt} = \rho C_{\mathrm{p}}\, dV \frac{\partial T}{\partial t}, \tag{14}$$

the final equation for the temperature distribution $T(\mathbf{r}, t)$ is given by

$$\frac{\partial T(\mathbf{r}, t)}{\partial t} = D \nabla^2 T(\mathbf{r}, t) - \mathbf{v} \cdot \nabla T(\mathbf{r}, t) + \frac{1}{\rho C_{\mathrm{p}}} Q(\mathbf{r}, t). \tag{15}$$

The constant $D = k/\rho C_{\mathrm{p}}$ is called the thermal diffusivity of the medium.

C. Solution of the Differential Equation for a Pulsed Source

A rigorous solution of the differential equation, Eq. (15), will now be attempted in order to determine the temperature distribution in the medium. For the conditions described earlier in Section I.A, that is, for the laser beam propagating through the medium in the z-direction and the flow velocity confined to the x-direction, the differential equation for $T(x, y, t)$

is

$$\frac{\partial T(x, y, t)}{\partial t} = D\nabla^2 T(x, y, t) - v_x \frac{\partial T(x, y, t)}{\partial x} + \frac{1}{\rho C_p} Q(x, y, t). \tag{16}$$

The first, second, and third terms on the right in Eq. (16) represent, respectively, the effects of thermal diffusion, fluid flow, and the heating due to pump-beam absorption. The two-dimensional differential equation implies that any inhomogeneities in the medium along the pump beam are negligible. The heat source is assumed to be again given by Eq. (1), i.e., we again assume the medium to be weakly absorbing, with the pump beam having a Gaussian spatial and rectangular temporal profile, etc. In other words, all assumptions leading to Eq. (1) are still valid. The assumption of short laser pulse is now relaxed, however, and the laser pulse width is now arbitrary.

Equation (16) is essentially the equation of conservation of energy. For a general solution to the problem at hand, one must also consider the conservation of mass and momentum. It is, however, not necessary to consider the latter two if the effect of the pressure pulse (photoacoustic effect) accompanying the photothermal signal (Rose, Salamo, and Gupta, 1984) is ignored. The pressure pulse occurs on a much shorter time scale than the thermal diffusion (Heritier, 1983), and the energy carried off by the pressure pulse is negligible (Tam and Coufal, 1983). Therefore, the effect of the pressure pulse on the photothermal signal will be neglected and the conservation of mass and momentum will not be considered.

The following boundary conditions are assumed to hold:

$$\begin{aligned} T(x, y, t)|_{t=0} &= 0 \qquad T'(x, y, t)|_{t=0} = 0, \\ T(x, y, t)|_{x=\pm\infty} &= 0 \qquad T(x, y, t)|_{y=\pm\infty} = 0. \end{aligned} \tag{17}$$

Here T' is the gradient of the temperature. These boundary conditions imply that we have taken the ambient temperature to be zero, and therefore, $T(x, y, t)$ in Eq. (16) may be considered to be the temperature above the ambient.

We shall use the Green's function method (used earlier by Rose, Vyas, and Gupta, 1986) for a solution to Eq. (16). The solution is given by (see for example, Arfken, 1985)

$$T(x, y, t) = \int_{-\infty}^{+\infty}\int_{-\infty}^{+\infty}\int_0^{\infty} Q(\xi, \eta, \tau) G(x/\xi; y/\eta; t/\tau)\, d\xi\, d\eta\, d\tau, \tag{18}$$

where the functional form of $Q(\xi, \eta, \tau)$ is given by Eq. (1). The Green's function satisfies the differential equation

$$-D\nabla_{xy}^2 G + v_x \frac{\partial G}{\partial x} + \frac{\partial G}{\partial t} = \frac{1}{\rho C_p}\,\delta(x-\xi)\,\delta(y-\eta)\,\delta(t-\tau), \quad (19)$$

with the boundary conditions

$$\begin{aligned} G(\pm\infty/\xi;\ y/\eta;\ t/\tau) &= 0, \\ G(x/\xi;\ \pm\infty/\eta;\ t/\tau) &= 0, \\ G(x/\xi;\ y/\eta;0/\tau) &= 0. \end{aligned} \quad (20)$$

The solution to Eq. (19) can be found conveniently by taking the Fourier transform of this equation (Arfken, 1985). The resulting equation in the ω_x, ω_y space is

$$\left(\omega_x^2 + \omega_y^2\right) DG_{\mathrm{F}} - i\omega_x v_x G_{\mathrm{F}} + \frac{\partial G_{\mathrm{F}}}{\partial t} = \frac{1}{2\pi\rho C_p} e^{i(\omega_x \xi + \omega_y \eta)} \delta(t-\tau). \quad (21)$$

Here G_{F} is the Fourier transform of the Green's function. An examination of Eq. (21) shows the simplification afforded by this transformation: Differential operators have been reduced to algebraic factors. Equation (21) can be further simplified by taking its Laplace transform from t-space to its complementary s-space. The result is

$$\left(\omega_x^2 + \omega_y^2\right) DG_{\mathrm{FL}} - i\omega_x v_x G_{\mathrm{FL}} + sG_{\mathrm{FL}} = \frac{1}{2\pi\rho C_p} e^{i(\omega_x \xi + \omega_y \eta)} e^{-s\tau}, \quad (22)$$

where G_{FL} is the Laplace transform of G_{F}. Eq. (22) is simply an algebraic equation with the solution,

$$G_{\mathrm{FL}} = \frac{e^{i\omega_x \xi} e^{i\omega_y \eta} e^{-s\tau}}{2\pi\rho C_p\left[D\left(\omega_x^2 + \omega_y^2\right) - i\omega_x v_x + s\right]}. \quad (23)$$

Now G_{F} can be obtained by taking the inverse Laplace transform of Eq. (23). This can be accomplished conveniently by recognizing that G_{FL} is of the form

$$C\frac{e^{-s\tau}}{s-k}, \quad (24)$$

where both C and k are complex functions, and that (Arfken, 1985)

$$L(e^{kt}) = \frac{1}{s-k}. \tag{25}$$

By using Eqs. (24) and (25), we get

$$\begin{aligned} G_{\mathrm{F}} &= L^{-1}(G_{\mathrm{FL}}) = L^{-1}\left[Ce^{-s\tau}L(e^{kt})\right] \\ &= CH_{\tau}(t)e^{k(t-\tau)}, \end{aligned} \tag{26}$$

where $H_{\tau}(t)$ is the unit step function:

$$H_{\tau}(t) = \begin{cases} 0 & \text{for} \quad 0 \leq t < \tau \\ 1 & \text{for} \quad t \geq \tau. \end{cases} \tag{27}$$

By inserting the values of C and k in Eq. (26), we obtain

$$G_{\mathrm{F}} = \frac{e^{i\omega_x \xi} e^{i\omega_y \eta} H_{\tau}(t)}{2\pi\rho C_{\mathrm{p}}} e^{[i\omega_x v_x - (\omega_x^2 + \omega_y^2)D](t-\tau)}. \tag{28}$$

The original Green's function is now obtained by taking the inverse Fourier transform of Eq. (28):

$$\begin{aligned} G = {} & \frac{H_{\tau}(t)}{4\pi^2\rho C_{\mathrm{p}}} \int_{-\infty}^{\infty} e^{i\omega_x(\xi + v_x(t-\tau))} e^{-\omega_x^2 D(t-\tau)} e^{-i\omega_x x}\, d\omega_x \\ & \times \int_{-\infty}^{\infty} e^{i\omega_y \eta} e^{-\omega_y^2 D(t-\tau)} e^{-i\omega_y y}\, d\omega_y. \end{aligned} \tag{29}$$

Both integrals in Eq. (29) can be evaluated conveniently by use of the convolution theorem (Arfken, 1985):

$$\int_{-\infty}^{\infty} F(\omega_x) G(\omega_x) e^{-i\omega_x x}\, d\omega_x = \int_{-\infty}^{\infty} f(\lambda) g(x-\lambda)\, d\lambda, \tag{30}$$

where functions f and g are the inverse Fourier transforms of the functions F and G, respectively. We obtain

$$G = \frac{H_{\tau}(t)}{4\pi\rho C_{\mathrm{p}} D(t-\tau)} e^{-[x-(\xi+v_x(t-\tau))]^2/4D(t-\tau)} e^{-[y-\eta]^2/4D(t-\tau)}, \tag{31}$$

where we have used the following two relationships,

$$\frac{1}{2\pi}\int_{-\infty}^{\infty} e^{i\omega_x(t-\lambda)}\, d\omega_x = \delta(t-\lambda)$$

and

$$\frac{1}{\sqrt{2\pi}} \int_{-\infty}^{\infty} \left(\frac{1}{b\sqrt{2}} e^{-\omega_x^2/4b^2} \right) e^{-i\omega_x \lambda} \, d\omega_x = e^{-b^2\lambda^2}.$$

Finally, the temperature distribution $T(x, y, t)$ is obtained by substituting the Green's function, Eq. (31), in Eq. (18). Integrations over ξ and η can be performed in a straightforward manner, with the result

$$T(x, y, t) = \frac{2\alpha E_0}{\pi t_0 \rho C_p} \int_0^{t_0} \frac{e^{-2\{[x - v_x(t-\tau)]^2 + y^2\}/\{8D(t-\tau) + a^2\}}}{\left[8D(t - \tau) + a^2\right]} \, d\tau. \tag{32}$$

for $t > t_0$.

Unfortunately, the integration over τ cannot be performed analytically and must be done numerically. This integration may be conveniently performed by using one of the commonly available subroutines in IMSL (International Mathematical and Statistical Library).

One very useful case in which the temperature distribution $T(x, y, t)$ can be written in a closed form is when the laser pulse is very short. In this case,

$$\lim_{t_0 \to 0} \int_0^{t_0} f(\tau) \, d\tau = f(0) t_0, \tag{33}$$

and

$$T(x, y, t) = \frac{2\alpha E_0}{\pi \rho C_p (8Dt + a^2)} e^{-2[(x - v_x t)^2 + y^2]/[a^2 + 8Dt]}. \tag{34}$$

This result is in agreement with that of a simple model, Eq. (5), and the results are shown in Fig. 1. Rose, Vyas, and Gupta (1986) have investigated the range of validity of Eq. (34) by comparing the prediction of Eq. (34) with that of Eq. (32) for different laser pulse widths t_0. They conclude that for pulse widths $t_0 \leq 10\ \mu s$, Eq. (34) gives almost exact agreement with Eq. (32), but caution must be used for longer pulse widths.

D. Solution of the Differential Equation for a CW Source

In many situations, it is more desirable to use a CW laser source for the pump beam. The photothermal signal can be measured conveniently if the CW laser is amplitude-modulated at some frequency ω, because then a phase-sensitive detection can be used, that is, a signal at frequency ω and having a definite phase with respect to the source can be detected. The source term $Q(x, y, t)$ in this case may be written as

$$Q(x, y, t) = \frac{2\alpha P_{av}}{\pi a^2}\left[e^{-2(x^2+y^2)/a^2}\right](1 + \cos \omega t), \tag{35}$$

where P_{av} is the *average* power of the pump beam in the presence of the modulation. The pump laser is sinusoidally modulated with the degree of modulation being 100%. The spatial profile of the pump beam is again assumed to be Gaussian with $1/e^2$-width a. The temperature distribution is again given by the solution of Eq. (16), that is, it is given by Eq. (18) with the functional form of $Q(\xi, \eta, \tau)$ given by Eq. (35). The Green's function is independent of the source term and it is still given by Eq. (31). As a matter of fact, the strength of the Green's function method lies precisely in the fact that a solution for $T(x, y, t)$ can be found for any source term that can be expressed analytically (although the final result may not be in the closed form). In the present case, the result is

$$T(x, y, t) = \frac{2\alpha P_{av}}{\pi \rho C_p} \int_0^t \frac{(1 + \cos \omega\tau)}{[a^2 + 8D(t-\tau)]} \times e^{-2\{[x - v_x(t-\tau)]^2 + y^2\}/[a^2 + 8D(t-\tau)]}\, d\tau. \tag{36}$$

The integration over τ must again be performed numerically.

A few typical temperature distributions produced by a CW laser beam are shown in Fig. 3 (Vyas *et al.*, 1988). The integration in Eq. (36) was performed by the method of 64-point Gaussian quadrature (Vyas *et al.*, 1988). In this calculation, the modulation frequency ω was set equal to zero. The temperature $T(x, y)$ has been plotted as a function of the distance x from the center of the pump beam. This distance has been expressed in units of the $1/e^2$-radius a. Positive x is measured in the direction of the gas flow. Because we have assumed our medium to have no boundaries, the temperature of the medium never attains a true steady state. However, for points close to the pump beam, a quasi-steady state is established, that is, the rate of temperature increase becomes negligible after a certain time (a few seconds for a stationary medium). The approach of this quasi-steady state becomes faster as the flow velocity of the medium increases.

Monson, Vyas, and Gupta (1988) have discussed the temporal evolution of the temperature in more detail. The curves shown in Fig. 3 have been computed for $t = 5$ sec, when a quasi-steady state has been attained. The bottom curve is for a stationary medium ($v_x = 0$), and the other curves are for $v_x = 1$ cm/s, 10 cm/s, and 1 m/s, as labeled. All other relevant parameters used in this calculation are given in the figure caption. For $v_x = 0$, the temperature distribution extends far outside the laser beam due to thermal diffusion. As the flow velocity increases, the temperature distribution becomes more and more asymmetric, and the peak value of the temperature rise becomes smaller because the heat is carried by the medium in the direction of the flow.

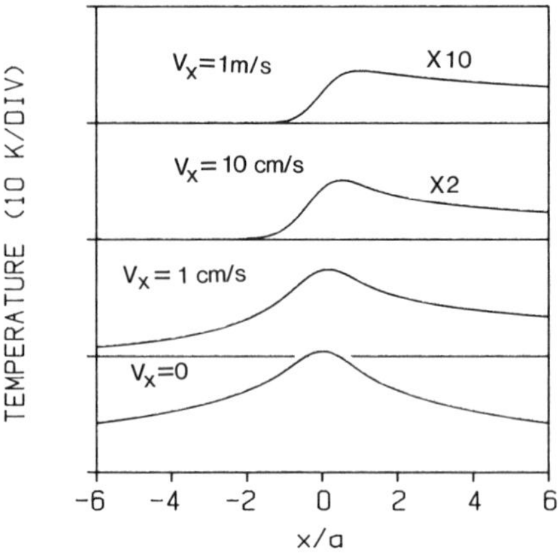

FIG. 3. Temperature distribution in a medium for zero-modulation frequency and for $v_x = 0$, 1 cm/sec, 10 cm/sec, and 1 m/sec, as noted on the diagram. In this computation, the medium was assumed to be N_2 seeded with 1000 ppm of NO_2 ($\alpha = 0.39\ m^{-1}$), laser power was 1 W, and the pump-beam radius a was taken to be 0.5 mm. Abscissa is plotted in units of a (0.5 mm). (From Vyas *et al.*, 1988.)

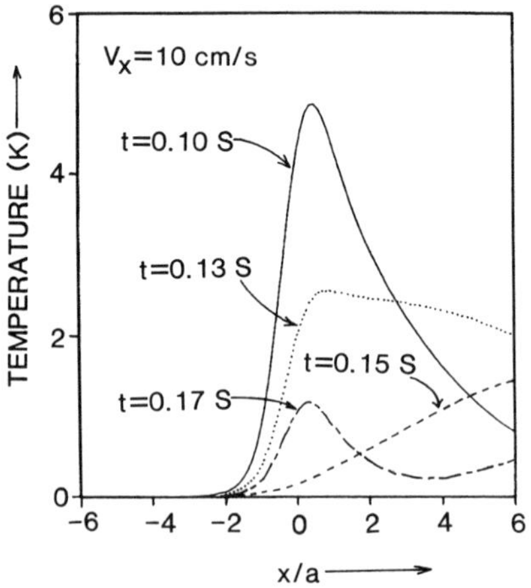

FIG. 4. Variation of the temperature distribution with time in a medium with flow velocity $v_x = 10$ cm/s, when the pump beam is modulated at 10 Hz. The four curves represent five different times in the modulation cycle of the pump beam as noted on the diagram. Time $t = 0.10$ sec corresponds to the peak of the laser power, while $t = 0.15$ sec corresponds to the minimum of laser power. All parameters used in this computation are given in the caption to Fig. 3. (From Vyas *et al.*, 1988.)

Figure 4 shows the effect of the modulation on the temperature distribution (Vyas *et al.*, 1988). Four curves are shown in the figure for $f = 10$ Hz and $v_x = 10$ cm/s. The four curves correspond to $t = 0.10$, 0.13, 0.15, and 0.17 sec. Time $t = 0.10$ corresponds to the peak of the laser power, whereas $t = 0.15$ corresponds to the minimum of laser power. As can be seen from these curves, the temperature distribution goes through drastic changes in shape as the laser power oscillates.

II. Detection of the Thermal Image

The change in the refractive index of the medium produced by the absorption of the pump beam may be detected in a number of different ways. For example, a probe-laser beam passing through the sample will experience a change in the optical path length. The change in the optical path length may conveniently be detected as a fringe shift in an interferometer (Stone, 1972; Stone, 1973; Davis, 1980; Davis and Petuchowski, 1981). In this method of detection, the signal is proportional to the change in refractive index $n(\mathbf{r}, t)$, and we shall refer to this technique as photothermal phase-shift spectroscopy (PTPS). PTPS will be discussed in Section II.A. Another convenient method for the detection of the thermal image relies on the measurement of the gradient of the refractive index, $\partial n(\mathbf{r}, t)/\partial r$. The change in the refractive index of the medium due to the absorption of the pump beam follows the spatial profile of the pump beam (which is generally assumed to be a Gaussian). A probe-laser beam passing through the inhomogeneous refractive index suffers a deflection that can easily be measured by a position-sensitive detector (Boccara, Fournier, and Badoz, 1980; Boccara *et al.*, 1980; Fournier *et al.*, 1980; Jackson *et al.*, 1981). This technique is called photothermal deflection spectroscopy (PTDS) and will be discussed in Section II.B. The nonuniform change in the refractive index may also produce a lensing effect in the medium. A probe beam passing through the medium may change shape (the sample may act as a lens) resulting in a change in the intensity of the probe beam passing through a pinhole (Gordon *et al.*, 1965; Hu and Whinnery, 1973). In this technique, the signal is proportional to the second derivative of the refractive index, $\partial^2 n(\mathbf{r}, t)/\partial r^2$. This technique is called the photothermal lensing spectroscopy (PTLS) and is discussed in Section II.C.

For each of the three methods mentioned above, we shall consider two cases, collinear and transverse. Figure 5 shows the typical pump- and probe-beam configuration. The pump beam propagates in the z-direction, and the flow velocity of the medium is in the x-direction. For the transverse case, the probe beam propagates in the y-direction, whereas for the collinear case, the probe beam propagates in the z-direction. Pump- and probe-

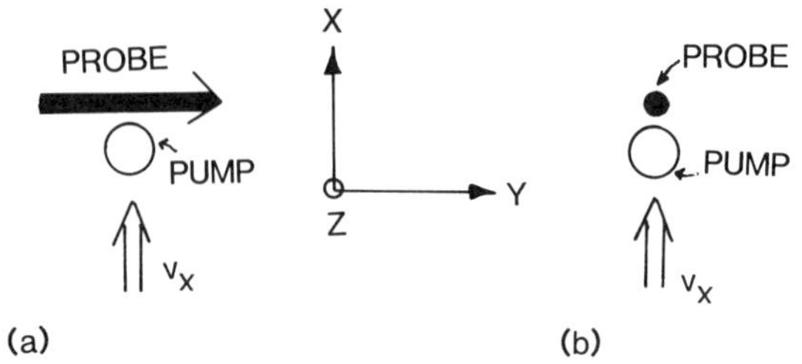

FIG. 5. Pump- and probe-beam configurations for (a) transverse and (b) collinear geometries.

beam axes may or may not intersect. In all cases, the probe-beam radius is assumed to be much smaller than the pump-beam radius.

A. Photothermal Phase - Shift Spectroscopy

In this section, expressions for PTPS signal size and shape for a sample placed in one arm of a Michelson interferometer will be derived. Similar expressions for other types of interferometers (e.g., Fabry–Perot interferometer) can be derived analogously (Campillo *et al.*, 1982). A typical experimental arrangement is shown in Fig. 6(a). M4 and M5 are dielectric mirrors that totally transmit the probe beam and totally reflect the pump beam. Broken lines show transverse PTPS, whereas the solid lines show the collinear PTPS. A general expression for the PTPS signal in terms of the change in refractive index $\Delta n(\mathbf{r}, t)$ is derived in Section II.A.1, and explicit expressions for pulsed and CW PTPS signals are derived in Sections II.A.2 and II.A.3, respectively (Monson, Vyas, and Gupta, 1988).

1. Thermally Induced Phase-Shift and Intensity Change in an Interferometer

Let us assume that the two interfering waves have amplitudes A and B and phases φ_A and φ_B. An additional phase difference $\gamma(t)$ is introduced into one of the waves by the photothermal effect. The interfering waves may be written as

$$a = A\sin(kx - \omega t + \varphi_A)$$

and

$$b = B\sin(kx - \omega t + \varphi_B + \gamma(t)). \tag{37}$$

The resultant wave is then

$$c = a + b = R\sin(kx - \omega t + \delta(t)),$$

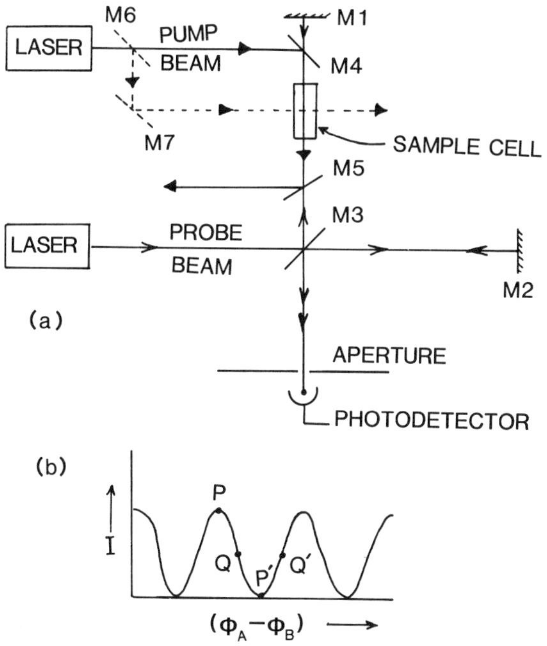

FIG. 6. (a) Schematic illustration of a PTPS experiment. The sample cell is placed in one arm of a Michelson interferometer. The pump beam passes through the cell either collinearly (solid line) or transversely (dotted line). (b) Intensity variation observed at the photodetector as a function of the phase difference $(\varphi_A - \varphi_B)$.

where R and $\delta(t)$ are given by

$$R \cos \delta(t) = A \cos \varphi_A + B \cos(\varphi_B + \gamma(t)),$$

and

$$R \sin \delta(t) = A \sin \varphi_A + B \sin(\varphi_B + \gamma(t)).$$

The intensity of resultant wave is then given by

$$I(t) = R^2 = A^2 + B^2 + 2AB \cos[(\varphi_A - \varphi_B) - \gamma(t)]. \qquad (38)$$

Here $(\varphi_A - \varphi_B)$ is the constant phase difference between the two waves (in the absence of the photothermal signal). The "operating point" on the intensity curve is determined by $(\varphi_A - \varphi_B)$, as shown in Fig. 6(b). If $(\phi_B - \varphi_B) = m\pi$, the operating point is either P or P', and the intensity is quite insensitive to small changes in $\gamma(t)$. This is the least desirable situation. On the other hand, if $(\varphi_A - \varphi_B) = (m + 1/2)\pi$, the operating point is either Q or Q', and the intensity is quite sensitive to small changes in $\gamma(t)$. This is the most desirable situation. Let us now determine $\gamma(t)$ in terms of the temperature change T produced by the absorption of the pump beam:

$$\gamma(x, y, t) = \frac{4\pi}{\lambda} \int_{\text{path}} \Delta n\,(x, y, t)\, ds, \tag{39}$$

where the integration has been carried out over the path of the probe beam, and we have assumed the probe beam to be infinitesimally thin. Since

$$\Delta n(x, y, t) = (n_0 - 1)\frac{T(x, y, t)}{T_\text{A}}, \tag{40}$$

where T_A is the ambient temperature and n_0 is the refractive index at the ambient temperature, Eq. (39) becomes

$$\gamma(x, y, t) = \frac{4\pi}{\lambda}\frac{(n_0 - 1)}{T_\text{A}} \int_{\text{path}} T(x, y, t)\, ds. \tag{41}$$

Assuming that we have chosen our operating point to be such that $(\varphi_A - \varphi_B) = (m + 1/2)\pi$, we get for the signal:

$$\delta V(x, y, t) \propto 2AB \sin\left(\frac{4\pi}{\lambda}\frac{(n_0 - 1)}{T_\text{A}} \int_{\text{path}} T(x, y, t)\, ds\right). \tag{42}$$

One convenient way to calibrate the detector is to simply note the difference between the maximum and the minimum voltage at the detector, V, as $(\varphi_A - \phi_B)$ is taken through a change of π (in the absence of the thermal signal). From Eq. (38), V is given by

$$V = V_{\text{max}} - V_{\text{min}} \propto 4AB. \tag{43}$$

Equation (42), along with Eq. (43), then gives the desired result,

$$\delta V(x, y, t) = \tfrac{1}{2} V \sin\left(\frac{4\pi}{\lambda}\frac{(n_0 - 1)}{T_\text{A}} \int_{\text{path}} T(x, y, t)\, ds\right). \tag{44}$$

2. Pulsed Photothermal Phase-Shift Spectroscopy

The collinear PTPS signal is given by

$$\begin{aligned} \delta V_\text{L}(x, y, t) &= \tfrac{1}{2} V \sin\left(\frac{4\pi}{\lambda}\frac{(n_0 - 1)}{T_\text{A}} \int_{\text{path}} T(x, y, t)\, dz\right) \\ &= \tfrac{1}{2} V \sin\left(\frac{4\pi}{\lambda}\frac{(n_0 - 1)l}{T_\text{A}} T(x, y, t)\right), \end{aligned} \tag{45}$$

where l is the length of interaction between the pump and the probe beams. By substituting for $T(x, y, t)$ from Eq. (32), we get

$$\delta V_{\mathrm{L}}(x, y, t) = \tfrac{1}{2} V \sin\left(\frac{4\pi}{\lambda} \frac{(n_0 - 1)l}{T_{\mathrm{A}}} \frac{2\alpha E_0}{\pi t_0 \rho C_{\mathrm{p}}} \times \int_0^{t_0} \frac{e^{-2[(x - v_x(t-\tau))^2 + y^2]/[a^2 + 8D(t-\tau)]}}{[a^2 + 8D(t-\tau)]} \, d\tau \right). \tag{46}$$

One very useful special case where one can write the result in closed form is when the laser pulse length t_0 is very short. In this case, by using Eq. (33), we get

$$\delta V_{\mathrm{L}}(x, y, t) = \tfrac{1}{2} V \sin\left(\frac{4\pi}{\lambda} \frac{(n_0 - 1)l}{T_{\mathrm{A}}} \frac{2\alpha E_0}{\pi \rho C_{\mathrm{p}}} \frac{e^{-2[(x - v_x t)^2 + y^2]/(a^2 + 8Dt)}}{(a^2 + 8Dt)} \right). \tag{47}$$

The transverse PTPS signal is given by

$$\delta V_{\mathrm{T}}(x, t) = \tfrac{1}{2} V \sin\left(\frac{4\pi}{\lambda} \frac{(n_0 - 1)}{T_{\mathrm{A}}} \int T(x, y, t) \, dy \right), \tag{48}$$

where $T(x, y, t)$ is given by Eq. (32). The y-integration is to be carried out over the interaction length of the pump and the probe beams. However, since the y-integrand is nearly zero outside this interaction length, the integration may be performed from $-\infty$ to $+\infty$. The result is

$$\delta V_{\mathrm{T}}(x, t) = \tfrac{1}{2} V \sin\left(\frac{4\pi}{\lambda} \frac{(n_0 - 1)}{T_{\mathrm{A}}} \frac{2\alpha E_0}{\sqrt{2\pi}\, t_0 \rho C_{\mathrm{p}}} \times \int_0^{t_0} \frac{e^{-2[(x - v_x(t-\tau))^2]/[a^2 + 8D(t-\tau)]}}{[a^2 + 8D(t-\tau)]^{1/2}} \, d\tau \right). \tag{49}$$

As in the collinear case, for a short laser pulse, this result may be written in a closed form using Eq. (33) as

$$\delta V_{\mathrm{T}}(x, t) = \tfrac{1}{2} V \sin\left(\frac{4\pi}{\lambda} \frac{(n_0 - 1)}{T_{\mathrm{A}}} \frac{2\alpha E_0}{\sqrt{2\pi}\, \rho C_{\mathrm{p}}} \frac{e^{-2[x - v_x t]^2/[a^2 + 8Dt]}}{[a^2 + 8Dt]^{1/2}} \right). \tag{50}$$

Figure 7 shows the PTPS signals for the collinear case calculated by using Eq. (46) for a stationary medium ($v_x = 0$) for the parameters given in the

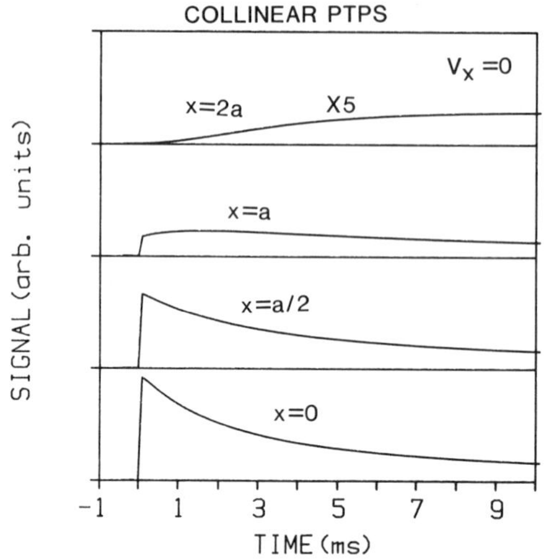

FIG. 7. Typical pulsed collinear PTPS signals in a stationary medium for several pump–probe separations (expressed in units of a); $\sin\gamma(t)$ has been plotted as a function of time, and one division corresponds to $\sin\gamma = 1$. The top curve has been expanded by a factor of 5 for clarity. Parameters used in this computation are $t_0 = 1\ \mu s$, $E_0 = 6$ mJ, $\alpha = 0.39\ m^{-1}$, $a = 0.5$ mm, $l = 1$ cm, $T_A = 300$ K, and $\lambda = 490$ nm. (From Monson, Vyas, and Gupta, 1988.)

figure caption (Monson, Vyas, and Gupta, 1988). Four curves are shown, for $x = 0$, $a/2$, a, and $2a$. The signal has the largest amplitude for $x = 0$, since the change in the refractive index is the maximum at this position. The intensity of light at the detector suffers a transient change when the pump laser is fired and returns to its original value as the heat diffuses out of the region. As the distance between the pump and the probe beams is increased, the signal becomes smaller and broader. When the probe beam is outside the pump beam ($x = 2a$), the heat arrives at the probe-beam position via thermal diffusion, and therefore the peak signal occurs later in time and is weak and broad. PTPS signals for the transverse case in a stationary medium have the same general shape as those in Fig. 7, except that they are broader and, of course, are smaller in amplitude due to smaller interaction length (Monson, Vyas, and Gupta, 1988).

Signal shapes for a flowing medium (collinear case) are shown in Fig. 8. The flow velocity of the medium is assumed to be $v_x = 2$ m/s, and all other parameters used in this calculation are the same as those used for Fig. 7. A positive value of x corresponds to the probe being downstream from the pump beam. Signal shape is essentially the spatial profile of the pump beam broadened by thermal diffusion. Transverse PTDS signals have essentially the same shape as the collinear ones, except for a smaller amplitude.

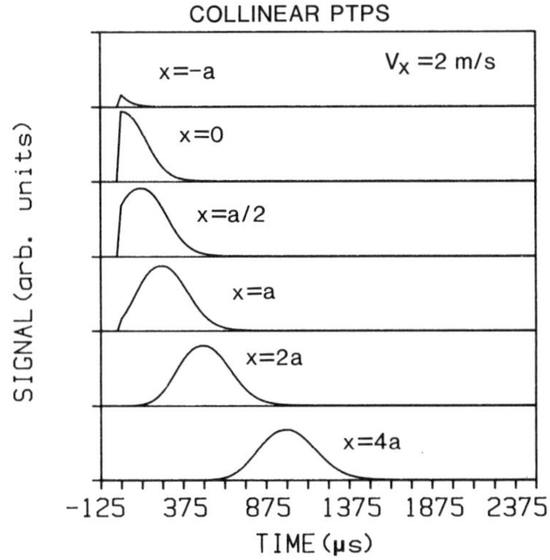

FIG. 8. Pulsed collinear PTPS signals in a flowing medium ($v_x = 2$ m/s). One division corresponds to $\sin \gamma(t) = 1$. All parameters used in this computation are given in the caption to Fig. 7. (From Monson, Vyas, and Gupta, 1988.)

3. CW Photothermal Phase-Shift Spectroscopy

Substitution of the equation for temperature distribution, Eq. (36), in Eq. (45) leads to the expression for collinear CW PTPS:

$$\delta V_L(x, y, t) = \tfrac{1}{2} V \sin\left(\frac{4\pi}{\lambda} \frac{(n_0 - 1)l}{T_A} \frac{2\alpha P_{av}}{\pi \rho C_p} \int_0^t \frac{(1 + \cos \omega\tau)}{[a^2 + 8D(t - \tau)]} \times e^{-2\{[x - v_x(t-\tau)]^2 + y^2\}/[a^2 + 8D(t-\tau)]} \, d\tau \right). \tag{51}$$

Substitution of Eq. (36) into Eq. (48) and integration over y leads to the expression for transverse CW PTPS signal:

$$\delta V_T(x, t) = \tfrac{1}{2} V \sin\left(\frac{4\pi}{\lambda} \frac{(n_0 - 1)}{T_A} \frac{2\alpha P_{av}}{\sqrt{2\pi}\, \rho C_p} \int_0^t \frac{(1 + \cos \omega\tau)}{[a^2 + 8D(t - \tau)]^{1/2}} \times e^{-2[x - v_x(t-\tau)]^2/[a^2 + 8D(t-\tau)]} \, d\tau \right). \tag{52}$$

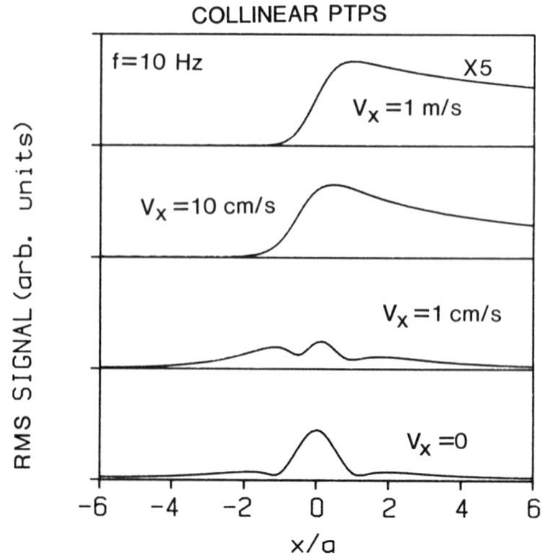

FIG. 9. CW collinear PTPS signals as a function of the pump–probe distance for four different flow velocities of the medium. RMS values of $\sin \gamma(t)$ have been plotted for a modulation frequency of 10 Hz and an interaction length of 1 cm. All other parameters are given in the caption of Fig. 3. One division corresponds to the rms value of $\sin \gamma = 0.5$. The top curve has been expanded by a factor of 5. (From Monson, Vyas, and Gupta, 1988.)

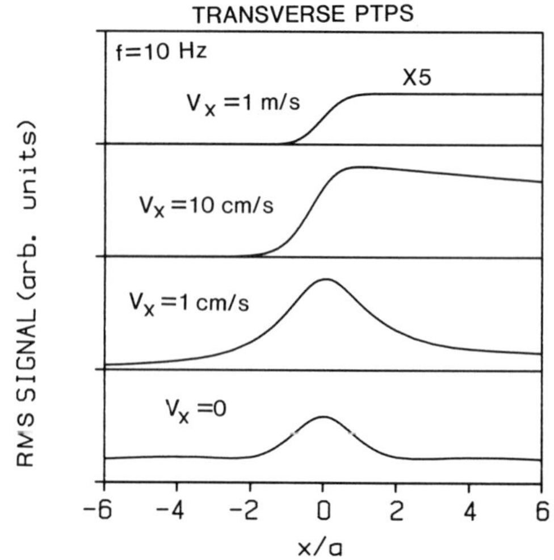

FIG. 10. CW transverse PTPS signals computed with the same parameters as in Fig. 9. One division corresponds to the rms value of $\sin \gamma = 0.05$. The top curve has been expanded by a factor of 5 for clarity. (From Monson, Vyas, and Gupta, 1988.)

Some of the predictions of CW collinear PTPS are shown in Fig. 9. RMS values of the PTPS signals have been plotted against the distance between the pump and the probe beams, x, for a modulation frequency of 10 Hz. This distance has been expressed in units of the $1/e^2$-radius of the pump beam. The distance x is taken to be positive downstream. Four curves are shown for $v_x = 0$, 1 cm/sec, 10 cm/sec, and 1 m/sec. As the velocity increases, these curves become more and more asymmetric, with the signal extending far to the right side (downstream). Low- and zero-velocity curves show interesting undulations. These undulations are a direct consequence of the change in temperature distribution as the pump-beam intensity oscillates (see Fig. 4 and Monson, Vyas, and Gupta, 1988). Figure 10 shows curves analogous to those of Fig. 9 for transverse PTPS, that is, predictions of Eq. (52). These curves have no undulations, and of course, they are smaller in amplitude.

B. Photothermal Deflection Spectroscopy

In this section, expressions for the deflection of a probe-laser beam when passing through a medium with an inhomogeneous refractive index created by the absorption of the pump-laser beam will be derived. A general ray equation describing the propagation of light in an inhomogeneous medium will be derived in Section II.B.1, explicit expressions for the pulsed photothermal signal (Rose, Vyas, and Gupta, 1986) will be derived in Section II.B.2, and those for the CW photothermal signals (Vyas *et al.*, 1988) will be derived in Section II.B.3.

1. The Ray Equation

We wish to derive an equation for the propagation of an optical ray between points A and B if the (inhomogeneous) refractive index of the medium in this region is $n(\mathbf{r})$. From Fermat's principle, the ray path $\mathbf{s}$ would be such that the optical path length between A and B would be an extremum, that is,

$$\delta \int_A^B n(\mathbf{r})\, ds = 0. \tag{53}$$

Here,

$$ds = \left(dx^2 + dy^2 + dz^2\right)^{1/2} = \left[1 + \dot{x}^2 + \dot{y}^2\right]^{1/2} dz, \tag{54}$$

where the dots represent differentiation with respect to z. Therefore Eq. (53) becomes

$$\delta \int_A^B n(x, y, z)(1 + \dot{x}^2 + \dot{y}^2)\, dz = 0. \tag{55}$$

Eq. (55) will be solved using Lagrangian mechanics (Ghatak and Thyagarajan, 1978). According to Hamilton's principle, the trajectory of a particle between times t_1 and t_2 is given by (Goldstein, 1959)

$$\delta \int_{t_1}^{t_2} L\, dt = 0, \tag{56}$$

where L is the Lagrangian of the particle. Comparison of Eq. (56) with Eq. (55) shows that the Lagrangian in this case is

$$L = n(x, y, z)(1 + \dot{x}^2 + \dot{y}^2)^{1/2}, \tag{57}$$

and z plays the role played by t in Eq. (56). The Lagrangian satisfies the equations

$$\frac{d}{dz}\left(\frac{\partial L}{\partial \dot{x}}\right) = \frac{\partial L}{\partial x} \tag{58a}$$

and

$$\frac{d}{dz}\left(\frac{\partial L}{\partial \dot{y}}\right) = \frac{\partial L}{\partial y}. \tag{58b}$$

By substituting Eq. (57) in Eq. (58a), we get

$$\frac{1}{(1 + \dot{x}^2 + \dot{y}^2)^{1/2}} \frac{d}{dz}\left(\frac{n}{(1 + \dot{x}^2 + \dot{y}^2)^{1/2}} \frac{dx}{dz}\right) = \frac{\partial n}{\partial x}.$$

By using Eq. (54), this can be written as

$$\frac{d}{ds}\left(n \frac{dx}{ds}\right) = \frac{\partial n}{\partial x}. \tag{59a}$$

Similarly, by using (58b), one gets

$$\frac{d}{ds}\left(n \frac{dy}{ds}\right) = \frac{\partial n}{\partial y}. \tag{59b}$$

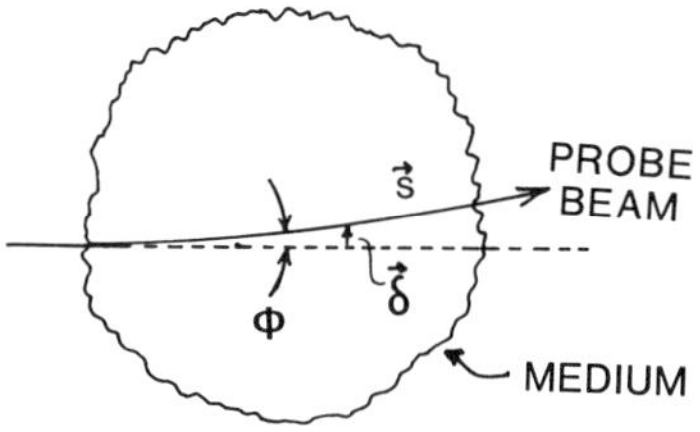

FIG. 11. Diagram showing the relationship between the probe-beam path s, perpendicular displacement δ, and the deflection angle φ.

By taking the z-direction to be *approximately* the ray direction (see Fig. 11), Eq. (59) may be written as

$$\frac{d}{ds}\left(n\frac{d\boldsymbol{\delta}}{ds}\right) = \nabla_{\perp} n, \tag{60}$$

where $\boldsymbol{\delta}$ is the perpendicular displacement of the beam from its original direction as shown in Fig. 11, and $\nabla_{\perp} n$ is the gradient of the refractive index perpendicular to the beam path. Equation (60) is the ray equation we had set out to derive.

The refractive index $n(\mathbf{r}, t)$ is a function of both space and time. We may replace $n(\mathbf{r}, t)$ by n_0, where n_0 is the uniform (unperturbed) refractive index of the medium on the left-hand side of Eq. (60):

$$\frac{d}{ds}\left(n_0\frac{d\boldsymbol{\delta}}{ds}\right) = \nabla_{\perp} n(\mathbf{r}, t). \tag{61}$$

This approximation is justified because any variations of n along s will not deflect the beam. The refractive index $\mathbf{n}(r, t)$ is related to the unperturbed refractive index n_0 by

$$n(\mathbf{r}, t) = n_0 + \left.\frac{\partial n}{\partial T}\right|_{T_A} T(\mathbf{r}, t), \tag{62}$$

where T_A stands for the ambient temperature. Using Eq. (62) in Eq. (61), we get

$$\frac{d\boldsymbol{\delta}}{ds} = \frac{1}{n_0}\frac{\partial n}{\partial T}\int_{\text{path}} \nabla_{\perp} T(\mathbf{r}, t)\, ds, \tag{63}$$

where the integration is carried out over the path of the probe beam. For our geometry (Fig. 5), and for small deflections, the deflection angle φ in the x-direction may be written as

$$\varphi(x, y, t) = \frac{1}{n_0} \frac{\partial n}{\partial T} \int_{\text{path}} \frac{\partial T(x, y, t)}{\partial x} \, ds. \tag{64}$$

Again, two cases will be considered: the probe beam propagating in the y-direction shown in Fig. 5a (transverse PTDS), and the probe beam propagating in the z-direction shown in Fig. 5b (collinear PTDS). In either case, deflection of the probe beam in the x-direction, φ, is observed. The general case, where the pump and probe beams make an arbitrary angle θ with respect to each other, has been considered by Rose, Vyas, and Gupta (1986) and will not be considered here. The probe beam is again considered to be infinitesimally thin. For transverse and collinear PTDS, Eq. (64) reduces, respectively, to

$$\varphi_{\mathrm{T}}(x, t) = \frac{1}{n_0} \frac{\partial n}{\partial T} \int \frac{\partial T(x, y, t)}{\partial x} \, dy \tag{65}$$

and

$$\varphi_{\mathrm{L}}(x, y, t) = \frac{1}{n_0} \frac{\partial n}{\partial T} \int \frac{\partial T(x, y, t)}{\partial x} \, dz. \tag{66}$$

2. Pulsed Photothermal Deflection

In this section, explicit expressions for pulsed photothermal deflection will be derived by using Eq. (32) in Eqs. (65) and (66). By differentiating Eq. (32) with respect to x inside the τ-integral, we obtain

$$\frac{\partial T(x, y, t)}{\partial x} = -\frac{8\alpha E_0}{\pi t_0 \rho C_{\mathrm{p}}} \int_0^{t_0} \frac{x - v_x(t-\tau)}{[8D(t-\tau) + a^2]^2} \times e^{-2\{[x - v_x(t-\tau)]^2 + y^2\}/\{8D(t-\tau) + a^2\}} \, d\tau \quad \text{for} \quad t > t_0. \tag{67}$$

Collinear PTDS is then given by

$$\varphi_{\mathrm{L}}(x, y, t) = -\frac{l}{n_0} \frac{\partial n}{\partial T} \frac{8\alpha E_0}{\pi t_0 \rho C_{\mathrm{p}}} \int_0^{t_0} \frac{\{x - v_x(t-\tau)\}}{\{8D(t-\tau) + a^2\}^2} \times e^{-2\{[x - v_x(t-\tau)]^2 + y^2\}/[8D(t-\tau) + a^2]} \, d\tau, \tag{68}$$

where l is the interaction length. Transverse PTDS is given by

$$\varphi_{\mathrm{T}}(x,t) = -\frac{1}{n_0}\frac{\partial n}{\partial T}\frac{8\alpha E_0}{\pi t_0 \rho C_{\mathrm{p}}}\int_0^{t_0} d\tau\left[\int_{-\infty}^{\infty} e^{-2y^2/[8D(t-\tau)+a^2]}\,dy\right]$$
$$\times \frac{x - v_x(t-\tau)}{[8D(t-\tau)+a^2]^2} e^{-2[x-v_x(t-\tau)]^2/[8D(t-\tau)+a^2]}. \quad (69)$$

The integrand in the y-integral vanishes outside an interaction length, which is of the order of $(a^2 + 8Dt)^{1/2}$. It is therefore permissible to choose the limits as $\pm\infty$ on this integral. The integral can be solved analytically, with the result

$$\varphi_{\mathrm{T}}(x,t) = -\frac{1}{n_0}\frac{\partial n}{\partial T}\frac{8\alpha E_0}{\sqrt{2\pi}\, t_0 \rho C_{\mathrm{p}}}\int_0^{t_0} \frac{[x - v_x(t-\tau)]}{[8D(t-\tau)+a^2]^{3/2}}$$
$$\times e^{-2[x-v_x(t-\tau)]^2/[8D(t-\tau)+a^2]}\,d\tau. \quad (70)$$

The τ-integral in Eqs. (68) and (70) must be evaluated numerically, which can be done conveniently by using the IMSL subroutine DCADRE. However, in a few special cases, $\partial T/\partial x$ may be written in closed form. In the following, we shall consider these cases.

a. *PTDS Signal in the Absence of a Flow ($v_x = 0$)*

In this case, Eq. (67) may be integrated to yield:

$$\frac{\partial T(x,y,t)}{\partial x} = \frac{\alpha E_0 x}{2\pi t_0 \rho C_{\mathrm{p}} D(x^2+y^2)}\left\{e^{-2(x^2+y^2)/[a^2+8D(t-t_0)]}\right.$$
$$\left. - e^{-2(x^2+y^2)/[a^2+8Dt]}\right\}. \quad (71)$$

Explicit expressions for φ_{T} and φ_{L} may now be written as:

$$\varphi_{\mathrm{T}}(x,t) = -\frac{1}{n_0}\left(\frac{\partial n}{\partial T}\right)\frac{\alpha E_0}{2 t_0 \rho C_{\mathrm{p}} D}\left\{\mathrm{erf}\left[\left\{\frac{2x^2}{a^2 + 8D(t-t_0)}\right\}^{1/2}\right]\right.$$
$$\left. - \mathrm{erf}\left[\left\{\frac{2x^2}{a^2+8Dt}\right\}^{1/2}\right]\right\} \quad (72)$$

and

$$\varphi_{\mathrm{L}}(x, y, t) = \frac{l}{n_0}\left(\frac{\partial n}{\partial T}\right)\frac{\alpha E_0 x}{2\pi t_0 \rho C_{\mathrm{p}} D(x^2 + y^2)}\left[e^{-2(x^2+y^2)/[a^2+8D(t-t_0)]} - e^{-2(x^2+y^2)/[a^2+8Dt]}\right]. \tag{73}$$

b. *PTDS Signals in Flow-Dominated Conditions*

If the flow velocity is very high and PTDS signal is observed downstream on such a short time scale that no appreciable thermal diffusion takes place, we may set $D = 0$ in Eq. (67). This equation can then be integrated analytically, with the result:

$$\frac{\partial T(x, y, t)}{\partial x} = \frac{2\alpha E_0 e^{-2y^2/a^2}}{\pi t_0 \rho C_{\mathrm{p}} a^2 v_x}\left[e^{-2[x-v_x(t-t_0)]^2/a^2} - e^{-2[x-v_x t]^2/a^2}\right]. \tag{74}$$

Explicit expressions for φ_{T} and φ_{L} are then given by

$$\varphi_{\mathrm{T}}(x, t) = \frac{1}{n_0}\left(\frac{\partial n}{\partial T}\right)\frac{2\alpha E_0}{\sqrt{2\pi}\, t_0 \rho C_{\mathrm{p}} a v_x}\left[e^{-2[x-v_x(t-t_0)]^2/a^2} - e^{-2[x-v_x t]^2/a^2}\right] \tag{75}$$

and

$$\varphi_{\mathrm{L}}(x, y, t) = \frac{l}{n_0}\frac{\partial n}{\partial T}\frac{2\alpha E_0 e^{-2y^2/a^2}}{\pi t_0 \rho C_{\mathrm{p}} a^2 v_x}\left[e^{-2[x-v_x(t-t_0)]^2/a^2} - e^{-2[x-v_x t]^2/a^2}\right]. \tag{76}$$

Equations (75) and (76) are most useful in liquids where the diffusion rates are low.

c. *PTDS Signal in the Impulse Approximation*

If the laser pulse duration t_0 is very short, Eq. (34) may be used to determine $\partial T/\partial x$. The expressions for φ_T and φ_L are then found to be

$$\varphi_T(x, t) = -\frac{1}{n_0}\frac{\partial n}{\partial T}\frac{8\alpha E_0}{\sqrt{2\pi}\,\rho C_p}\frac{(x - v_x t)}{(8Dt + a^2)^{3/2}} e^{-2(x - v_x t)^2/[a^2 + 8Dt]} \quad (77)$$

and

$$\varphi_L(x, y, t) = -\frac{l}{n_0}\frac{\partial n}{\partial T}\frac{8\alpha E_0}{\pi\rho C_p}\frac{(x - v_x t)}{(8Dt + a^2)^2} \times e^{-2[(x - v_x t)^2 + y^2]/[a^2 + 8Dt]}. \quad (78)$$

Rose, Vyas, and Gupta (1986) have discussed the range of validity of impulse approximation.

Figure 12 shows some of the results predicted by Eq. (70). The photothermal deflection has been plotted as a function of time for different positions of the probe beam. The center-to-center distance between the pump and the probe beams is represented by x. A negative value of x corresponds to the probe beam being upstream from the pump beam, whereas a positive value of x corresponds to the probe beam being downstream. We note that as the probe beam is moved upstream, the signal quickly disappears because the heat is unable to diffuse against the gas flow. As the probe beam is moved downstream, the signal gets stronger at first and then acquires a shape that is essentially the derivative of the spatial profile of the pump beam (which is assumed to be a Gaussian here). This shape, of course, can be understood easily. As the heat pulse passes by, the probe beam at first experiences a positive gradient of the heat, followed by a negative gradient. As x is increased further, the signal becomes broader and smaller due to thermal diffusion. Flow velocity of the medium has been measured in several different ways by using these signals. The simplest method is based on the measurement of the time-of-flight of the heat pulse between two probe-beam positions downstream from the pump beam (Sontag and Tam, 1985; Rose and Gupta, 1985; Sell, 1985). One may also determine the flow velocity by fitting the shape of the PTDS signal (Sell, 1985; Weimer and Dovichi, 1985). Yet another way is to use the amplitude of the signal that depends on the flow velocity as described below (Sell, 1984; Nie, Hane, and Gupta, 1986).

Figure 13 shows results similar to that of Fig. 12 but in a stationary medium ($v_x = 0$ in Eq. (70)). In this case, the signal is zero for $x = 0$,

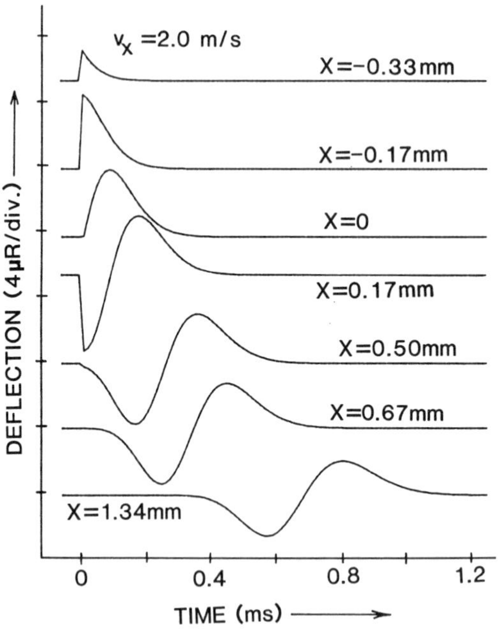

FIG. 12. Pulsed transverse PTDS signal shapes in a flowing medium for several different pump-to-probe-beam distances. $x = 0$ corresponds to the case when the axes of the two beams intersect. Negative x corresponds to the probe beam being upstream from the pump beam, and positive x corresponds to it being downstream. The deflection is expressed in microradians. Flow velocity of the medium (N_2 seeded with 1000-ppm NO_2) was assumed to be 2.0 m/s. Laser pulse energy was assumed to be 1.65 mJ, beam radius a was 0.33 mm, and all the other parameters used in this computation are given in the caption for Fig. 1. (From Rose, Vyas, and Gupta, 1986.)

because $\partial T/\partial x$ is zero at this point, and reverses sign as x changes sign, as expected. For small values of x ($0 \leq x \leq a$), the signal consists of a sharp deflection of the probe beam shortly after the pump-laser firing, followed by a gradual return of the probe beam to its original position on the time scale of the diffusion time of the heat out of the probe region. The signal attains its maximum value for $x = a/2$, where the gradient of $T(x, y, t)$ is a maximum. For larger values of x, the peak of the signal occurs later in time, and the signal is broader and weaker, as expected.

For collinear PTDS (Eq. (68)), the signal shapes are similar to those shown in Figs. 12 and 13. However, the amplitude of φ_L is larger than that of φ_T and it depends on the interaction length. Moreover, $\varphi_T(t)$ in a stationary medium is broader than $\varphi_L(t)$ (Rose, Vyas, and Gupta, 1986). Physical reason for the difference in the width is as follows: In the case of

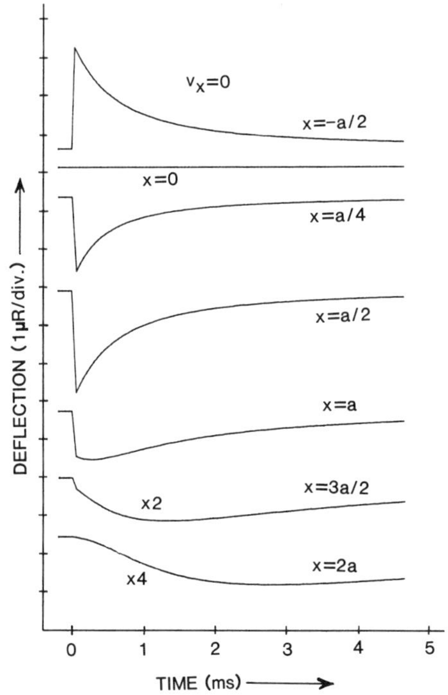

FIG. 13. Pulsed transverse PTDS signal in a stationary medium ($v_x = 0$) for several probe-to-pump-beam distances as indicated above. Laser energy was assumed to be $E_0 = 1$ mJ, and all the other parameters used in this calculation are the same as for Fig. 12. Two curves on the bottom have been expanded by the indicated factors for clarity. (From Rose, Vyas, and Gupta, 1986.)

collinear PTDS, the width of the pump beam in the y-direction (for an infinitely thin probe beam) does not contribute significantly to the width of $\varphi_L(t)$. However, in the case of transverse PTDS, the probe beam samples heat from different parts of the heat source in the y-direction. Since it takes longer for the heat to diffuse out of the probe beam for $y \neq 0$ than for $y = 0$, the $\varphi_T(t)$ signal is broader. For a flowing medium, the difference in the widths of $\varphi_T(t)$ and $\varphi_L(t)$ is very small.

Figure 14 shows the peak value of the probe-beam deflection $\varphi_m(x)$ plotted as a function of distance x between the pump and the probe beams for a stationary medium (solid curve) and for two different values of the flow velocity. As before, the gas flow is along positive x, that is, a positive value of x corresponds to the probe beam being downstream from the pump beam. For $v_x = 0$, as expected, the PTDS signal is zero at $x = 0$, and

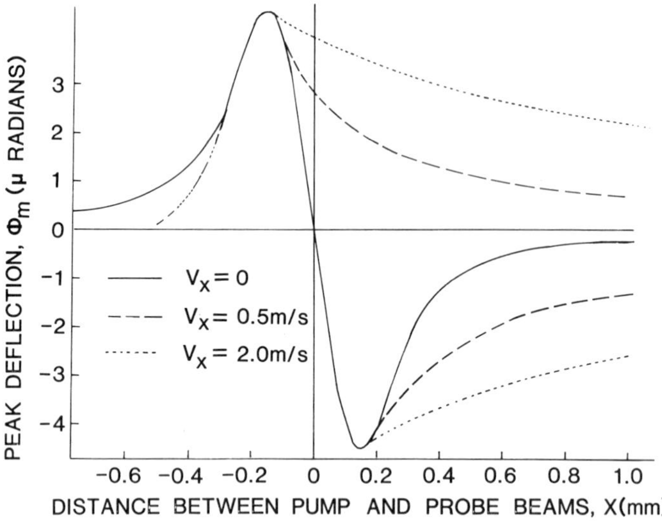

FIG. 14. Peak value of the probe-beam deflection φ_m plotted as a function of the pump–probe distance x for $v_x = 0$, $v_x = 0.5$ m/s, and $v_x = 2.0$ m/s. The pump laser was assumed to have a pulse length of 1 μs and a pulse energy equal to 1.4 mJ, and the radius of the pump beam was assumed to be 0.3 mm. The medium was assumed to be N_2 at atmospheric pressure with 1000-ppm NO_2 (absorption coefficient $\alpha = 0.39$ m^{-1}). (From Nie, Hane, and Gupta, 1986.)

it is antisymmetric about $x = 0$. For larger values of v_x, we find that the symmetry of the curve about $x = 0$ is broken, as expected, and a nonzero deflection signal is obtained at $x = 0$. As a matter of fact, the signal at $x = 0$ rises monotonically with v_x until it reaches saturation (Nie *et al.*, 1986). This fact can be used to measure very small flow velocities, that is, velocities so low that the transit time method is inappropriate due to thermal diffusion broadening of the signal. We also note in Fig. 14 that the curve splits into two branches for $v_x > 0$ and $x > 0$. This can be understood with the aid of Fig. 12. The negative and positive peaks of the signal in Fig. 12 correspond to the negative and positive branches in Fig. 14 for $v_x > 0$ and $x > 0$.

3. CW Photothermal Deflection Spectroscopy

In this section, explicit expressions for CW photothermal deflection will be derived by using Eq. (36) in Eqs. (65) and (66). The expressions for φ_L

and φ_T are found to be

$$\varphi_L = -\frac{l}{n_0}\frac{\partial n}{\partial T}\frac{8\alpha P_{av}}{\pi\rho C_p}\int_0^t \frac{(1+\cos\omega\tau)(x-v_x(t-\tau))}{[a^2+8D(t-\tau)]^2} \times e^{-2\{(x-v_x(t-\tau))^2+y^2\}/[a^2+8D(t-\tau)]}\,d\tau, \tag{79}$$

and

$$\varphi_T = -\frac{1}{n_0}\frac{\partial n}{\partial T}\frac{8\alpha P_{av}}{\sqrt{2\pi}\,\rho C_p}\int_0^t \frac{(1+\cos\omega\tau)(x-v_x(t-\tau))}{[a^2+8D(t-\tau)]^{3/2}} \times e^{-2\{(x-v_x(t-\tau))^2\}/[a^2+8D(t-\tau)]}\,d\tau. \tag{80}$$

Figures 15 and 16 show CW PTDS signal shapes for collinear and transverse cases evaluated using Eqs. (79) and (80), respectively (Vyas *et al.*, 1988). The method of 64-point Gaussian quadrature was used to perform the integration. These figures show the equilibrium values of the PTDS signals. Even though the temperature distribution for cw excitation does not reach an equilibrium value, the PTDS signals do reach equilibrium (Vyas, *et al.*, 1988). RMS values of the deflection have been plotted as a function

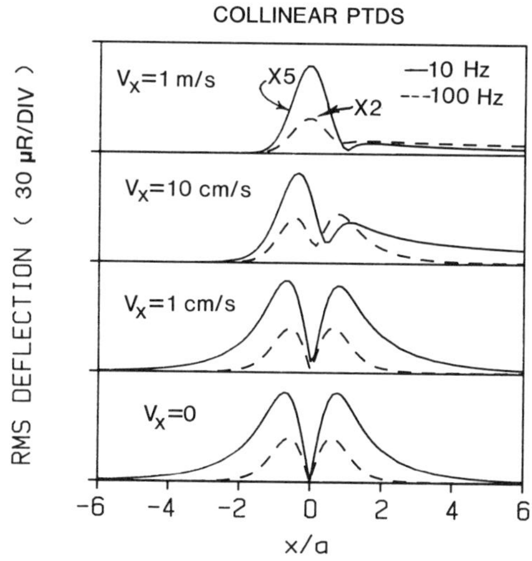

FIG. 15. CW collinear PTDS signals for modulation frequencies of 10 Hz (solid lines) and 100 Hz (broken lines) for four different flow velocities. RMS values of the deflection in μ radians have been plotted as a function of the pump–probe distance (in units of a). The top curves have been expanded by the indicated factors. The interaction length is assumed to be 1 cm, and all other parameters used in this calculation are given in the caption to Fig. 3. (From Vyas *et al.*, 1988.)

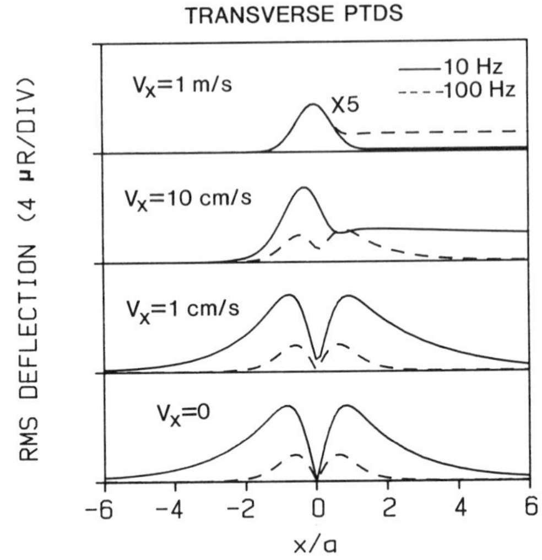

FIG. 16. CW PTDS signals, similar to those of Fig. 15, except for the transverse geometry. (From Vyas *et al.*, 1988.)

of the distance x between the pump and the probe beams for four velocities $v_x = 0$, $v_x = 1$ cm/sec, $v_x = 10$ cm/sec, and $v_x = 1$ m/sec. Solid lines are for a modulation frequency of 10 Hz, whereas the dotted lines are for 100-Hz modulation frequency. Consider Fig. 15 first. As the flow velocity of the medium increases, the curves become more and more asymmetric, as expected. Note that for large velocities, the signal downstream becomes very small even though a significant temperature distribution (above ambient) exists (see Fig. 3). This is because the *gradient* of the temperature in this region is very small. In order to understand the dip in the signal at $x \simeq a(v_x = 1$ m/s), one must examine the change in the temperature gradient as the pump-laser intensity oscillates (Vyas *et al.*, 1988). We also note that as the modulation frequency increases, the signal amplitude in general decreases, and for $v_x = 0$, the signal peaks occur closer to $x = a/2$ because the heat is able to diffuse only a shorter distance during the modulation cycle. The transverse PTDS signals shown in Fig. 16 are similar to those of collinear PTDS, except that their magnitudes are smaller.

C. Photothermal Lensing Spectroscopy

In this section, expressions for the change in intensity of a probe laser passing through a thermal lens will be derived (Fang and Swofford, 1983;

Harris and Dovichi, 1980; Dovichi, Nolan, and Weimer, 1984). The thermal lens, of course, is created by the nonuniform refractive index, which is caused by the absorption of the pump beam. An expression for the signal strength in terms of the focal length of the thermal lens is derived in Section II.C.1. General expressions for the focal length of the thermal lens are derived in Section II.C.2. Explicit expressions for the pulsed and CW lensing signals are derived in Sections II.C.3 and II.C.4, respectively.

1. Detection of the Thermal Lens

Figure 17 shows a typical configuration for the detection of a thermal lens (Fang and Swofford, 1983). The thermal lens is placed a distance z_1 in front of the probe-beam waist. A screen with a pinhole is placed at a distance z_2 in front of thermal lens. Intensity of the probe beam passing through the pinhole of radius b is observed by a photodetector; w_0 is the $1/e^2$-radius of the probe beam at its waist; w_1 and w_2 are the radii at the position of the thermal lens and the screen, respectively. When the thermal lens is activated (by turning on the pump beam), w_2 changes, resulting in a change of intensity at the detector. Our aim is to derive an expression for the signal $s(t)$ in terms of the focal length $f(t)$ of the thermal lens, where $s(t)$ is defined as

$$s(t) = \frac{P_{\text{det}}(t) - P_{\text{det}}(t=0)}{P_{\text{det}}(t=0)}, \tag{81}$$

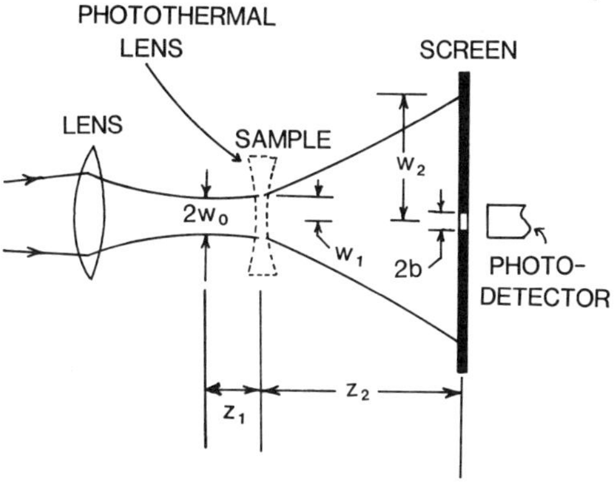

FIG. 17. Detection of the photothermal lensing effect by observation of the change in intensity of the probe beam passing through an aperture. (Adapted from Swofford and Morrell, 1978.)

where $P_{\text{det}}(t)$ is the power at the detector at time t, with $t = 0$ meaning an instant before the laser is turned on. For the CW case, the time t at which observation is made is generally large compared to the thermal diffusion time. For the pulsed case, generally the observation is made at $t > t_0$. The radial intensity distribution of the probe beam at the screen is given by

$$I(r) = \frac{2P}{\pi w_2^2} e^{-2r^2/w_2^2}, \tag{82}$$

where P is the power of the probe beam. Then

$$\begin{aligned} P_{\text{det}} &= \int_0^b I(r) 2\pi r\, dr \\ &\simeq 2P \frac{\pi b^2}{\pi w_2^2}. \end{aligned} \tag{83}$$

The signal $s(t)$ is then

$$\begin{aligned} s(t) &= \frac{w_2^2(0) - w_2^2(t)}{w_2^2(t)} \\ &\simeq \frac{w_2^2(0) - w_2^2(t)}{w_2^2(0)}, \end{aligned} \tag{84}$$

where we have replaced $w_2(t)$ by $w_2(0)$ in the denominator because the change in the radius of the beam is small.

We can find $w_2(t)$ by using the $ABCD$ law (Yariv, 1976). According to the $ABCD$ law, the complex beam parameter q_2 at the position of the screen is given in terms of the parameter q_0 at the beam waist by

$$q_2 = \frac{Aq_0 + B}{Cq_0 + D}, \tag{85}$$

where A, B, C, and D are the elements of the transformation matrix representing translation by a distance z_1, focusing by a lens of focal length f, and a translation by distance z_2, that is,

$$\begin{pmatrix} A & B \\ C & D \end{pmatrix} = \begin{pmatrix} 1 & z_2 \\ 0 & 1 \end{pmatrix} \begin{pmatrix} 1 & 0 \\ -\dfrac{1}{f} & 1 \end{pmatrix} \begin{pmatrix} 1 & z_1 \\ 0 & 1 \end{pmatrix}. \tag{86}$$

Explicitly, the matrix elements are

$$A = 1 - z_2/f,$$
$$B = z_1 + z_2 - z_1 z_2/f,$$
$$C = -\frac{1}{f},$$

and

$$D = 1 - z_1/f. \tag{87}$$

The complex beam parameter $q(z)$ is defined by

$$\frac{1}{q(z)} = \frac{1}{R(z)} - i\frac{\lambda}{\pi n w^2(z)}, \tag{88}$$

where $R(z)$ is the radius of the phase front, $w(z)$ is the $1/e^2$-radius of the intensity, n is refractive index, and λ is the wavelength of the probe radiation. At the waist of the beam, the radius $R_0 = \infty$ and q_0 is simply given by

$$q_0 = iz_0, \tag{89}$$

where $z_0 \equiv \pi n w_0^2/\lambda$ is the confocal distance (distance in which the beam diameter increases by a factor of $\sqrt{2}$). By substituting Eq. (89) into Eq. (85) and by using Eq. (88), we find that

$$-\frac{\lambda}{\pi n w_2^2} = \mathrm{Im}\left(\frac{1}{q_2}\right) = \frac{z_0(BC - AD)}{z_0^2 A^2 + B^2}.$$

But $AD - BC = 1$ from Eq. (87). Therefore,

$$w_2^2(t) = w_0^2\left(A^2 + \frac{B^2}{z_0^2}\right). \tag{90}$$

By substituting for A and B from Eq. (87) into Eq. (90), and by making the approximation $z_2 \gg z_1$ (which can generally be arranged), we get

$$w_2^2(t) = w_0^2\left\{\left(1 \frac{z_2}{f(t)}\right)^2 + \frac{z_2^2}{z_0^2}\left(1 - \frac{z_1}{f(t)}\right)^2\right\} \tag{91}$$

Equation (91) can be further simplified if $z_0 \ll z_2$. Then

$$w_2(t) = \frac{w_0 z_2}{z_0}\left(1 - \frac{z_1}{f(t)}\right). \tag{92}$$

Remembering that $f(t = 0) = \infty$, substitution of Eq. (92) into Eq. (84) yields the signal $s(t)$:

$$s(t) = \frac{-z_1^2}{f^2(t)} + \frac{2z_1}{f(t)}. \tag{93}$$

Generally, $f(t) \gg z_1$. In that case, Eq. (93) simplifies to

$$s(t) = \frac{2z_1}{f(t)}, \tag{94}$$

which is the relationship we had set out to derive. A general equation for $s(t)$ without the approximation is given by Vyas and Gupta, (1988).

In many situations, the thermal lens may be astigmatic. In these cases, the beam radius w_2 is different in two orthogonal directions. In these cases, Eq. (83) modifies to

$$P_{\text{det}} = 2P\frac{\pi b^2}{\pi w_x w_y}, \tag{95}$$

where w_x and w_y are the beam radii in the x- and the y-directions, respectively. The signal $s(t)$ is then given by

$$s(t) = \frac{w_x(0)w_y(0) - w_x(t)w_y(t)}{w_x(0)w_y(0)}. \tag{96}$$

A straightforward calculation similar to the one leading to Eq. (94) gives

$$s(t) = \frac{z_1}{f_x(t)} + \frac{z_1}{f_y(t)}. \tag{97}$$

2. Focal Length of the Thermal Lens

In order to evaluate Eq. (94) or Eq. (97), we need to know the focal length of the thermal lens. An expression for the focal length can be derived easily with the aid of Fig. 18 (Fang and Swofford, 1983). We shall follow Fang and Swofford's treatment for the cylindrically symmetric configuration, and will generalize it later to include the astigmatic thermal lens. A more sophisticated treatment of this subject is given by Vyas and Gupta (1988). Consider a medium of length l with refractive index $n(r)$ that increases radially. We shall make the thin-lens approximation, that is, we shall assume that l is very small compared with focal length f. Consider an

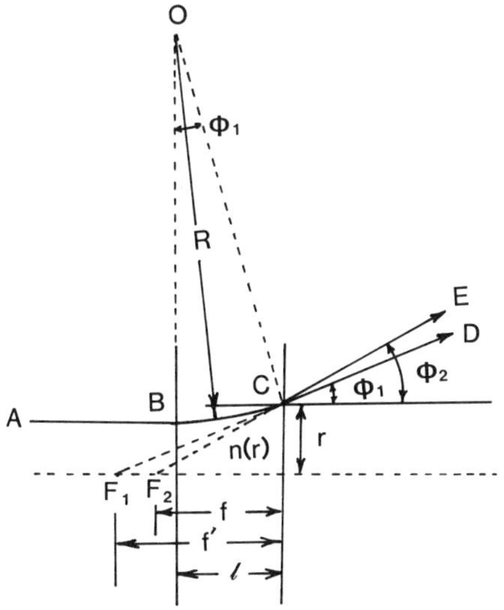

FIG. 18. Diagram showing the focal length of the thermal lens. (Adapted from Fang and Swofford, 1983.)

optical ray AB incident normally at the left interface of the medium. Inside the medium, the ray follows a curved trajectory BC of radius R. The ray arrives at the right interface making an angle φ_1 with the normal to the interface. Due to the refraction at the right interface, the ray actually travels along CE, making an angle φ_2 with the normal. The effective focal length of this medium then is f, as shown in the diagram. In the derivation below, small angle approximation will be made throughout. From the geometry,

$$\varphi_1 = \frac{BC}{R} \simeq \frac{l}{R}.$$

Therefore,

$$f' = -\frac{r}{\varphi_1} = -\frac{rR}{l}, \tag{98}$$

but

$$\frac{f}{f'} = \frac{\varphi_1}{\varphi_2} = \frac{1}{n}, \tag{99}$$

where the last step is given by Snell's law. By using Eq. (98) in Eq. (99), we get

$$f = -\frac{r}{n}\frac{R}{l}; \tag{100}$$

R can easily be found from the ray equation, Eq. (61),

$$\frac{d}{ds}\left(n_0 \frac{d\delta}{ds}\right) = \nabla_\perp n(\mathbf{r}, t).$$

For small deflection, $d\delta/ds = \varphi_1$, and

$$\frac{d\varphi_1}{ds} = \frac{1}{n_0}\nabla_\perp n.$$

But $ds = R\,d\varphi_1$, and therefore

$$\frac{1}{R} = \frac{1}{n_0}\nabla_\perp n. \tag{101}$$

In many cases, the refractive index is cylindrically symmetric, i.e., the refractive index depends only on the radial distance from the axis (for example, if the pump-laser beam has a Gaussian spatial profile and the medium is stationary). Moreover, we assume the probe beam to propagate collinearly at $r = 0$. In this case, it is convenient to expand the refractive index $n(r)$ in a MacLaurin series,

$$n(r) = n(0) + r\left(\frac{\partial n}{\partial r}\right)_{r=0} + \tfrac{1}{2}r^2\left(\frac{\partial^2 n}{\partial r^2}\right)_{r=0} + \cdots . \tag{102}$$

Because of the cylindrical symmetry, $(\partial n/\partial r)_{r=0} = 0$. Therefore the leading nonuniform term is a quadratic term that makes the medium act like a lens. By using Eq. (102) in Eq. (101), we get

$$\frac{1}{R} = \frac{1}{n_0}\frac{\partial n(r)}{\partial r} = \frac{r}{n_0}\left(\frac{\partial^2 n}{\partial r^2}\right)_{r=0}. \tag{103}$$

The focal length may now be found by using Eq. (103) in Eq. (100):

$$\frac{1}{f} = -l\left(\frac{\partial^2 n}{\partial r^2}\right)_{r=0}, \tag{104}$$

where we have set $n \simeq n_0$ in Eq. (100). If $(\partial^2 n/\partial r^2)_{r=0}$ is not constant over the path of the probe beam, Eq. (104) may be generalized as follows:

$$\frac{1}{f} = -\int_{\text{path}}\left(\frac{\partial^2 n}{\partial r^2}\right)_{r=0} ds. \tag{105}$$

By using Eq. (62), Eq. (105) may finally be written as

$$\frac{1}{f} = -\frac{\partial n}{\partial T}\int_{\text{path}}\left(\frac{\partial^2 T}{\partial r^2}\right)_{r=0} ds. \tag{106}$$

This is the desired equation. We find $(\partial^2 T/\partial r^2)_{r=0}$ from Eq. (32) or Eq. (36), as appropriate. Equation (106), when substituted in Eq. (94), yields the desired signal $s(t)$.

Now, we shall consider both transverse and collinear beam configurations, as we have done before in connection with PTPS and PTDS. For the beam configurations shown in Fig. 5, the probe beam propagates along the z-axis for the collinear case, whereas it propagates along the y-axis for the transverse case. It is obvious that a *cylindrical* thermal lens is experienced by the probe beam in the case of transverse PTLS. Even in the case of collinear PTLS, if the medium is not stationary, ($v_x \neq 0$), and/or the probe beam is not placed at $x = 0$ and $y = 0$, an astigmatic lens is experienced by the probe beam. Therefore, we must distinguish between focal lengths f_x and f_y. Thus, it is necessary to expand the refractive index $n(x, y)$ in a Taylor series about a point x', y' (a point on the axis of the probe beam):

$$\begin{aligned} n(x, y) &= n(x', y') + (x - x')\left(\frac{\partial n}{\partial x}\right)_{\substack{x=x'\\y=y'}} + (y - y')\left(\frac{\partial n}{\partial y}\right)_{\substack{x=x'\\y=y'}} \\ &+ \frac{(x - x')^2}{2}\left(\frac{\partial^2 n}{\partial x^2}\right)_{\substack{x=x'\\y=y'}} + \frac{(y - y')^2}{2}\left(\frac{\partial^2 n}{\partial y^2}\right)_{\substack{x=x'\\y=y'}} \\ &+ (x - x')(y - y')\left(\frac{\partial^2 n}{\partial x \partial y}\right)_{\substack{x=x'\\y=y'}} + \cdots. \end{aligned} \tag{107}$$

By following the steps leading from Eq. (102) to Eq. (106), we get

$$\frac{1}{f_x} = -\frac{\partial n}{\partial T}\int_{\text{path}}\left(\frac{\partial^2 T}{\partial x^2}\right)_{\substack{x=x'\\y=y'}} ds, \tag{108}$$

$$\frac{1}{f_y} = -\frac{\partial n}{\partial T}\int_{\text{path}}\left(\frac{\partial^2 T}{\partial y^2}\right)_{\substack{x=x'\\y=y'}} ds, \tag{109}$$

and

$$\frac{1}{f_z} = 0. \tag{110}$$

Note that in this case, the first derivatives of the refractive index $(\partial n/\partial x)$ and $(\partial n/\partial y)$ in Eq. (107) are not necessarily zero. These terms result in the *deflection* of the probe beam, as discussed in Section II.B (see Vyas and Gupta, 1988). For the collinear case, Fig. 5b, we may simplify Eqs. (108) and (109) as

$$\frac{1}{f_x} = -\frac{\partial n}{\partial T} l \left(\frac{\partial^2 T}{\partial x^2} \right)_{\substack{x=x' \\ y=0}} \tag{111}$$

and

$$\frac{1}{f_y} = -\frac{\partial n}{\partial T} l \left(\frac{\partial^2 T}{\partial y^2} \right)_{\substack{x=x' \\ y=0}}. \tag{112}$$

Similarly, for the transverse case, Fig. 5a, we have

$$\frac{1}{f_x} = -\frac{\partial n}{\partial T} \int \left(\frac{\partial^2 T}{\partial x^2} \right)_{x=x'} dy \tag{113}$$

and

$$\frac{1}{f_z} = 0. \tag{114}$$

3. Pulsed Photothermal Lensing Spectroscopy

In this section, explicit expressions for pulsed PTLS signals will be derived (Vyas and Gupta, 1988). We start with the collinear case. Substitution of Eq. (32) for $T(x, y, t)$ in Eqs. (111) and (112) leads to

$$\frac{1}{f_x} = \frac{8\alpha E_0 l}{\pi \rho C_p t_0} \left(\frac{\partial n}{\partial T} \right) \int_0^{t_0} \frac{1}{[a^2 + 8D(t-\tau)]^2} \left[1 - \frac{4(x - v_x(t-\tau))^2}{[a^2 + 8D(t-\tau)]} \right] \times e^{-2[x - v_x(t-\tau)]^2/[a^2 + 8D(t-\tau)]} \, d\tau \tag{115}$$

and

$$\frac{1}{f_y} = \frac{8\alpha E_0 l}{\pi \rho C_p t_0} \left(\frac{\partial n}{\partial T} \right) \int_0^{t_0} \frac{e^{-2[(x - v_x(t-\tau))^2]/[a^2 + 8D(t-\tau)]}}{[a^2 + 8D(t-\tau)]^2} \, d\tau, \tag{116}$$

where we have dropped the primes on x for convenience. Substitution into

Eq. (97) gives

$$s_{\mathrm{L}}(t) = \frac{8\alpha E_0 l z_1}{\pi \rho C_{\mathrm{p}} t_0}\left(\frac{\partial n}{\partial T}\right)\int_0^{t_0} \frac{1}{\left[a^2 + 8D(t-\tau)\right]^2}\left\{2 - \frac{4\left[x - v_x(t-\tau)\right]^2}{\left[a^2 + 8D(t-\tau)\right]}\right\} \times e^{-2[x - v_x(t-\tau)]^2/[a^2 + 8D(t-\tau)]}\, d\tau. \tag{117}$$

Similarly, using Eq. (32) in Eq. (113), and using Eqs. (113) and (114) in Eq. (97) gives the transverse PTLS:

$$s_{\mathrm{T}}(t) = \frac{8\alpha E_0 z_1}{\sqrt{2\pi}\, \rho C_{\mathrm{p}} t_0}\left(\frac{\partial n}{\partial T}\right)\int_0^{t_0} \frac{1}{\left[a^2 + 8D(t-\tau)\right]^{3/2}} \times \left\{1 - \frac{4\left[x - v_x(t-\tau)\right]^2}{\left[a^2 + 8D(t-\tau)\right]}\right\} \times e^{-2[x - v_x(t-\tau)]^2/[a^2 + 8D(t-\tau)]}\, d\tau. \tag{118}$$

These are the final equations that we wanted to derive. For a short laser pulse, Eqs. (117) and (118) can be written in a closed form, by using Eq. (33), as

$$s_{\mathrm{L}}(t) = \frac{8\alpha E_0 l z_1}{\pi \rho C_{\mathrm{p}}}\left(\frac{\partial n}{\partial T}\right)\frac{1}{\left[a^2 + 8Dt\right]^2}\left\{2 - \frac{4(x - v_x t)^2}{(a^2 + 8Dt)}\right\} \times e^{-2(x - v_x t)^2/(a^2 + 8Dt)} \tag{119}$$

and

$$s_{\mathrm{T}}(t) = \frac{8\alpha E_0 z_1}{\sqrt{2\pi}\, \rho C_{\mathrm{p}}}\left(\frac{\partial n}{\partial T}\right)\frac{1}{\left[a^2 + 8Dt\right]^{3/2}}\left\{1 - \frac{4(x - v_x t)^2}{(a^2 + 8Dt)}\right\} \times e^{-2(x - v_x t)^2/(a^2 + 8Dt)}. \tag{120}$$

Figure 19 shows a few typical pulsed PTLS signals in a stationary medium for the transverse case. The fractional change in the power at the detector, $s_{\mathrm{T}}(t)$, has been plotted against the time. A negative value of the signal corresponds to a diverging lens, whereas a positive value corresponds to a

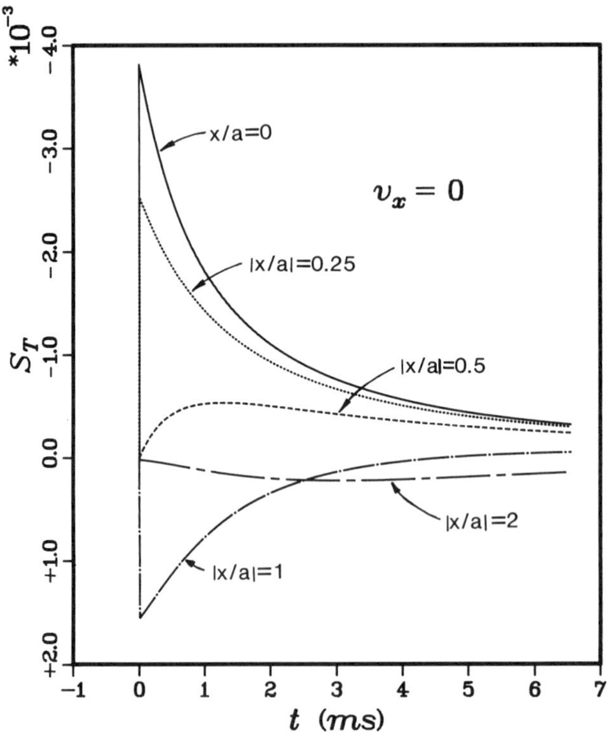

FIG. 19. Transverse photothermal lensing signals for a pulsed laser and a stationary medium. The five curves are for five different pump–probe separations, as labeled. The curves have been computed for the following parameters: E_0 = 6 mJ, t_0 = 1 μs, a = 0.5 mm, and α = 0.39 m^{-1}, corresponding to 1000-ppm NO_2 in N_2. (From Vyas and Gupta, 1988).

converging lens. For $|x| < 0.5a$, the photothermal lens behaves like a diverging lens. For $|x| = 0.5a$ the signal vanishes at $t = t_0$ (because the second derivative of temperature is zero at this place). However, for $t > t_0$, some signal gets generated by the heat diffusing out from the inside of the pump beam. For $|x| > 0.5a$, the photothermal lens behaves like a converging lens. Figure 20 shows pulsed PTLS signals in a flowing medium for the transverse case. In this case, the thermal lens travels downstream with the flow of the medium. The signal at $x = 3a$ shows the shape of the signal due to the entire lens. The thermal lens is a diverging lens in the center, whereas it is a converging lens in the wings. The wings are asymmetric due to the thermal diffusion. Collinear PTLS signal shapes are similar to those of Figs. 19 and 20; however, the signal amplitudes are larger (by about a factor of 30), and the zeros of signal occur at $x = \pm 0.707a$ (see Vyas and Gupta, 1988).

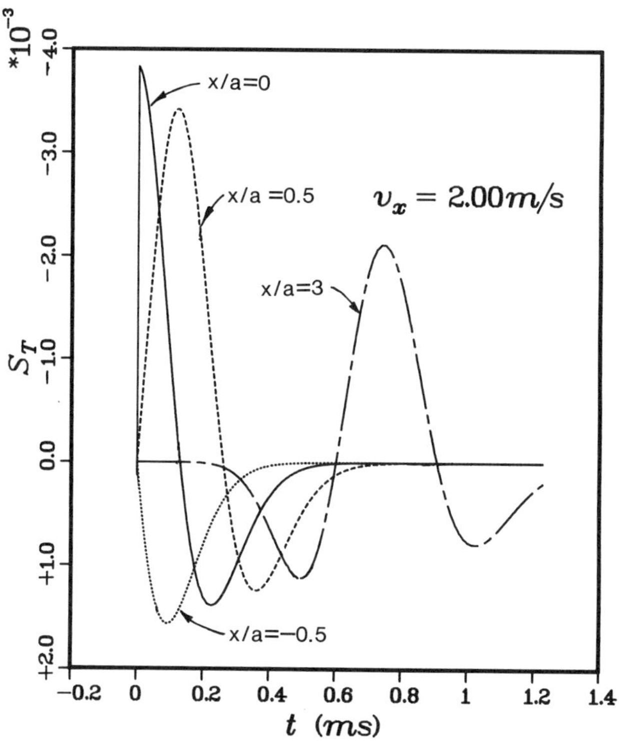

FIG. 20. Transverse photothermal lensing signals for a pulsed laser and a medium flowing with a velocity $v_x = 2$ m/s. All parameters used in this computation are the same as those for Fig. 19. (From Vyas and Gupta, 1988).

4. CW PHOTOTHERMAL LENSING SPECTROSCOPY

Explicit expressions for CW PTLS signals (Vyas and Gupta, 1988) are obtained in a manner completely analogous to that of the last section by using Eqs. (36) in Eqs. (111)–(114) and by using Eq. (97). The results are

$$s_L(t) = \frac{8\alpha P_{av} l z_1}{\pi\rho C_p}\left(\frac{\partial n}{\partial T}\right)\int_0^t \frac{(1+\cos\omega\tau)}{[a^2+8D(t-\tau)]^2} \times\left[2 - \frac{4[x-v_x(t-\tau)]^2}{[a^2+8D(t-\tau)]}\right] e^{-2[x-v_x(t-\tau)]^2/[a^2+8D(t-\tau)]}\, d\tau \tag{121}$$

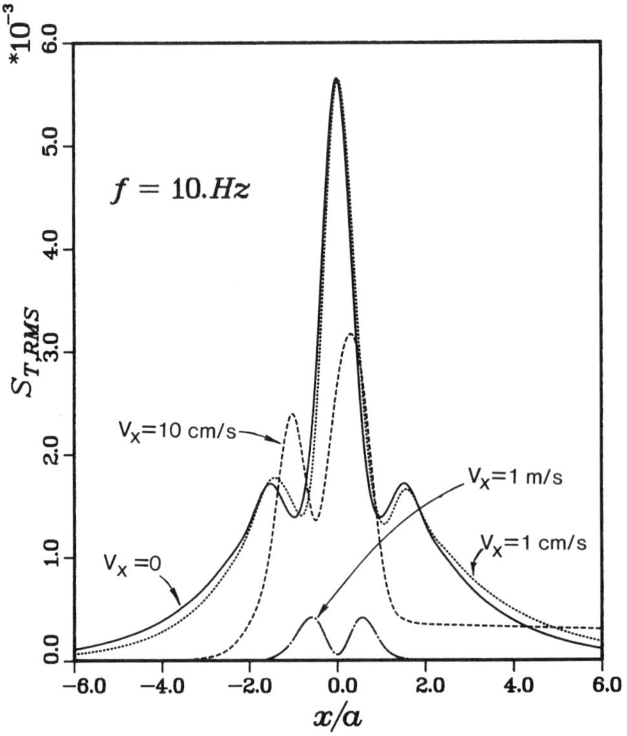

FIG. 21. RMS values of the CW PTLS signals for the transverse geometry plotted against the pump–probe distance for four different flow velocities. The modulation frequency was assumed to be 10 Hz, and all other parameters used in this computation are given in the caption to Fig. 3. (From Vyas and Gupta, 1988).

and

$$
\begin{aligned}
s_{\mathrm{T}}(t) = {} & \frac{8\alpha P_{\mathrm{av}} z_1}{\sqrt{2\pi}\,\rho C_{\mathrm{p}}}\left(\frac{\partial n}{\partial T}\right)\int_0^t \frac{[(1+\cos\omega\tau)]}{\left[a^2+8D(t-\tau)\right]^{3/2}} \\
& \times\left[1-\frac{4\left[x-v_x(t-\tau)\right]^2}{\left[a^2+8D(t-\tau)\right]}\right] e^{-2[x-v_x(t-\tau)]^2/[a^2+8D(t-\tau)]}\,d\tau .
\end{aligned}
\tag{122}
$$

Figure 21 shows the CW PTLS signals for the transverse case. The equilibrium values of the RMS signals are plotted against the pump–probe distance for modulation frequency of 10 Hz. Whereas the temperature of the medium for CW excitation does not reach equilibrium, the PTLS signals

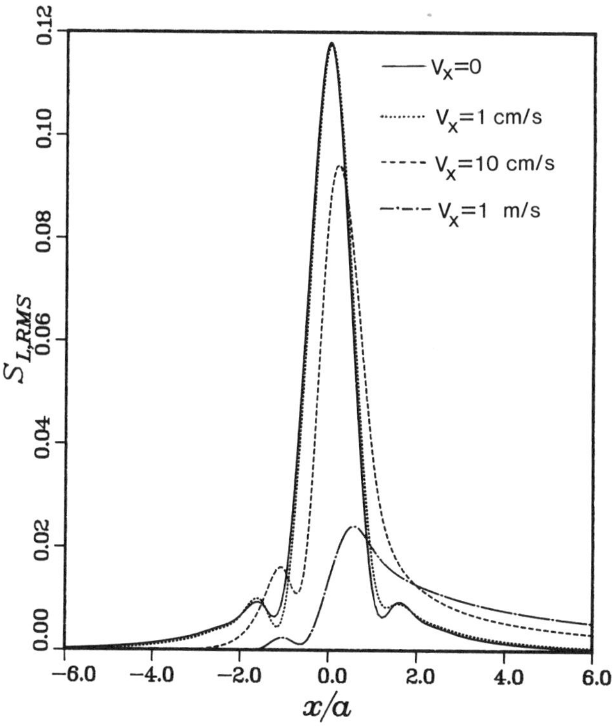

FIG. 22. RMS values of the CW PTLS signals for the collinear case. The interaction length was assumed to be 1 cm, and all other parameters are the same as for Fig. 21. (From Vyas and Gupta, 1988).

do reach equilibrium. A stationary medium, and flow velocities of 1 cm/s, 10 cm/s, and 1 m/s are considered. The signal is symmetric about $x = 0$ for $v_x = 0$, and this symmetry is lost as v_x is increased, as expected. The apparent symmetry of the signal for $v_x = 1$ m/s is fortuitous. For $v_x = 0$, the signal consists of three peaks; the central peak corresponds to a diverging lens, whereas the side peaks correspond to a converging lens. For higher velocities, ($v_x \geq 10$ cm/s), the peaks on the left correspond to converging lenses, whereas those on the right correspond to diverging lenses. When RMS values are measured, this information is difficult to obtain except by measuring the phase of the signal. Figure 22 shows the corresponding signals for the collinear case. As expected, these signals are larger in magnitude and have a less pronounced structure.

Vyas and Gupta (1988) have also investigated the dependence of the PTLS signals on modulation frequency and they find that very significant changes in signal shapes can occur as the modulation frequency is varied.

ACKNOWLEDGMENTS

Contents of this chapter are based on work done in collaboration with Reeta Vyas, Allen Rose, and Brian Monson, which was supported in part by Air Force Wright Aeronautical Laboratories.

REFERENCES

Arfken, G. (1985). *Mathematical Methods for Physicists*, Academic Press, New York.

Boccara, A. C., Fournier, D., Badoz, J. (1980). *Appl. Phys. Lett.* **36**, 130.

Boccara, A. C., Fournier, D., Jackson, W. B., and Amer, N. M. (1980). *Opt. Lett.* **5**, 377.

Campillo, A. J., Petuchowski, S. J., Davis, C. C., and Lin, H.-B. (1982). *Appl. Phys. Lett.* **41**, 327.

Davis, C. C. (1980). *Appl. Phys. Lett.* **36**, 515.

Davis, C. C., and Petuchowski, S. J. (1981). *Appl. Opt.* **20**, 2539.

Dovichi, N. J., Nolan, T. G., and Weimer, W. A. (1984). *Anal. Chem.* **56**, 1700.

Fang, H. L., and Swofford, R. L. (1983). In *Ultrasensitive Laser Spectroscopy* (D. S. Kliger, Ed.). Academic Press, New York.

Fournier, D., Boccara, A. C., Amer, N. M., and Gerlach, R. (1980). *Appl. Phys. Lett.* **37**, 519.

Ghatak, A. K., and Thyagarajan, K. (1978). *Contemporary Optics*. Plenum Press, New York.

Goldstein, H. (1959). *Classical Mechanics*. Addison-Wesley, Reading, Mass.

Gordon, J. P., Leite, R. C. C., Moore, R. S., Porto, S. P. S., and Whinnery, J. R. (1965). *J. Appl. Phys.* **36**, 3.

Gupta, R. (1987). In *Proc. IX Int. Conf. on Lasers and Applications* (*Lasers '86*). STS Press, McLean, VA.

Harris, J. M., and Dovichi, N. J. (1980). *Anal. Chem.* **52**, 695A.

Heritier, J. M. (1983). *Opt. Commun.* **44**, 267.

Hu, C., and Whinnery, J. R. (1973). *Appl. Opt.* **12**, 72.

Jackson, W. B., Amer, N. M., Boccara, A. C., and Fournier, D. (1981). *Appl. Opt.* **20**, 1333.

Monson, B., Vyas, R., and Gupta, R. (1988). *Appl. Opt.* (submitted.)

Nie, Y.-X., Hane, K., and Gupta, R. (1986). *Appl. Opt.* **25**, 3247.

Rose, A., and Gupta, R. (1985). *Opt. Lett.* **10**, 532.

Rose, A., Salamo, G. J., and Gupta, R. (1984). *Appl. Opt.* **23**, 781.

Rose, A., Vyas, R., and Gupta, R. (1986). *Appl. Opt.* **25**, 4626.

Sell, J. A. (1984). *Appl. Opt.* **23**, 1586.

Sell, J. A. (1985). *Appl. Opt.* **24**, 3725.

Sontag, H., and Tam, A. C. (1985). *Opt. Lett.* **10**, 436.

Stone, J. (1972). *J. Opt. Soc. Am.* **62**, 327.

Stone, J. (1973). *Appl. Opt.* **12**, 1828.

Swofford, R. L., and Morrell, J. A. (1978). *J. Appl. Phys.* **49**, 3667.

Tam, A. C., and Coufal, H. (1983). *J. Phys. Colloq.* C6, **44**, 9.

Vyas, R., and Gupta, R. (1988). *Appl. Opt.* (To be published.)

Vyas, R., Monson, B., Nie, Y.-X., and Gupta, R. (1988). *Appl. Opt.* (September 15, 1988).

Weimer, W. A., and Dovichi, N. J. (1985). *Appl. Opt.* **24**, 2981.

Yariv, A. (1976). *Introduction to Optical Electronics*. Holt, Rinehart and Winston, New York.

CHAPTER 4

PHOTOTHERMAL SPECTROSCOPY: APPLICATIONS IN CHROMATOGRAPHY AND ELECTROPHORESIS

MICHAEL D. MORRIS
FOTIOS K. FOTIOU
Department of Chemistry
University of Michigan
Ann Arbor, MI

I. General Introduction to Chromatography and Electrophoresis 127
A. Chromatography 127
B. Electrophoresis 129
II. Photothermal Gas-Chromatography Detectors 131
III. Photothermal Liquid-Chromatography Detectors 133
IV. Photothermal Flow-Injection Analysis Detectors 140
V. Photothermal Electrophoresis Detectors 141
VI. Photothermal Detectors for Thin-Layer Chromatography 145
A. Single-Point Measurements 145
B. Spatial Multiplexing 147
VII. Conclusions 153
References 153

I. General Introduction to Chromatography and Electrophoresis

A. Chromatography

Contemporary chemical and biological research often require the separation and quantification of complex mixtures of compounds. Samples may contain fewer than a dozen constituents, or they may contain several hundred; complex biological samples may contain well over a thousand different proteins. Qualitative and quantitative measurement is important, but the actual isolation of the components of a mixture is needed infrequently. Therefore, separation techniques can use total sample sizes of a few nanograms to a few milligrams. In fact, the most efficient separation procedures require small samples. Even where the highest resolution is

ISBN 0-12-636345-5

unnecessary, the use of small samples is desirable to minimize reagent consumption and apparatus size.

For analytical separations, chemists rely almost exclusively on chromatographic techniques, principally gas chromatography and reverse-phase liquid chromatography. Thin-layer chromatography is employed less frequently. It is used, principally in industrial laboratories, as a rapid and inexpensive semiquantitative technique. Recent monographs (Heftmann, 1984; Poole and Scheutte, 1984; Perry, 1981) provide an introduction to general principles and instrumentation, and survey broad classes of applications.

Liquid chromatography is increasingly important in biological practice. However, workers in the life sciences still rely heavily on gel electrophoresis because existing chromatographic techniques fail with large proteins and nucleic acids. Polyacrylamide and agarose gels have pore sizes large enough to accommodate these macromolecules. Electrophoresis in these gels (Andrews, 1986; Jorgenson, 1986; Hames and Rickwood, 1983) is widely employed.

Electrophoresis offers more than temporary technological advantages. Separation by protein isoelectric point, for example, has no chromatographic analog. Accommodation of ultrahigh-resolution two-dimensional techniques is relatively straightforward.

The chromatographic separation process dilutes a sample; at the end of the separation, the sample occupies a volume 10–100 times larger than its starting volume. To preserve the resolution achieved in the separation, the detector active- and dead-volume must be small compared to the volume occupied by the eluted sample. A good working rule is that these volumes must be less than or equal to the volume of sample originally introduced. In practice, liquid and gas chromatography require detectors that operate with microliter or submicroliter volumes.

Similar considerations apply in thin-layer chromatography and gel electrophoresis. In these cases, spatial resolution along the direction of development is required. Spatial averaging transverse to the direction of development is necessary because inhomogeneities in the medium and nonidealities in the separation process cause nonuniform transverse sample distribution. In thin-layer chromatography, an ideal probe would sample a rectangle about 0.1×10 mm. In electrophoresis, the longitudinal dimension should be 0.025–0.05 mm, but a 10-mm transverse dimension is still acceptable.

High-resolution gas-chromatographic separations are usually performed on capillary columns, typically 0.53 mm i.d. Commonly, the stationary phase is present as a thin coating on the wall itself, rather than on a support material. Capillaries having their internal surface covered with a porous support, which is coated with the stationary phase, are also used.

Most liquid chromatography employs monolayer stationary phases bonded to 2–5-μm diameter silica supports. The common stationary phases are C-18 and C-8 hydrocarbons. The most common mobile phase is a methanol/water mixture, typically about 60% methanol. For historical reasons, liquid chromatography with a hydrocarbon stationary phase and a polar mobile phase is called reverse-phase liquid chromatography.

Two types of liquid-chromatography columns are in common use. The most common configuration is a column 10–15 cm long by 4.6 mm i.d. This column geometry provides reasonable resolution combined with high speed. Mobile phase flow rate is typically about 1 ml/min. For high resolution, microbore columns, which are typically 25 cm by 1 mm i.d., are used. Mobile phase flow rate is about 25–50 μl/min., and instrument time is usually 10–45 minutes. Both column types operate with sample sizes of 0.1–1 μl.

The stationary phase in modern (Fenimore and Davis, 1981) thin-layer chromatography is a uniform layer of silica or alumina firmly bound to a supporting glass plate. Microliter samples are applied with a syringe to one end of the plate. Ten to twenty samples can be placed on the same plate and separated simultaneously. This feature facilitates direct comparison among samples and makes the per-sample time surprisingly short.

B. Electrophoresis

Gel electrophoresis functions well with proteins and polynucleotides too large to be separated by liquid chromatography (Andrews, 1986; Hames and Rickwood, 1983). Electrophoresis is used both qualitatively, with simple visual indication of the presence of sample constituents, and quantitatively, with densitometric evaluation of the gels.

Electrophoresis is the general name for a family of techniques in which ions migrate in an electric field (Andrews, 1986; Hames and Rickwood, 1983). In biological and biochemical practice, migration occurs in a gel formed by polymerizing acrylamide or agarose in an aqueous buffer. The gel is usually cast in the form of a thin slab. In proteins, for example, the most common gel size is 1 mm thick, 75 mm wide, and 100 mm long. The wide gel allows separation of 10–15 samples simultaneously. The multiple sample channels compensate for the long separation time (45 minutes to several hours) and the large number of manual operations usually required.

The samples to be separated are placed at one end of the gel. The ends of the gel are in contact with buffer reservoirs, which hold the electrodes for the driving electric field. These electrodes located near the sample and at the opposite end of the gel define the direction in which the proteins or nucleic acids migrate.

Proteins are usually separated in polyacrylamide gels. The gels contain 5%–20% polymer by weight, depending on the protein size range expected. The most common technique, SDS-PAGE, is separation by molecular weight of proteins that have been denatured with sodium dodecyl sulfate (SDS). Proteins may be sorted by isoelectric point in gels containing a pH buffer gradient. This technique is called isoelectric focusing. High-resolution two-dimensional electrophoresis usually is isoelectric focusing followed by SDS-PAGE. The two-dimensional technique allows separation of mixtures containing hundreds or even thousands of proteins.

Nucleic acids are often too large to be accommodated in polyacrylamide gels, and are separated on agarose or mixed agarose/polyacrylamide gels. The separation is essentially by molecular weight. If the direction of the electric field is changed periodically to allow relaxation of the molecules, nucleic acids up to several million base pairs long can be separated by this technique (Chu, Vollrath, and Davis, 1986).

In an interesting variant on electrophoresis, the separated proteins or nucleic acids are transferred to sheets of modified cellulose nitrate or similar material. On the surface of the transfer medium, they are available for biological reactions, which can improve both the sensitivity and selectivity of the separation. The general procedure is called blotting (Gershoni and Palade, 1983) because the earliest versions employed transfer by adsorption onto filter paper.

Direct photometric detection of proteins or nucleic acids is possible, but with poor sensitivity. Detection of separated proteins or nucleic acids is performed by staining them with dyes or colloidal metals, by autoradiography, or by detection with antigen–antibody binding reactions performed on blots.

Two staining procedures are commonly employed for proteins. For moderate sensitivity, the fabric dye Coomassie Brilliant Blue R250 is employed. The molecule is bound electrostatically to NH_3^+ moieties and through noncovalent bonds to nonpolar regions. About 100 ng of protein can be detected by Coomassie staining by conventional densitometry on 1-mm-thick gels (Jorgenson, 1986).

Silver staining provides 50–100 times greater sensitivity, but at the expense of a more tedious and less reproducible procedure. The polyacrylamide gel is immersed in silver nitrate solution. Proteins bind the silver and form sites at which silver is reduced to the metal. Further reduction is used to form colloidal silver around these nuclei.

For nucleic acids, the fluorescent aminoacridine dye ethidium bromide provides excellent sensitivity and selectivity. Silver staining is also commonly used with nucleic acids.

Autoradiography, the fogging of photographic film by radioisotopes in a gel, provides ultrasensitive detection of proteins labeled with isotopes sulfur or iodine. Autoradiography is 100–1000 times more sensitive than staining with Coomassie Brilliant Blue R250 (Andrews, 1986). However, several days of exposure may be necessary for high sensitivity. The technique is actually used to follow the fate of specific proteins, rather than as a general method of detection.

II. Photothermal Gas-Chromatography Detectors

Photothermal detection has been applied infrequently in gas chromatography. The technique is most useful where the special combination of ultralow detection limits and specificity for a few compounds is needed at a capital cost lower than that of mass spectrometry. The typical effluent is transparent in the visible and ultraviolet, so that only infrared (IR) lasers are of broad practical use in detectors. Vibrational absorption coefficients are an order of magnitude weaker than electronic absorption coefficients, however. Of the common infrared systems, only carbon dioxide lasers provide adequate pulse energy or continuous wave (CW) power in a useful region of the spectrum.

Nickolaisen and Bialkowski (1985) have studied thermal lens formation with pulsed lasers at gas-chromatographic flow rates. At flow rates ranging from about 3 ml/min. to 100 ml/min., the photothermal signal is independent of flow rate because the rise time of the pulsed photothermal signal is shorter than the time scale of flow or mixing. By contrast in liquid-phase measurements, sensitivity loss and increased noise are easily observed at chromatographic flow rates, both with CW lasers (Dovichi and Harris, 1981) and, less severely, with pulsed lasers.

In addition, the enhancement factor for the pulsed thermal lens relative to the CW lens is greater than unity under chromatographic conditions. For this reason, also, the use of a pulsed laser should be favored. Using typical experimental parameters, the enhancement factor for the pulsed thermal lens calculated for a CO_2 laser ranges from 3.6×10^6 in He, the most common carrier gas, to 2.8×10^7 in Ar (Nickolaisen and Bialkowski, 1985).

The same group has used pulsed carbon dioxide laser photothermal deflection for selective detection of chlorofluorocarbons in chromatographic effluents. As shown in Fig. 1, the carbon dioxide laser can be tuned to monitor Freon-11 (trichlorofluoromethane), Freon-12 (dichlorodifluoromethane) and Freon-22 (chlorodifluoromethane and chlorotrifluoroethylene) simultaneously or to function as a selective detector for any one of them.

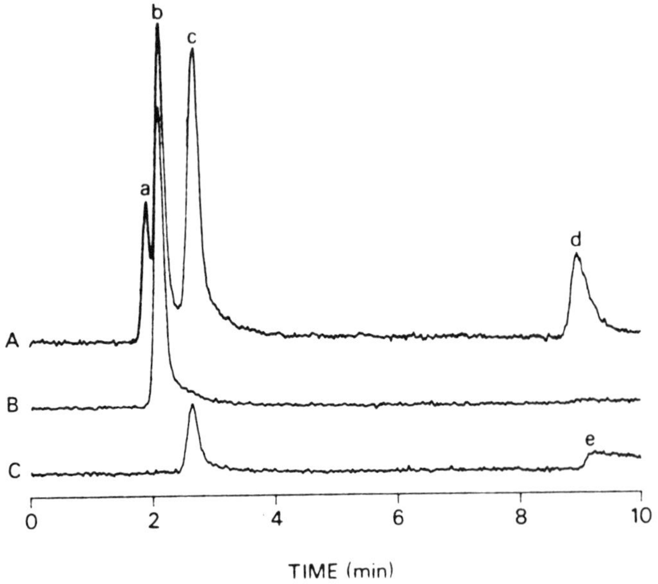

FIG. 1. Gas chromatograms of freon mixture, detected at carbon dioxide laser wavelengths as indicated. (A) 9.282 μM. (B) 9.473 μM. (C) 10.719 μM. The peaks are (a) Freon-22; (b) Freon-12; (c) chlorotrifluoroethylene; (d) Freon-11; (e) dichloromethane (Nickolaisen and Bialkowski, 1986a).

With 20-mJ pulse energies, the mass detection limits for Freons are in the nanogram to high-picogram range. These values are about two orders of magnitude higher than are obtained by electron-capture detection (ECD). However, experiments in static cells suggest that an optimized experimental configuration would probably compare favorably to ECD.

The sensitivity problem of CW laser photothermal detection in the infrared has been addressed by Fung and Gaffney (1986). They employed an intracavity detection scheme, using a modulated CO_2 laser as the heating beam and a He–Ne probe, as shown in Fig. 2. The system has been demonstrated as a detector for permanent gases, using a column packed with Porapack N and operated at $-15°C$ with N_2 carrier gas at flow rates from 20 ml/min to 40 ml/min.

The CW intracavity system has been shown to perform well with both ethylene and sulfur hexafluoride. Linearity and reproducibility are both good. The detection limit for SF_6 is 60 parts per trillion, equivalent to a change of refractive index of 5×10^{-8}. That value is within a factor of 2 of the detection limit obtained under similar experimental conditions using phase-fluctuation optical heterodyne spectroscopy (Lin, Gaffney, and

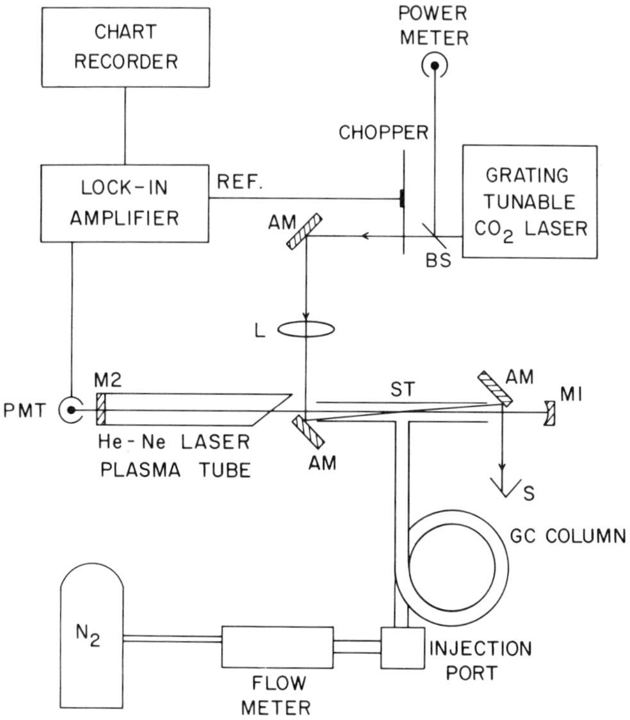

FIG. 2. Experimental configuration for gas chromatography with intracavity photothermal deflection detection. (AM) aluminum mirror; (BS) beam splitter; (L) focusing lens; (M1) and (M2) laser resonator mirrors; (PMT) photomultiplier tube; (S) beam stop; (ST) tubular sample cell (Fung and Gaffney, 1986.)

Campillo, 1981). However, the photothermal deflection system is far simpler to construct and maintain.

III. Photothermal Liquid-Chromatography Detectors

Photothermal detectors for liquid chromatography are more highly developed than systems for other separation techniques. Applications to liquid chromatography have been recently reviewed (Morris, 1986). High-performance liquid chromatography is currently the most widely used analytical separation technique. Although a wide variety of detector systems are in use, few offer low detection limits at moderate cost combined with ready adaptability to submicroliter flow cells.

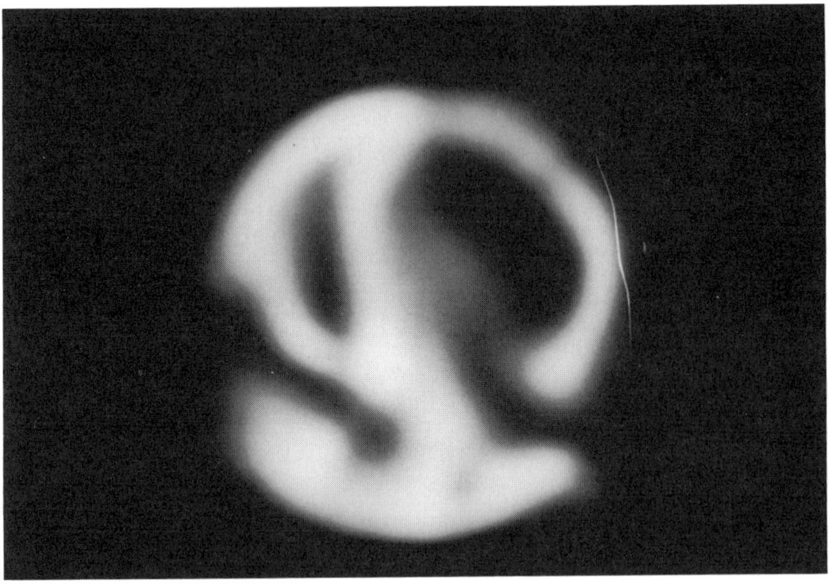

FIG. 3. Shadowgraph of flow pattern in liquid-chromatography flow cell. The sample is water-injected into 50% methanol. The dark areas are regions of rapidly changing refractive index (Peck and Morris, 1988a).

Absorbance cells for liquid chromatography deliberately include provision for introducing some nonuniformity into mobile phase flow. Typically, the liquid entering the cell is directed onto one window. The exit flow is at an angle to the other window. This "Z"-configuration insures that there are no stagnant regions of liquid in the cell to degrade resolution.

There are adverse consequences to the conventional cell design. The sudden changes in flow direction introduce high-frequency components in several directions into the flow velocity to induce mixing. The mixing process breaks up thermal lenses and becomes a source of excess noise.

Sample mixing is not actually complete during the residence time of the cell (Peck and Morris, 1988a). The shadowgraph photograph in Fig. 3 shows typical behavior of a commercial 1-mm-path-length cell. The sample flow bifurcates as a result of contact with the cell window. The mixing action flushes the cell more rapidly than unperturbed laminar flow, but makes local concentration a strong function of position. This behavior can lead to systematic errors in thermal lens measurements, which probe a small fraction of the cell volume.

These problems were recognized early in the development of photothermal and photoacoustic detectors. Many early reports in both fields are devoted to attempts to describe and circumvent the problems. Oda and Sawada (1981) studied the modulation frequency dependence of photoacoustic signals and signal–noise ratio in a liquid-chromatographic detector. Their cell, which employed a "U"-flow geometry, was noisy at low frequencies, but quiet above 2 kHz, when pumped with a piston pump.

The problem can be alleviated with pumps that produce smooth rather than pulsating flow. Buffett and Morris (1983) found that the combination of a Kratos SFA-234 flow cell, which employs a "Z"-flow geometry, and a dual-piston pump with pulse dampener could not be successfully employed in thermal lens detectors. The same cell performed well when used with a syringe pump, which produces pulseless flow. Because they minimize problems with flow fluctuations, syringe pumps are most commonly employed with thermal lens detectors.

Pang and Morris (1985) demonstrated that flow-related noise could also be reduced by using matched-sample and mobile phase blank flows in a series differential thermal lens configuration. To the extent that flow patterns in the cells are matched, they can be made to cancel. Imprecise measures of flow matching were used, but noise reduction by a factor of 2.5–4 was observed. Further improvement might be obtained by careful matching of system components and careful flow balancing. However, this approach doubles solvent consumption, and would only be practical for use with microbore or capillary columns, in which flow rates are 50 μl/min or less.

Mobile phase flow is equivalent to an increase in thermal conductivity of the system (Dovichi and Harris, 1981). Over the limited flow range encountered in liquid-chromatographic practice, the change in thermal lens enhancement factor with flow velocity, dE/dv, is constant. Flow-rate fluctuations are a source of noise. At low frequencies, the noise in a flowing thermal lens system can be described as the sum of a static contribution and an independent contribution that depends on flow-rate fluctuations (Eq. 1).

$$\sigma^2 = \sigma_0^2 + A^2\left(\frac{dE}{dv}\right)^2 \sigma_v^2. \tag{1}$$

The standard deviation of the thermal-lens strength in the flow system is σ, the component from the stationary thermal lens is σ_0, and from flow-rate fluctuations is σ_v; E is the thermal lens enhancement factor, A is the absorbance, and v is the mobile phase flow velocity. Equation (1) is an incomplete description of the thermal lens in a typical liquid chromatogra-

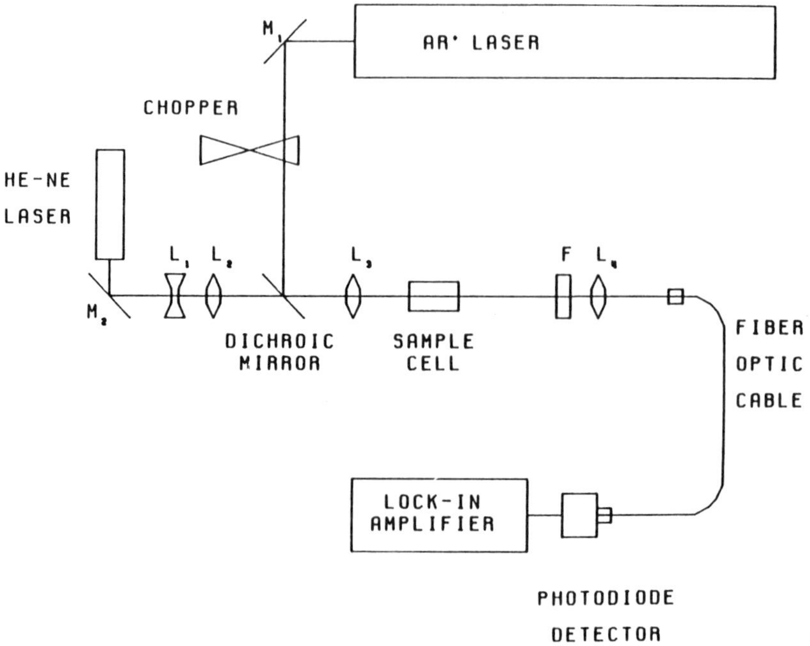

FIG. 4. Thermal lens detector for liquid chromatography. (L_1) h −40 mm f.l.; (L_2) 125 mm f.l.; (L_3) and (L_4) 100 mm f.l.; (F) Corning 3–69; (M_1) aluminum mirror (Buffett and Morris, 1982. Reprinted by permission of the American Chemical Society.)

phy, since it accounts for pumping fluctuations only, and not the perturbations caused by the flow patterns within the cell.

Leach and Harris (1981) were the first to use the thermal lens effect for liquid chromatography detection. They employed a time-resolved single-beam measurement with an argon-ion laser (190 mW) and an 18-μl 10-mm-path-length flow cell. Using a reverse-phase system (50 : 50 methanol–water) and a 250 × 4.6-mm column packed with ODS on 5-μm silica, they demonstrated noise levels of about 1.5×10^{-5} absorbance/cm, with a 5-sec time constant. These early results were about an order of magnitude better than absorbance detectors of the same period.

Buffett and Morris (1982) introduced the two-laser (pump–probe) configuration (Fig. 4) to liquid chromatography (LC) detection. They also employed a conventional reverse-phase system (LiChrosorb RP-18 on 10-μm silica, 80 : 20 methanol) and a 10-mm-path flow cell. As shown in Fig. 5, they were able to obtain absorbance noise of about 2×10^{-6} with a 1-sec time constant. The performance improvement over the Leach and Harris system can be attributed to two factors. First, the use of an 80 : 20 methanol

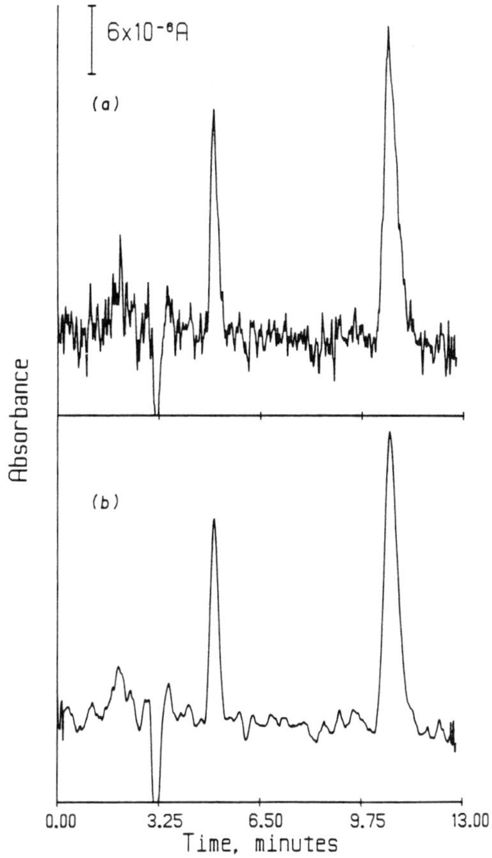

FIG. 5. Chromatograms of o-nitroaniline (0.56 ng) and N,N-dimethyl-3-nitroaniline (3.4 ng): (a) unsmoothed; (b) 25-point smooth. (Buffett and Morris, 1982. Reprinted by permission of the American Chemical Society.)

mobile phase increased the sensitivity of the experiment. Second, with the pump/probe system, they were able to use synchronous demodulation to recover the thermal lens signal. Leach and Harris were forced to use a simplified data extraction algorithm that achieved real-time performance in their time-resolved system at the expense of signal–noise ratio.

Buffett and Morris (1983) demonstrated that high laser power was not needed for low-noise liquid-chromatography detection. They were able to reduce the pump-beam power from 100 mW to 7.5 mW with no degradation in signal-to-noise ratio, if they maintained a pump–probe ratio of 30 : 1 or higher.

Since the demonstration of the advantages of the pump–probe configuration for liquid-chromatographic detection, most workers have employed variants on this technique. In addition to thermal lens measurements, photothermal deflection with pump and probe beams crossed at a small angle (Collette *et al.*, 1986) or at right angles (Nolan, Hart, and Dovichi, 1985) have been employed in liquid-chromatographic measurements.

Several groups have attempted single-CW-laser variants on the configuration. Typically, a small fraction of the laser beam is used as the probe. Polarization encoding (Yang, 1984; Pang and Morris, 1985) and counter-propagating beams (Yang, Hall, and De La Cruz, 1986) are the most common versions. Alternatively, the thermal lens signal can be detected on the modulated pump laser itself at the second harmonic of modulation frequency (Pang and Morris, 1984). These configurations eliminate the alignment instability problems of the two-laser designs. Unless the beams are crossed or counter-propagating, most of these schemes place extraordinary demands on the rejection properties of polarizing prisms or lock-in amplifiers. They provide somewhat worse performance than the traditional configurations.

Skogerboe and Yeung (1986) have used high-frequency (150 kHz) acousto-optic modulation to direct a single laser beam alternately through and around a flow cell to the same photodiode to achieve a pseudo-steady-state measurement of the thermal lens (Fig. 6). This approach allows direct differential measurement by lock-in detection, without the disadvantages associated with some of the earlier single-laser designs.

Early photothermal detectors achieved noise levels equivalent to 10^{-5}–10^{-6} absorbance (Leach and Harris, 1981; Buffett and Morris, 1983).

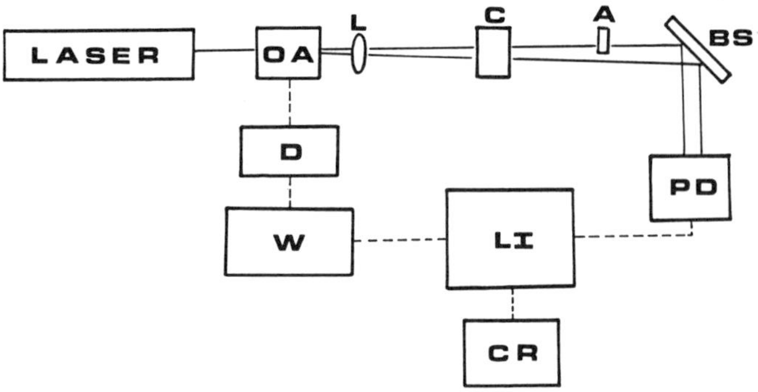

FIG. 6. Experimental apparatus for high-frequency modulation thermal lens detector. (OA) Bragg cell; (D) driver; (L) lens; (C) chromatographic flow cell; (A) aperture; (BS) beam splitter; (PD) photodiode; (W) square wave generator; (LI) lock-in amplifier; (CR) chart recorder. (Skogerboe and Yeung, 1986. Reprinted by permission of the American Chemical Society.)

More recent designs have demonstrated performance approaching 10^{-7} absorbance (Nolan, Hart, and Dovichi, 1985; Collette *et al.*, 1986). The improvement results largely from careful choice of pump and flow cells to minimize noise associated with these components, rather than from improvements in the photothermal experiment itself.

Most research on photothermal detectors has focused on detector design rather than application. Nitroanilines have been common substrates for these studies (Leach and Harris, 1981; Buffett and Morris, 1983). Nitroanilines absorb fairly strongly at the blue wavelengths of argon-ion and helium–cadmium lasers, although their absorption maxima are in the 350–400-nm range.

Nolan, Hart, and Dovichi (1985) have shown that DABSYL derivatives of amino acids can be detected at the femtomole level, using a microbore (1 mm i.d.) C-18 column and 60% methanol as the mobile phase, and argon-ion 488 nm to generate the thermal lens. Detection limits range from about 5 femtomoles for glycine to 300 femtomoles for methionine. The wide range of detection limits results from differences in the reactivity of the amino acids to the chromogenic reagent and differences in extraction

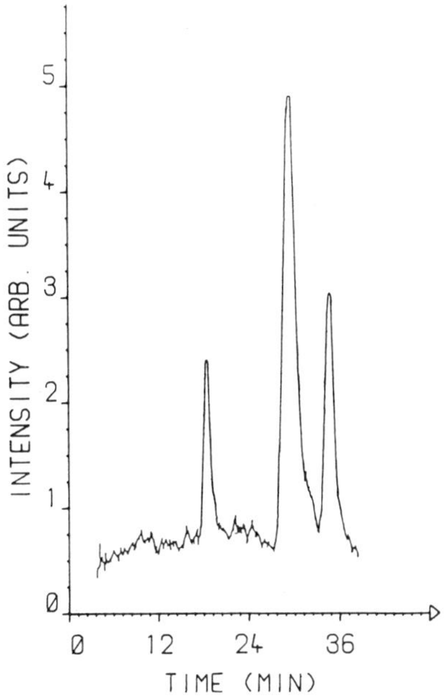

FIG. 7. Chromatograms of sulfonate rug dyes AR337, AO128, and AY151 in 65% methanol, Ar^+ 458-nm pump beam. (Collette *et al.*, 1986.)

efficiency. The absorption coefficients of the amino acid DABSYL derivatives at 488 nm vary by only a factor of 2.

Collette *et al.* (1986) have used photothermal detection to quantify acid sulfonate rug dyes (Fig. 7) with about 500-femtogram (1–2 femtomoles) detection limits. A C-18 microbore column and 65% methanol were employed. They also successfully employed gradient elution to improve the separation. At 488 nm, the solvent background is sufficiently low that the change in background is small compared to the size of the signals observed. However, detection limits are increased about an order of magnitude.

There have been no published studies of pulsed ultraviolet lasers as photothermal liquid-chromatography detectors. The ultraviolet region is more useful to practicing chromatographers than the visible. Fixed-wavelength photothermal detectors operating at 250 nm or lower, or tunable detectors in the 200–300-nm range would be of great interest to the chromatographic community. Nikolaisen and Bialkowski (1986b) have reported preliminary flow-cell studies using a pulsed nitrogen laser as the pump source. With a capillary tube, which provides unperturbed laminar flow, radial velocity dispersion causes deviation from the expected (milliseconds) time dependence. There appear to be no insurmountable obstacles to the use of low-energy pulsed ultraviolet lasers in photothermal detectors.

IV. Photothermal Flow-Injection Analysis Detectors

Flow-injection analysis (FIA) has certain operational similarities to liquid chromatography (Ruzicka and Hansen, 1981). FIA employs flow rates in the ml/min range. LC detectors and fittings are often used in FIA systems. However, because sample must be mixed with one or more reagent streams, FIA systems use ganged peristaltic pumps rather than piston or syringe pumps. Versatility and economy are gained at the expense of pulseless flow. Flow-related noise problems are more severe in photothermal FIA detectors than in LC detectors. However, FIA is used to automate colorimetric analysis. There is greater need for detectors operating in the mid-visible than in liquid chromatography.

An FIA detector based on a single-beam thermal lens measurement has been demonstrated (Leach and Harris, 1984). Colorimetric determination of iron as its 1,10-phenanthroline chelate yields detection limits of 37-pg iron in a 100-μl sample, with a flow rate of 0.5 ml/min. A crossed-beam thermal lens system with polarization encoding of pump and probe beams has been used to measure bromophenol blue (Yang and Hairrell, 1984). Detection limits of 2.5×10^{-7} M have been demonstrated. Better performance is obtained with the crossed-beam system than with collinear beams, because of spatial rejection of the pump-beam intensity.

Yang and Hairrell have investigated the flow-rate behavior of their detector over the range 0.1–10 ml/min. They observed optimum response at a flow rate of 0.3 ml/min. Above this rate, the response declines rapidly with increasing flow rate.

V. Photothermal Electrophoresis Detectors

Photothermal spectroscopy appears to be more closely matched to the detection needs of gel electrophoresis than to those of most other separation techniques. The most widely used protein and nucleic acid stains absorb at the visible output wavelengths of the common gas lasers. The Coomassie dyes are blue and are detectable with either a He–Ne laser or an argon-ion laser. The plasmon resonances of colloidal silver extend out to at least 500 nm, making detection with He–Cd or argon-ion lasers feasible. The more specialized staining techniques also employ reagents with strong visible transitions that allow detection by convenient gas lasers.

Peck and Morris (1986a, b; 1988b) have demonstrated ultrasensitive photothermal detection and quantification of Coomassie Brilliant Blue–stained proteins on polyacrylamide gels. They have simply employed mechanical scanning of the stained gels through a stationary optical system. Both He–Ne 633-nm and Ar^+ 528.7-nm radiation have been used to generate the thermal lens, usually with about 20 mW delivered to the sample. Both crossed-beam and collinear-beam pump and probe configurations have provided the same performance.

Preliminary studies using albumin stained with Coomassie Brilliant Blue G250 in a nondenaturing electrophoresis (Davis, 1967) on a 1-mm-thick gel demonstrated a detection limit of 0.95 ng. This detection limit is two orders of magnitude below the threshold of visibility and at least one order of magnitude better than conventional densitometry. Quite similar detection limits have been obtained for several proteins separated on denaturing gels (Laemmli, 1970) using either Coomassie Brilliant Blue R250 or G250.

Polyacrylamide electrophoresis gels usually contain 5–20% polymer in an aqueous buffer. Peck and Morris (1988b) have shown that the gels have the same thermal properties as aqueous solutions, but larger refractive-index temperature coefficients. They model the thermal lens in a polyacrylamide gel as an expansion of an aqueous solution through a rigid porous matrix. The presence of polyacrylamide modifies the optical properties of the

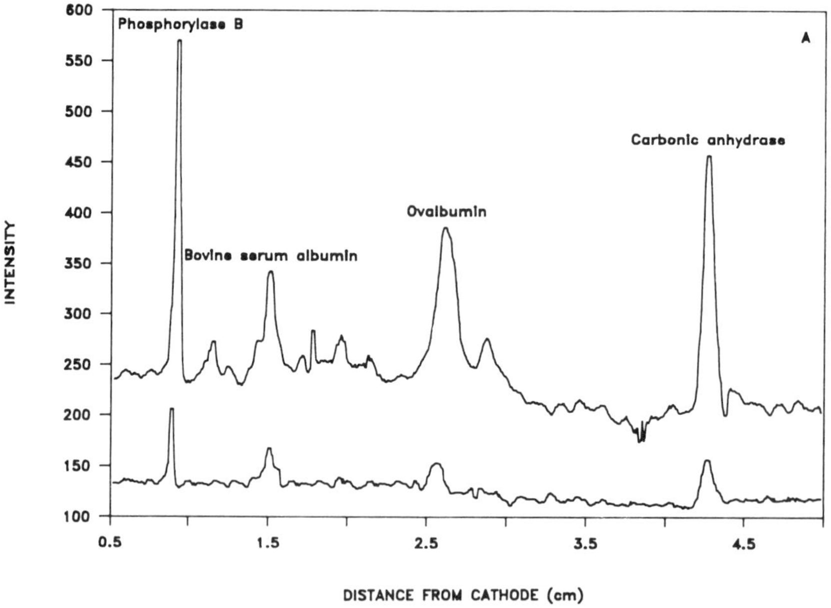

FIG. 8. Thermal lens densitometry of proteins stained with (A) Coomassie Brilliant Blue R250 and (B) Coomassie Brilliant Blue G250. Each graph shows both 70-ng and 10-ng quantities of proteins. (Peck and Morris, 1986b. Reprinted by permission of the American Chemical Society.)

system only, leading to a thermo-optic response that is 1.5 to 2 times greater than in an aqueous solution.

Although the thermo-optic properties of a polyacrylamide gel are no more than twice as favorable as those of water, the Peck–Morris densitometer provides detection limits close to 50 times better than are obtainable by conventional transmission densitometers. The improvement results in part because most of a photothermal signal is generated near the focal volume of the pump laser (Carter and Harris, 1983, 1984). Modern protein gel electrophoresis employs gels 0.5–1.0 mm thick. The short path length is more disadvantageous to transmission densitometry based directly on Beer's law than to photothermal densitometry.

A consequence of the use of tightly focused laser beams is that a photothermal densitometer interrogates only a small region of the protein sample. Co-addition of two scans of each lane, separated by about 1 mm, improves the reproducibility of the data. The small separation provides sufficient averaging of local variations in stain intensity.

Figure 8 shows typical thermal lens traces for 70-ng and 10-ng quantities of stained proteins separated on 10% denaturing polyacrylamide gels.

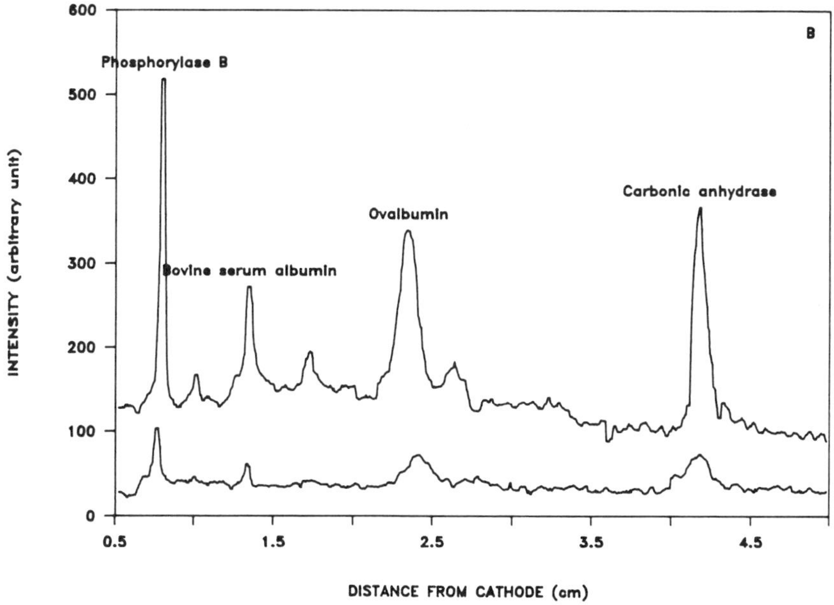

FIG. 8. (*continued*)

Although the base lines appear noisy, they actually contain many reproducible small peaks caused by the presence of minor impurities or degradation products in the protein mixtures used. Some of the more prominent minor peaks are easily visible in the traces from 70-ng samples.

Both Coomassie stains provide the same detection limits, about 0.5–1.5 ng, depending on the protein. Coomassie Brilliant Blue G250 stains are stable for several weeks. Historically, Coomassie Brilliant Blue R250 has been more widely used for this application, although R250 stains are stable for only two or three days, and the gels require more destaining than with G250.

The data in Fig. 8 have been subjected to software removal of noise spikes and a sloping baseline by robust smoothing (Bussian and Hardle, 1984). The sloping base line results from slight misalignment of the gel holder in the optical path. The spikes in the are caused by scattering of the probe laser beam. Scattering centers are formed by minor damage to the gels that occurs during the staining, destaining, and mounting procedures. Dust and impurities in the gel and dye are further minor contributors to this noise source.

Figure 9 shows calibration curves obtained for several proteins stained with Coomassie Brilliant Blue R250 and G250. The curves are linear over the 10–100-ng region but fall off steeply below 10 ng. The most probable

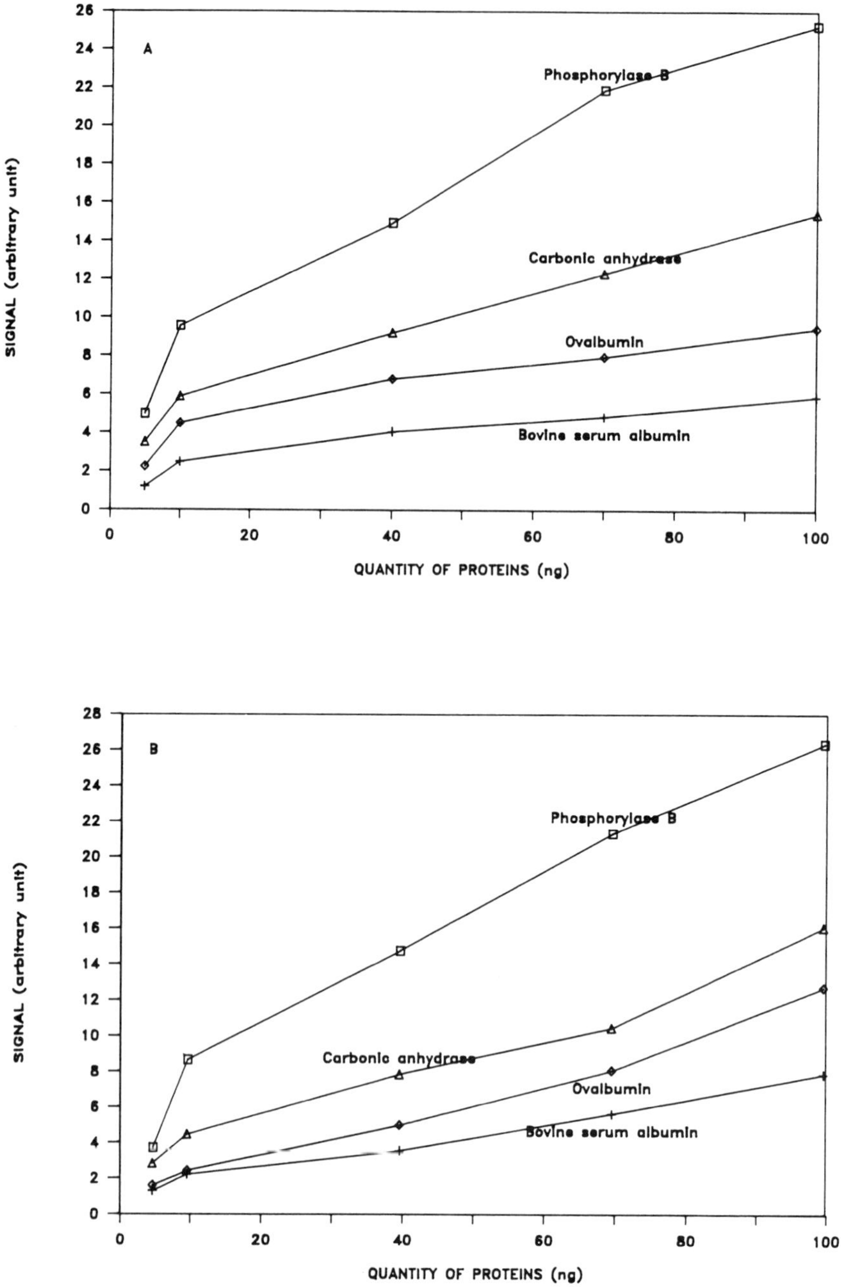

FIG. 9. Calibration curves for proteins stained with (A) Coomassie Brilliant Blue R250 and (B) Coomassie Brilliant Blue G250. (Peck and Morris, 1986b. Reprinted by permission of the American Chemical Society.)

explanation is that the dye–protein complexes dissociate at these low concentrations.

In early work, gels were scanned at 0.18 mm/sec. More recent designs (Peck and Morris, 1987b) allow an increase in scan rate to about 4 mm/sec with no decrease in signal–noise ratio. The increased scan rate is due largely to the use of higher-quality mechanical components and some refinement in the optical design.

Photothermal densitometry is not limited to Coomassie Blue–stained proteins. Other protein stains or silver staining can be used with this system.

VI. Photothermal Detectors for Thin-Layer Chromatography

A. Single-Point Measurements

Chen and Morris (1984a) have used transverse photothermal deflection to quantify compounds separated by thin-layer chromatography (TLC). Detection limits in the picogram range were achieved for intensely absorbing derivatives of several quinones. Their system employed an argon-ion laser to generate the mirage and mechanical scanning of the TLC plate in a stationary optical system. As with most mirage-effect systems, satisfactory operation required shielding from laboratory air currents, which degraded signal-to-noise ratios by a factor of 10.

Photothermal signals from typical chromatograms are shown in Fig. 10. The spatial resolution of the system shows the irregularity of the chromatographic process. In trace b, the α-ionone band is broken into several resolved sub-bands.

Pump- and probe-beam shaping has been shown to reduce detection limits by about one order of magnitude, by allowing integration over local irregularities in the chromatogram (Chen and Morris, 1984b). By using a one-dimensional $2\times$ beam expander to increase the probe area parallel to the plate and a defocused pump-laser beam to match the probe cross-section, Chen and Morris found a 1.3-picogram detection limit for phenanthrenequinone, compared to 31 picograms with an unexpanded beam. A similar improvement is obtained for 1,2-naphthoquinone. However, detection limits for α-ionone are unchanged, apparently because the limiting factor is partial decomposition of the ionone in the sulfuric acid reagent.

Peck *et al.* (1985) have shown that careful optimization of experimental parameters including the angle between the probe-laser beam and the TLC plate, probe-beam offset, and pump-beam modulation frequency are re-

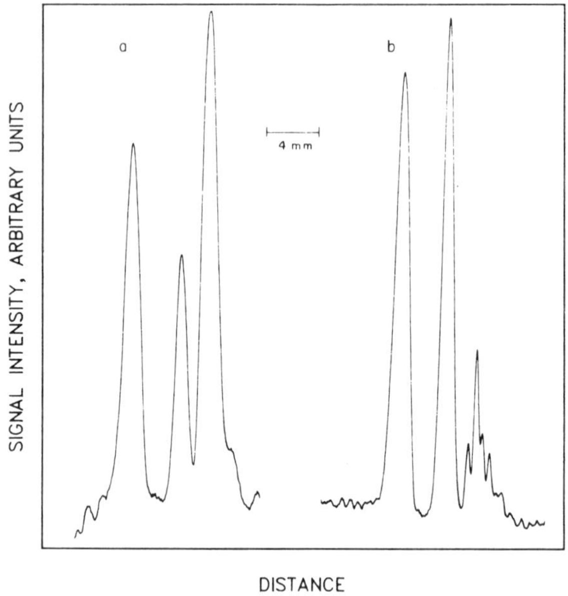

FIG. 10. Transverse photothermal deflection detection of derivatized quinones on thin-layer chromatography plates. (a) 144-ng 1,2-naphthoquinone, 93-ng phenanthrenequinone, 144-pg α-ionone. (b) 144-ng 1,2-naphthoquinone, 186-ng phenanthrenequinone, 14.9-pg α-ionone. (Chen and Morris, 1984a. Reprinted by permission of the American Chemical Society.)

quired to obtain maximum sensitivity from the photothermal densitometer. By using only 21-mW pump-beam power, they have demonstrated that their system gives 5 × lower detection limits than commercial reflectance or transmission densitometers. They have also identified wobble in the mechanical stage as a serious source of excess noise, and estimate that a further 4 × reduction in detection limits could be obtained with smooth stage travel.

Masujima *et al.* (1985) have described a mirage-effect densitometer using an intracavity probe arrangement. They obtained detection limits 20 × lower with the intracavity configuration than with a conventional extracavity design. Although the observed detection limits are similar to those of Peck *et. al.* (1985), the work suggests that a completely optimized intracavity probe design would produce extremely low detection limits.

Fotiou and Morris (1986a) have developed a densitometer based on a pulsed laser, which provides access to the ultraviolet. They used a 351-nm XeF excimer laser in the mirage-effect configuration, with gated integrator processing of the transient thermal refractive-index gradient.

To avoid thermal damage, the laser pulse energy delivered to the sample was limited to 1–2 mJ. With this system, they obtained a subnanogram detection limit and three-decade linearity for 2,4-dinitroaniline. In contrast to CW photothermal deflection, the pulsed system was found to be only moderately susceptible to laboratory air currents. Operation of the system without its enclosure increased noise signals by no more than $4\times$. Even with millijoule pulse energies, detection limits were set by absorption by residual impurities on the plate and possibly by the organic binder on the plate. The background from these sources obscured small signals.

Pulsed photothermal deflection spectroscopy (PDS) with UV lasers is sensitive and experimentally simple. The low power requirements demonstrate that the technique can be useful with relatively inexpensive pulsed lasers. Lower detection limits can be reached by careful pre-elution of plates to remove impurities. It may also be necessary to use plates with binders selected for lowest absorption at the wavelengths of the experiment. Even further improvements are possible by increasing the incident power to the sample. However, power increases must be made with caution, because photochemical transformations or photodecomposition are common. High average power is best achieved with low-pulse-energy/high-rep-rate lasers or by the use of distributed power, as described below.

B. Spatial Multiplexing

Pulsed-laser illumination is difficult to employ with solid samples. Gross thermal decomposition is common, especially with high-peak-power pulsed lasers. Less obviously, photochemical transformations, such as *cis–trans* isomerization, may be quite facile under laser illumination. Since photochemical reaction may cause absorbance changes without obvious changes in color, systematic errors from that source can go completely undetected.

It is important to reconcile the need for high laser power or energy to maximize signal-to-noise ratio with the need for gentle irradiation to minimize unwanted sample transformations. This problem has been recognized by several groups working in both photoacoustic and transverse photothermal deflection spectroscopies. Signal multiplexing is also described in Chapter 2 of this book.

Distribution of the laser energy over the total sample surface is necessary in order to minimize or to avoid the thermal damage and to retain the inherent sensitivity of PDS. Coufal, Moller, and Schneider (1982a, b) have demonstrated the feasibility of power distribution in photoacoustic spectroscopy using both Hadamard (Coufal *et al.*, 1982a) and Fourier

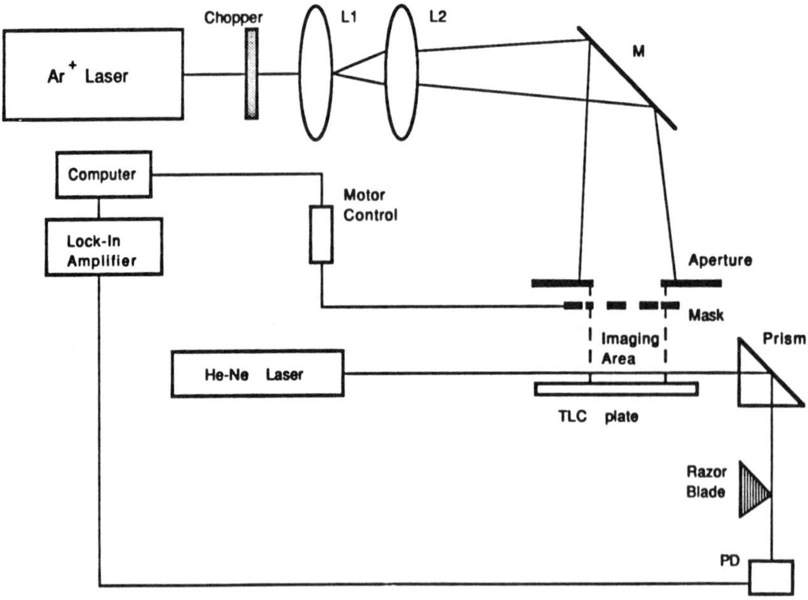

FIG. 11. Apparatus for Hadamard transform transverse photothermal deflection imaging. (PD) photodiode; (M) mirror; (L_1) and (L_2) lenses. (Fotiou and Morris, 1987a. Reprinted by permission of the American Chemical Society.)

(Coufal *et al.* 1982b) coding of the expanded excitation beam. Hadamard masks are simpler to construct, whereas Fourier (sinusoidal) masks allow variable resolution from a single mask.

Fotiou and Morris introduced the use of Hadamard coding for transverse photothermal densitometry employing both pulsed (1986b) and CW (1987a) lasers as light sources. In their work, an encoded laser beam was line-focused and used to irradiate samples on a stationary TLC plate to create refractive-index gradients for transverse photothermal deflection measurements. Their apparatus is shown as Fig. 11.

Hadamard transform photothermal deflection spectroscopy takes advantage of the fact that the photothermal deflection signal generated along a line is the sum of the incremental signals generated at each point along the line. In this multiplexing technique, the excitation beam is passed through a mask that encodes it. The encoded beam is line focused along the direction of the probe beam. Therefore, the signal is the sum of signals from each illuminated resolution element along the line of the probe beam. A resolution element is defined by the image of the unit aperture of the mask. If a suitable sequence of masks is chosen, it is possible to make a series of measurements from which the signals at each resolution element can be recovered.

When a mask consisting of opaque and transparent slits intercepts the pump beam, the measured signal generated by mask j is given by Eq. (2).

$$y_j = \sum_{i=1}^{n} s_{ij} x_i. \tag{2}$$

Here y_j is the signal from the jth mask, $s_j = (s_{1j}, s_{2j}, \ldots, s_{nj})$ is the matrix of mask elements of the jth mask, and x_i is the signal at the position i on the sample. The element s_{ij} has the value 1 if the corresponding ith resolution element is illuminated, and 0 if it is not illuminated. For n resolution elements, a sequence of at least n different masks is required. If the n masks are properly chosen, they define a system of n linear independent equations that completely describe the system:

$$y_1 = \sum_{i=1}^{n} s_{i1} x_i, \tag{3.1}$$

$$y_2 = \sum_{i=1}^{n} s_{i2} x_i, \tag{3.2}$$

$$\ldots$$

$$y_n = \sum_{i=1}^{n} s_{in} x_i. \tag{3.n}$$

In matrix notation, the set of equations (3) may be written as Eq. (4).

$$\mathbf{Y} = \mathbf{S} \cdot \mathbf{X}. \tag{4}$$

The system can be solved by calculating the inverse of the matrix $\mathbf{S}$, according to Eq. (5).

$$\mathbf{X} = \mathbf{S}^{-1} \cdot \mathbf{Y}. \tag{5}$$

The inverse $\mathbf{S}^{-1}$ is computationally simple to generate and is given by Eq. (6).

$$\mathbf{S}^{-1} = \frac{\mathbf{2}}{\mathbf{n+1}} \cdot \mathbf{W}. \tag{6}$$

Here $\mathbf{W}$ is a matrix that has -1's where $\mathbf{S}^{\mathrm{T}}$ has 0's, and $+1$'s where $\mathbf{S}^{\mathrm{T}}$ has $+1$'s (T stands for transpose).

There are three known constructions for S-matrices that yield cyclic matrices (Harwit and Sloane, 1979). A cyclic matrix has the property that

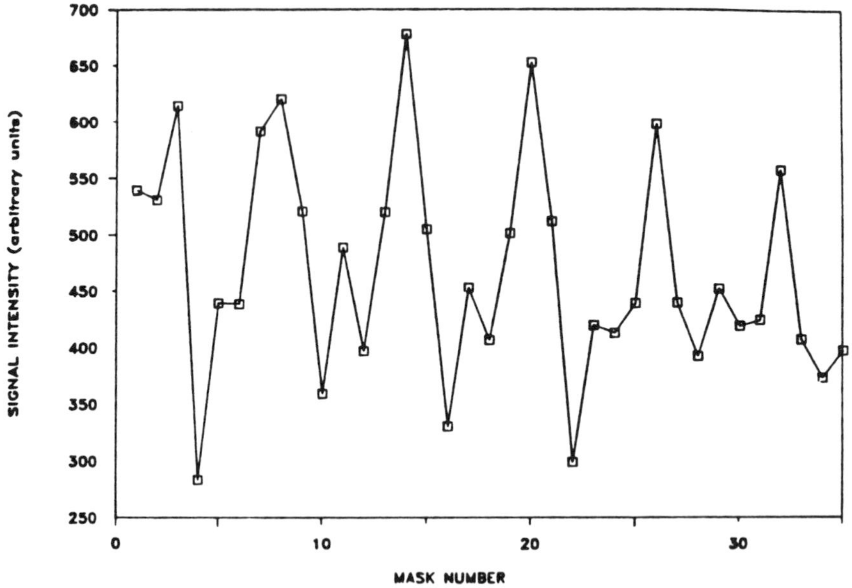

FIG. 12. Hadamard mask-encoded transverse photothermal deflection signals obtained with a 35-active-element mask. The data is obtained from a TLC plate spotted with six 100-ng quantities of *trans*-azobenzene on 2.5-mm spacings. (Fotiou and Morris, 1987a. Reprinted by permission of the American Chemical Society.)

each row is generated from the previous one by shifting its elements one position to the left (or right) and placing the overflow in the position of the element that was first shifted. The advantage of using a cyclic matrix is that a single mask of $2n - 1$ slits can be used to generate the configurations of all n individual masks. This mask is called a Hadamard mask. In operation, the Hadamard mask is shifted incrementally underneath a limiting aperture (or frame) a distance of one slit width. Each shift of the mask generates another row of the S-matrix. Shifting the mask $n - 1$ times generates the entire S-matrix sequentially.

Figure 12 shows experimental data for a Hadamard transform photothermal system based on a 69-element Hadamard mask with 406-μm element width and an Ar^+ laser as the pump beam (Fotiou and Morris, 1987a). The encoded data are from a thin-layer chromatography plate spotted with six 100-ng aliquots of *trans*-azobenzene deposited at 2.5-mm intervals along a straight line. The transformed data are plotted in Fig. 13A. The recovered spot separations are in good agreement with the actual spot separations. Peak areas do not correspond to the actual amounts of analyte.

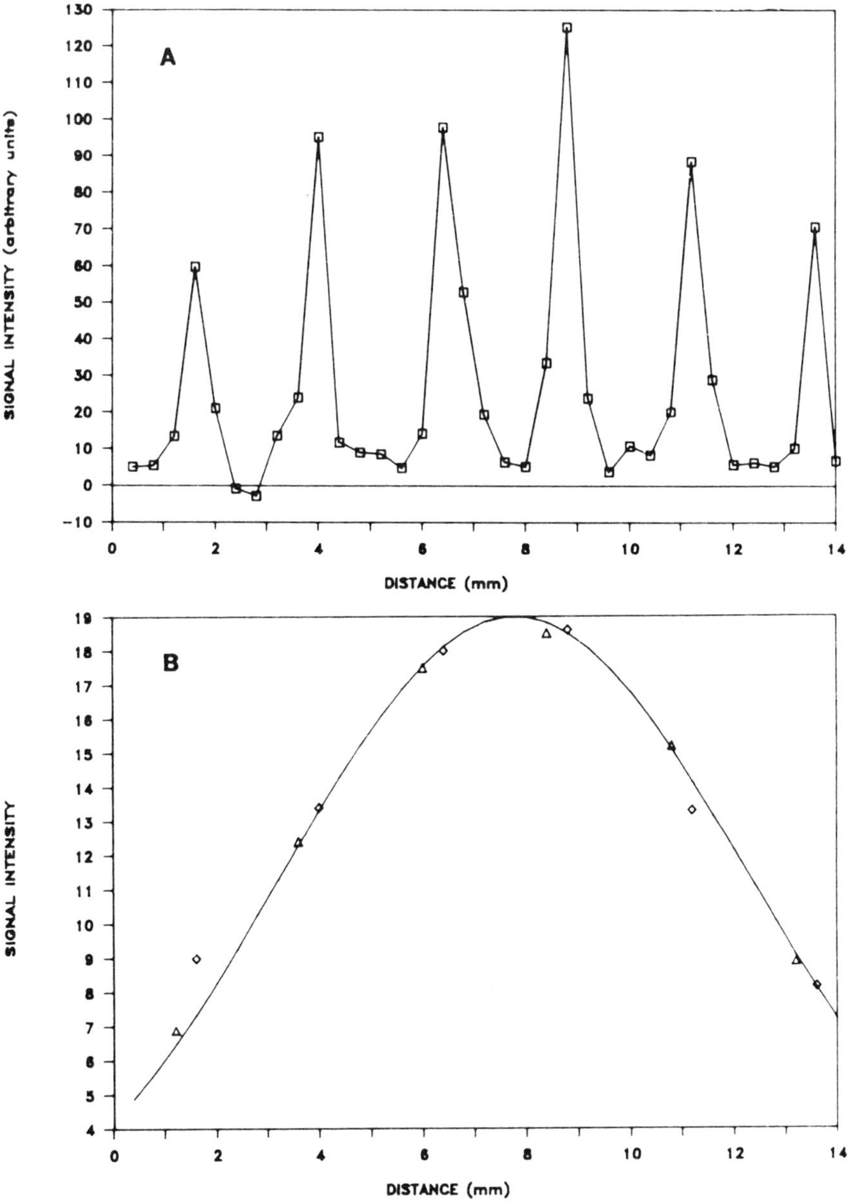

FIG. 13. (A) data of Figure 12 recovered by inverse Hadamard transformation. (B) areas under the peaks of the data of trace A. Open squares denote data from samples at original positions. Open triangles denote data from samples after plate is translated through one resolution element. Solid line shows regression of data to a Gaussian. (Fotiou and Morris, 1987a. Reprinted by permission of the American Chemical Society.)

The data of Fig. 13A have not been corrected for nonuniform photothermal response across the plate. The nonuniform response results largely from the pump-beam Gaussian intensity profile. There are small contributions from imperfect collimation of the probe beam and from imperfect alignment to the plane of the TLC plate. There is a further error because the absolute deflection of the laser beam is taken as a measure of the angular deflection, causing a systematic variation in sensitivity along the probed region. Several procedures for compensation for these factors have been suggested (Fotiou and Morris, 1987a). The simplest is normalization to a calibration curve obtained from identical spots of an absorbing compound. A calibration curve is shown in Fig. 13B.

The Hadamard transform approach is surprisingly simple to implement and causes only a small increase in detection limit (Fotiou and Morris, 1987a). Hadamard masks containing up to 509 elements and with element widths as small as 25 μm have been used successfully (Fotiou and Morris, 1987b). With adequately precise stepwise motion and properly fabricated masks, spatial resolution of a few microns should be readily achievable.

The modulation transfer function (MTF) of a Hadamard-encoded photothermal system has been investigated by Fotiou and Morris (1987b). They demonstrate that if the pump-laser beam is adequately collimated, photothermal measurements are made in the near field. In this case, a mask behaves as a system of independent apertures of unit width, rather than a far-field ensemble of apertures. For masks with coarse unit aperture ($d \geq 20$ μm), the largest contribution to the transfer function is convolution with the aperture width, which causes a linear decrease in contrast with spatial frequency. If the mask is continuously moved, rather than incrementally shifted, then a blur contribution to the MTF becomes important, and the transfer function becomes parabolic in spatial frequency. Fotiou and Morris were unable to observe thermal diffusion contributions to the MTF. Under their experimental conditions, thermal diffusion should be a small contributor to the MTF.

Like any multiplexing technique, Hadamard masking has the property that A/D converter resolution is divided between representation of multiple channels and representation of signal intensities. If a 2^j-element mask is employed, and the signal dynamic range in each channel is 2^k, and one bit is used to represent noise, then Eq. (7) describes b, the number of bits required to digitize the Hadamard-encoded photothermal image (Morris and Fotiou, 1987b):

$$b = j + k - 4. \tag{7}$$

Equation (7) has been verified for photothermal signals obtained from 63-, 127-, and 255-element mask systems. The A/D-converter resolution requirements of multiplexed Hadamard-encoded signals are surprisingly small. Dynamic range compression of the signals results from the fact that high resolution is required only to encode small changes in the shape of the signals. Limiting the A/D converter range usually smooths the signal somewhat and may lead to loss of resolution if the conditions of Eq. (6) are violated.

In practice, the signal dynamic range rarely exceeds 4 or 5 bits. A standard 12-bit converter is adequate to handle signals from a 4095-element mask. A 16-bit converter would be needed to handle masks up to 16 K elements with sufficient data dynamic range to be useful. It is unlikely that any transducer presently available has sufficient linear dynamic range to allow greater resolution than 16 bits.

VII. Conclusions

Experimental photothermal detectors have been constructed for use in most forms of chromatography and electrophoresis. Although these detectors provide excellent sensitivity, they are not yet in widespread use. The reasons are merely technological. Adequate power, wavelength coverage or tunability is usually found only in a large package, which is unsuited to the chromatographic laboratory. Recent advances in diode-pumped Nd:YAG lasers and the development of compact nitrogen pumped dye lasers suggests that the size and cost problems will be overcome in the next few years. As small and inexpensive lasers proliferate, the role of photothermal detectors will expand.

REFERENCES

Andrews, A. T. (1986). *Electrophoresis. Theory, Techniques and Biochemical and Clinical Applications*. Oxford Univ. Press, New York.

Bertsch, W., Hara, S., Kaiser, R. E., and Zlatkis, A. (1980). *Instrumental HPTLC*. Huthig, Heidelberg.

Buffett, C. E. and Morris, M. D. (1982). *Anal. Chem.* **54**, 1824–1828.

Buffett, C. E. and Morris, M. D. (1983). *Anal. Chem.* **55**, 376–378.

Bussian, B. M., and Hardle, W. (1984). *Appl. Spectrosc.* **38**, 309–313.

Carter, C. A., and Harris, J. M. (1983). *Appl. Spectrosc.* **37**, 166–172.

Carter, C. A., and Harris, J. M. (1984). *Anal. Chem.* **56**, 922–925.

Chen, T. I., and Morris, M. D. (1984a). *Anal. Chem.* **56**, 19–21.

Chen, T. I., and Morris, M. D. (1984b). *Anal. Chem.* **56**, 1674–1677.

Chu, G., Vollrath, D., and Davis, R. W. (1986). *Science* **234**, 1582–1585.

Collette, T. W., Parekh, N. J., Griffin, J. H., Carreira, L. A., and Rogers, L. B. (1986). *Appl. Spectrosc.* **40**, 164–169.

Coufal, H., Moller, U., and Schneider, S. (1982a). *Appl. Opt.* **21**, 116–120.

Coufal, H., Moller, U., and Schneider, S. (1982b). *Appl. Opt.* **21**, 2339–2343.
Davis, B. T. (1967). *Ann. N.Y. Acad. Sci.* **121**, 404.
Dovichi, N. J., and Harris, J. M. (1981). *Anal. Chem.* **53**, 689–692.
Fenimore, D. C., and Davis, C. M. (1981). *Anal. Chem.* **53**, 252A–266A.
Fotiou, F. K., and Morris, M. D. (1986a). *Appl. Spectrosc.* **40**, 700–704.
Fotiou, F. K., and Morris, M. D. (1986b). *Appl. Spectrosc.* **40**, 704–706.
Fotiou, F. K., and Morris, M. D. (1987a). *Anal. Chem.* **59**, 185–187.
Fotiou, F. K., and Morris, M. D. (1987b). *Anal. Chem.* **59**, 1446–1452.
Fung, K. H., and Gaffney, J. S. (1986). *J. Chromatogr.* **363**, 207–215.
Gershoni, J. M., and Palade, G. E. (1983). *Anal. Biochem.* **131**, 1–15.
Hames, B. D., and Rickwood, D. (1983). *Gel Electrophoresis of Proteins. A Practical Approach.* IRL Press, Oxford.
Harwit, M., and Sloane, N. A. (1979). *Hadamard Transform Optics.* Academic Press, New York.
Heftmann, E. (1984). *Chromatography. Fundamentals and Applications of Chromatographic and Electrophoretic Methods*, Vol. 22A. Elsevier, New York.
Horvath, C. (1986). *High-Performance Liquid Chromatography, Advances and Prospective*, Vol. I, pp. 1–65. Academic Press, New York.
Jorgenson, J. W. (1986). *Anal. Chem.* **58**, 743A–760A.
Laemmli, U. K. (1970). *Nature.* **227**, 680–685.
Leach, R. A., and Harris, J. M. (1981). *J. Chromatogr.* **218**, 15–19.
Leach, R. A., and Harris, J. M. (1984). *Anal. Chim. Acta* **164**, 91–101.
Lin, H. B., Gaffney, J. S., and Campillo, A. J. (1981). *J. Chromatogr.* **206**, 205.
Masujima, T., Sharda, A. N., Lloyd, L. B., Harris, J. M., and Eyring, E. M. (1985). *Anal. Chem.* **56**, 2975–2977.
Morris, M. D. (1986). In *Detectors for Liquid Chromatography* (Yeung, E. S., ed.), pp. 105–147. Wiley, New York.
Nikolaisen, S. L., and Bialkowski, S. E. (1985). *Anal. Chem.* **57**, 758–762.
Nickolaisen, S. L., and Bialkowski, S. E. (1986a). *J. Chromatogr.* **366**, 127–133.
Nickolaisen, S. L., and Bialkowski, S. E. (1986b). *Anal. Chem.* **58**, 215–220.
Nolan, T. G., Hart, B. K., and Dovichi, N. J. (1985). *Anal. Chem.* **57**, 2703–2705.
Oda, J., and Sawada, T. (1981). *Anal. Chem.* **53**, 471–477.
Pang, T.-K. J., and Morris, M. D. (1984). *Anal. Chem.* **56**, 1467–1469.
Pang, T.-K. J., and Morris, M. D. (1985). *Anal. Chem.* **57**, 2153–2155.
Peck, K., and Morris, M. D. (1986a). *Anal. Chem.* **58**, 506–507.
Peck, K., and Morris, M. D. (1986b). *Anal. Chem.* **58**, 2876–2879.
Peck, K., and Morris, M. D. (1988a). *J. Chromatogr.* **448**, 193–201.
Peck, K., and Morris, M. D. (1988b). *Appl. Spectrosc.*, **42**, 513–515.
Peck, K., Fotiou, F. K., and Morris, M. D. (1985). *Anal. Chem.* **57**, 1359–1362.
Perry, J. A. (1981). *Introduction to Analytical Gas Chromatography. History, Principles, and Practice.* Chromatographic Science Series, Vol. 14. M. Dekker, New York.
Poole, C. K. and Scheutte, S. (1984). *Contemporary Practice of Chromatography*, Elsevier, Amsterdam.
Roberts, T. R. (1978). *Radiochromatography. The Chromatography and Electrophoresis of Radiolabelled Compounds*, Vol. 14. Elsevier, Amsterdam.
Ruzicka, J., and Hansen, E. H. (1981). *Flow Injection Analysis*, Wiley, New York.
Sepaniak, M. J., Vargo, J. D., Kettler, C. N., and Maskarinec, M. P. (1984). *Anal. Chem.* **56**, 1252–1257.
Skogerboe, K. J., and Yeung, E. S. (1986). *Anal. Chem.* **58**, 1014–1018.
Yang, Y. (1984). *Anal. Chem.* **56**, 2336–2338.
Yang, Y., and Hairrell, R. E. (1984). *Anal. Chem.* **56**, 3002–3004.
Yang, Y., Hall, S. C., and De La Cruz, M. S. (1986). *Anal. Chem.* **58**, 758–761.

CHAPTER 5

PHOTOTHERMAL STUDIES OF ENERGY TRANSFER AND REACTION RATES

JOHN R. BARKER
BEATRIZ M. TOSELLI
Department of Atmospheric, Oceanic, and Space Sciences
Space Physics Research Laboratory
University of Michigan,
Ann Arbor, MI

I. Introduction 155
II. Time-Resolved Optoacoustics 156
III. Time-Dependent Thermal Lensing 157
IV. Timescales 160
V. TDTL and TROA Theory 162
A. TDTL Optics 162
B. Density and Pressure Perturbations 165
VI. Experimental Methods 169
VII. Observed and Calculated Results 170
A. Signals From Thermal Lensing 170
B. Signals From Acoustic Waves 181
VIII. Conclusions 188
Acknowledgements 188
References 189

I. Introduction

Many different analytical techniques have been used in chemical-kinetics and energy-transfer studies. Both chemical-kinetics and energy-transfer investigations require highly sensitive analytical methods that can be used in real time. Moreover, in energy-transfer studies, it is particularly useful to distinguish between various modes of energy exchange, because this detailed information can help to determine the mechanisms involved. Photothermal spectroscopic techniques are highly sensitive, they can be used in real time, and they only depend on energy deposition in the translations (and rotations), which makes them very specific for measuring energy release to the translational degrees of freedom.

In this chapter, two closely related photothermal techniques will be discussed: time-dependent thermal lensing (TDTL) and time-resolved optoacoustics (TROA). Both of these techniques rely on detection of the

ISBN 0-12-636345-5

density (TDTL), or pressure (TROA) fluctuations induced by a pulsed-light source, such as a pulsed laser. Both of these methods have "steady-state" versions as analytical tools, in which fast time resolution is not utilized (Whinnery, 1974; Pao, 1977). The "steady-state" applications usually involve chopped CW lasers and phase-sensitive detection using lock-in amplifiers to achieve extreme analytical sensitivity. These applications are discussed in other chapters of this volume. [A new photothermal method that will not be discussed here is based on atomic resonance line broadening, which can be used to monitor small time-dependent temperature changes (Braun *et al.*, 1986; Wallington *et al.*, 1987).]

The distinguishing characteristic of the techniques described here is that they require measurement of the actual time record of the photothermal signal produced by a pulsed laser. Extensive signal averaging may be used, but the result is a record of the actual time-dependent signal.

The TDTL and TROA methods have been applied mostly to energy-transfer studies, but there are a few isolated examples of chemical-reaction studies. In this paper, we endeavor to treat energy transfer and chemical reactions in the same way, because only the rate of energy deposition (or absorption) affects photothermal techniques, and it does not matter whether a chemical transformation has taken place.

II. Time-Resolved Optoacoustics

Optoacoustic and ultrasonic measurements have been used for many years as sensitive analytical tools and for the measurement of chemical-reaction rates (Herzfeld and Litovitz, 1959) and of energy transfer in gases (Cottrell and McCoubrey, 1961). With the advent of the laser, optoacoustic spectrophone measurements became practical. In these measurements, the laser beam is chopped, or modulated, and phase-sensitive detection is employed. More recently, individual time-resolved acoustic signals produced by pulsed lasers have been used as analytical tools and for energy deposition rate measurements (Pao, 1977; see, for example, Presser, Barker, and Gordon, 1983).

As discussed in Section V, an acoustic disturbance is produced when the warmer gas at the center of the laser beam expands initially. This disturbance controls the early-time behavior of the thermal lens as it propagates outwards. Once the disturbance is well away from the beam center, it has no influence on the TDTL signal, until it is reflected from the cell walls and returns once more to the center of the cell. If a microphone is placed outside the original laser beams and away from the cell wall, the acoustic disturbance is detected as the pressure fluctuation associated with a density perturbation. Any theory that produces an expression for the density perturbation can be used with the assumption of adiabatic expansion and compression of the acoustic waves (Herzfeld and Litovitz, 1959) to describe the time-resolved optoacoustic signal.

Wrobel and Vala (1978) and Bailey *et al.* (1983a) have developed theories of acoustic waves in cylindrical cells that neglect thermal conductivity and diffusion, but specifically include the rate constant for energy deposition. The main differences between the two theories is that the former workers used an eigenfunction expansion, whereas the latter workers used a Green's function approach. Moreover, the Vala and Wrobel results apply to cells with multiply-reflected acoustic waves, whereas the result of Bailey *et al.* (1983a) applies only to the incident wave.

Rohlfing, Gelfand, and Miles (1982) and Rohlfing *et al.* (1984) have considered the TROA technique for slow energy transfer and have considered conditions that minimize the effects of cell resonances. A simplified model of the TROA technique has been developed by Smith, Davis, and Smith (1984), who included the energy deposition rate and explained the gross shape of the signal. Their approach also is useful for investigating relatively slow energy transfer, but they chose to neglect the early-time portion of the signal, which is largely controlled by the pump-laser beam spatial profile.

Below, we shall describe the results obtained from a theory (Barker and Rothem, 1982) that includes energy transfer, thermal conductivity, and diffusion of excited species, although viscosity is neglected. In appropriate limits, this more complete theory reduces to those of the other workers. This theory was originally developed to describe the TDTL technique, but it also applies to the closely related TROA technique, as shown in this chapter.

III. Time-Dependent Thermal Lensing

Thermal lensing, or "blooming," of laser beams can be a severe problem in the propagation of high-energy laser beams through media that absorb some of the laser power (Gordon *et al.*, 1964, 1965; Hu and Whinnery, 1973). However, this problem has been turned to advantage in TDTL experiments that monitor transfer of energy from the laser-excited species to the translational motion of the bath gas (Grabiner, Siebert, and Flynn, 1972; Siebert, Grabiner, and Flynn, 1974). Electronically excited states can also lead to transient lensing, which can be used to monitor electronic-state lifetimes (Friedrich and Klem, 1979), but this effect will not be considered here (a straightforward extension of the TDTL theory could accommodate this effect).

In the TDTL experiment shown in Fig. 1 (Trevor, Rothem, and Barker, 1982), a pulsed-laser beam (pump laser) is passed through a gas that absorbs some of the energy. The excited molecules then undergo collisional energy transfer, and the absorbed laser energy is released to the translational and rotational degrees of freedom as heat. (Because translational–rotational (T–R) energy transfer is rapid, the rotational and translational degrees of freedom are strongly coupled, except at extremely low pressures, and we assume that they are in equilibrium with one another.) The gas will be warmest at the center, where the laser beam is most intense. Because gas molecules tend to move away from the warmer center of the laser beam path, the density becomes lower there, causing a variation in the refractive index. The nonhomogeneous refractive-index variation produces a "lens," which dissipates at longer times due to thermal conductivity.

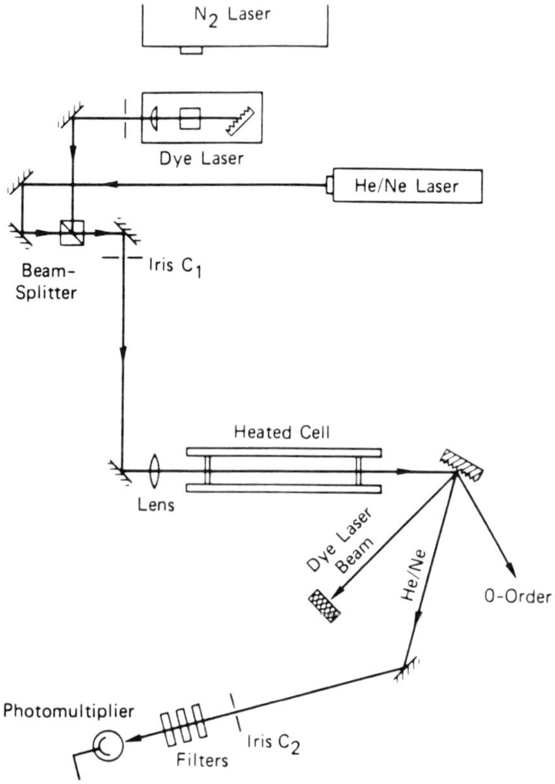

FIG. 1. Schematic of TDTL apparatus (Trevor, Rothem, and Barker, 1982).

The time-dependent lensing is sensed by a CW laser beam (probe laser) that is coaxial with the pump-laser beam. (In photothermal deflection experiments, the probe-laser beam is perpendicular to the pump-laser beam (Tam and Hess, 1987).) The ray paths of the CW laser are initially nearly parallel, but they are perturbed by the refractive-index variations. Divergent lensing of the probe laser is produced when heat is deposited at the center of the laser beam, while convergent lensing is observed when the translational temperature is lowered at the center of the beam. The sign and magnitude of the lensing are measured by monitoring the light intensity at the center of the probe-laser beam with a photomultiplier behind a pinhole.

Typically, a pulsed tunable dye laser is used to excite absorber molecules, which then undergo energy-transfer collisions with the bath gas in the cell. Since the pump laser may be more than 10 orders of magnitude more intense than the low-power He–Ne probe laser, various tricks must be used to reduce undesirable scattered light (Trevor, Rothem, and Barker, 1982). When the wavelengths of pump and probe laser are widely separated, such as with CO_2 pump and He–Ne probe lasers, less effort is needed to reduce scattered light.

The gas-density variations following the pump-laser pulse depend on at least five gas-dynamic factors: 1) energy-release rate, 2) acoustic waves generated by the heat deposition, 3) binary diffusion of the excited molecules through the bath gas, 4) thermal conductivity, and 5) viscosity. In the theory summarized here (Barker and Rothem, 1982), the effects of the first four factors are included; the effect of viscosity is small and has been neglected thus far, although it is included in an improved version of the theory (Jacobs, 1988).

A theoretical description of the TDTL technique was published by Flynn and coworkers (Grabiner, Siebert, and Flynn, 1972; Siebert, Grabiner, and Flynn, 1974), who were the first to apply the technique to energy-transfer studies; they considered only the energy-release rate and the acoustic waves. Earlier treatments of the theory were not concerned with the energy-release rate, but considered only the signal strength and the dissipation of the lens by thermal conductivity (Longaker and Litvak, 1969).

Bailey *et al.* (1980, 1983a, b) have included the energy-release rate and have treated the theory in various limits by applying Green's function techniques. They first considered the optics of the lens and the effects of the energy-release rate constant and thermal conductivity, while neglecting acoustic effects and diffusion (Bailey *et al.*, 1980). They later developed (Bailey *et al.*, 1983a, b) the limiting case considered earlier by Flynn and coworkers, where only acoustic effects and the energy-release rate are important. The strength of the approach by Bailey *et al.* is that the results can be obtained in convenient closed forms, because it assumes a cell of infinite radius with no reflected acoustic waves. Independently, Twarowski and Kliger (1977a, b) also used Green's function methods to treat the time development of thermal lenses produced by fast pulsed excitation and by step-function excitation; these authors treated the energy-release rate as instantaneous and made an assessment of sensitivity and the effects of thermal conductivity.

As mentioned above, Wrobel and Vala (1978) derived a theory of the acoustic waves that includes the energy-release rate, but neglects the other factors. Their development is based on the Bessel function expansion approach (Longaker and Litvak, 1969; also see Tough and Willetts, 1982) and is directly comparable to the Green's function approach of Bailey *et al.* (1980, 1983a, b), who neglected reflected acoustic waves, which are treated accurately by Wrobel and Vala. Although the Wrobel and Vala theory has not been applied to the TDTL effect, it is closely related to the other theories discussed here.

Our approach (Barker and Rothem, 1982) follows that of Flynn and coworkers (Grabiner, Siebert, and Flynn 1972; Siebert, 1973; Siebert, Grabiner, and Flynn, 1974), but it includes binary diffusion, thermal conductivity, and acoustic effects in addition to energy transfer. In the appropriate limits, our unified approach reproduces the other TDTL theories and shows that none of the four factors can be neglected in actual experiments, unless great care is taken in the selection of experimental conditions. The present version of the unified theory gives an exact treatment of the generation and propagation of the acoustic waves until they reach the vicinity of the cell wall, but reflected waves are not treated as accurately. An improved version of the theory treats reflected waves accurately, as well (Jacobs, 1988).

IV. Timescales

Before the discussion of the TDTL signals, it is worthwhile to consider the regimes in which acoustic effects, thermal conductivity, diffusion, and energy transfer may be important. For each of these processes, there is a characteristic time scale (Barker and Rothem, 1982).

For a two-level system, the energy-transfer process can be written as

$$A^* + M \rightarrow A + M, \tag{1}$$

where A^* is the excited molecule, M is the collider gas, and A is the unexcited molecule. The energy-release process is characterized by the bimolecular rate constant k_1, and the time constant for the process is

$$\tau_e = (k_1[M])^{-1} = k_e^{-1}, \tag{2}$$

where the square brackets denote concentrations and $[M]$ is assumed to be much greater than $[A^*]$; k_e is the first order rate constant.

When acoustic waves are produced by a Gaussian pump-laser beam of $1/e$-radius r_b is a gas with speed of sound c, the wavelength of the sound wave is approximately equal to r_b. The characteristic time associated with the sound wave is the acoustic transit time:

$$\tau_s = r_b/c \tag{3}$$

For two-dimensional diffusion in a cylindrical cell, the characteristic time scale for diffusion is

$$\tau_d = \frac{r_b^2}{4D_{12}}, \tag{4}$$

where D_{12} is the binary diffusion coefficient.

The time scale for thermal conductivity has been shown by Bailey *et al.* (1980) to be

$$\tau_t = \frac{PC_p r_b^2}{4RT\lambda}, \tag{5}$$

where P is the total pressure, C_p is the molar heat capacity at constant pressure, R is the gas law constant, T is temperature, and λ is the thermal conductivity coefficient.

For energy-transfer experiments the time scale should be as well separated as possible, so that the energy transfer is not obscured by other processes. As a rule of thumb, the following inequalities should hold:

$$\tau_t, \tau_D > 10\tau_e > 100\tau_s. \tag{6}$$

Under these conditions, the theories of Flynn and coworkers (Grabiner, Siebert, and Flynn, 1972; Siebert, 1973; Siebert, Grabiner, and Flynn, 1974) and of Bailey *et al.* (1983a) can be used.

For thermal conductivity experiments, the original theory of Bailey *et al.* (1980) can be used if the following inequalities hold:

$$\tau_e, \tau_s < \tau_t/100 \qquad \text{and} \qquad \tau_D > 10\tau_t. \tag{7}$$

These conditions often are satisfied at moderate gas pressures. Bailey *et al.*

(1981, 1982a, b, 1983b, 1984, 1987) have used this method to determine thermal conductivities of gases and organic vapors. They have also estimated the thermal conductivity cooling in infrared multiphoton experiments (Bailey *et al.*, 1982c).

These rules of thumb are useful for gaining some idea of the conditions under which energy transfer can be studied with minimal interference from other effects. For molecules that undergo slow energy transfer, it is sometimes possible to satisfy condition (6). However, most polyatomic gases exhibit such rapid energy transfer that the gas-dynamic effects cannot be disentangled. Under these conditions, the full unified theory must be used for quantitative interpretation of the data.

V. TDTL and TROA Theory

A. TDTL Optics

The optics of the time-dependent transient lens have been discussed in detail Bailey *et al.* (1978, 1980). For present purposes, it is assumed that 1) the cell radius b is much larger than the laser-beam radius r_b, 2) light absorption is negligible over the length Z of the cell, and 3) the lensing effect is detected by measuring the light transmitted by a pinhole located a distance L after the exit of the cell. As shown by Bailey *et al.* (1980) and by Twarowski and Kliger (1977a, b), the second condition is not necessary, but we shall retain it, because absorption is negligible in most experiments.

The transient lensing signal S is defined as

$$S = 1 - I/I_0, \tag{8}$$

where I_0 is the intensity at the center of the unperturbed probe laser beam, and I is the intensity when lensing is present. The intensity at the beam center may be monitored with a photomultiplier positioned behind a pinhole that is of a radius much smaller than the pump-laser beam radius.

Following Bailey *et al.* (1980), the thermal lensing signal is evaluated by starting with the differential ray equation and neglecting beam divergence, to produce an ordinary differential equation that describes the ray trajectory in terms of the index of refraction gradient. For small perturbations, the refractive index can be expanded in a Taylor series with retention of terms only to second order in the spatial coordinates. If the refractive index depends only on the radial cylindrical coordinate r, the differential equation is linear and is easily solved for the ray trajectory, where the boundary conditions are that $r(z=0)=r_0$, and the divergence of the beam at the cell entry is $(\partial r/\partial z)_0 \simeq 0$. The ray trajectory is given by:

$$r(z) = r_0 \cosh(\xi^{1/2} z) \qquad \xi > 0, \tag{9a}$$

$$r(z) = r_0 \cos(|\xi|^{1/2} z) \qquad \xi < 0, \tag{9b}$$

where

$$\xi = \frac{1}{n_0}\left(\frac{\partial^2 n}{\partial r^2}\right)_{r=0}, \tag{9c}$$

n_0 and n are the refractive indices of the unperturbed and perturbed gas, respectively, and r is the radial distance from the optical (cell) axis.

After exiting the cell, the ray trajectory is a straight line, and the radius at the detector is given by

$$r(Z+L) = r(Z) + L(\partial r/\partial z)_Z. \tag{10}$$

The intensity of the transmitted beam is inversely proportional to the square of the radius, giving the following expressions for the thermal lensing signal S:

$$S = 1 - \left[\cosh(\xi^{1/2} Z) + L\sinh(\xi^{1/2} Z)\right]^{-2} \qquad \xi > 0, \tag{11a}$$

$$S = 1 - \left[\cos(|\xi^{1/2}| Z) - L\sin(|\xi^{1/2}| Z)\right]^{-2} \qquad \xi < 0. \tag{11b}$$

If we assume that L can be neglected, these expressions reduce to

$$S = \left[\tanh(\xi^{1/2} Z)\right]^2 \qquad \xi > 0, \tag{12a}$$

$$S = \left[\tan(|\xi^{1/2}| Z)\right]^2 \qquad \xi < 0. \tag{12b}$$

In both cases, the small argument limit gives

$$S = \left(\frac{Z^2}{n_0}\right)\left(\frac{\partial^2 n}{\partial r^2}\right)_{r=0}. \tag{13}$$

This is the approximate expression we shall use in this chapter for illustrating the properties of the thermal lens; note that it differs by a factor of 2 from that used earlier (Siebert, 1973; Barker and Rothem, 1982).

The refractive index n_0 is related by the Lorenz–Lorentz law (Born and Wolf, 1975) to the gas density ρ_0 through the molar refractivity A:

$$A = (M/\rho_0)\left[(n_0^2 - 1)/(n_0^2 + 2)\right], \tag{14}$$

where M is the molecular weight of the gas. According to this equation, the variation in the refractive index due to density fluctuations is given by:

$$\frac{\partial n}{\partial \rho} = \frac{-A(n_0^2 + 2)}{2n_0(M + \rho_0 A)}. \tag{15}$$

In this and all subsequent equations, variables with "naught" subscripts refer to the unperturbed case, whereas variables without subscripts refer to the *perturbations*; for example, the density at any time is given by the sum $(\rho_0 + \rho)$.

For pressures of a few atmospheres or less, $n_0 \simeq 1$, and $M \gg \rho_0 A$, giving

$$\frac{\partial n}{\partial \rho} \simeq \frac{3}{2}\frac{A}{M}$$

and

$$n \simeq \frac{3}{2}\frac{A}{M}\rho. \tag{16}$$

Using this expression, the TDTL signal is written in terms of the density fluctuation, which can be obtained by solving the hydrodynamic equations:

$$S \simeq \frac{3Z^2 A}{2M}\left(\frac{\partial^2 \rho}{\partial r^2}\right)_{r=0}. \tag{17}$$

Note that this expression differs by a factor of 2 from that used in previous work (Barker and Rothem, 1982).

Instead of measuring the intensity at beam center, an alternative approach is to monitor the actual spot size of the laser beam (Jansen and Harris, 1985). This can be accomplished through use of an appropriate array detector, although the computer-processing requirements are large and time-consuming. Another approach to spot-size measurements is to use "optical computation" of the spot size by directing the probe laser through a transmission mask that has an optical density proportional to $1/r^2$, as has been done in steady-state thermal lensing experiments (Jansen and Harris, 1985). The total transmission is proportional to the laser-beam spot size. This method has several potential advantages: It uses more of the probe-laser beam, thus making it less susceptible to spatial noise; more light is delivered to the detector, reducing photon statistics noise; and it is just as rapid a method as the ordinary pinhole method. To our knowledge, this approach has not yet been applied to time-dependent thermal lensing, but it seems to have potential benefits.

A cautionary note is in order when considering thermal lens optics. Throughout the above ray analysis, diffraction effects have been neglected. If small apertures are used to confine the pump-and/or probe-laser beams, Fresnel and Fraunhofer diffraction may become very significant. The Taylor expansion of the refractive index may still be valid near the beam center, but both the sign and magnitude of the curvature can be affected by diffraction leading to unpredictable TDTL behavior. This anomalous behavior has been documented by Bailey *et al.* (1978) and can be easily observed by using an iris to narrow the laser beams. For small iris diameters, Fraunhofer interference rings are easily seen visually, and Fresnel diffraction effects are manifest in the intensity variations of the probe laser at beam center when the iris diameter is changed. Similarly, Fresnel diffraction will also affect the pump laser, leading to the anomalous TDTL behavior. The solution to this problem is to avoid small apertures; instead, use beam telescopes to control beam diameters and divergences (Bailey *et al.*, 1978).

B. Density and Pressure Perturbations *(Barker and Rothem, 1982; Jacobs, 1988)*

Excited molecules are produced by absorption of energy from the pump-laser pulse. Their initial spatial distribution depends on that of the pump-laser beam. Subsequently, the excited molecules undergo collisional deactivation even as they diffuse through the bath gas, which is in great excess. The concentration of excited molecules $[N_1(r, t)]$ is a function of both position and time, and it can be expressed as an expansion in terms of zero-order Bessel functions:

$$N_1(r, t) = \sum_{i=0}^{\infty} G_i \exp(\alpha_i t) J_0(\nu_i r), \tag{18}$$

$$\alpha_i = -\left[D_{12}(X_i/b)^2 + k_e\right]. \tag{19}$$

In these expressions, k_e is the first order rate constant for energy transfer $\nu_i = X_i/b$, where X_i are the zeros of $J_0(X_i) = 0$; and the coefficients G_i depend on the pump-laser beam spatial profile, which determines the initial concentration of excited molecules:

$$G_i = \frac{2}{b^2[J_1(\nu_i b)]^2} \int_0^b N_1(r, 0) J_1(\nu_i r) r\, dr, \tag{20}$$

where $J_1(x)$ is the Bessel function of order 1. The boundary condition selected here requires that $N_1(b, t) = 0$, which is appropriate if the excited molecules are annhilated at the cell wall, as expected for energy transfer.

For a Gaussian laser beam with $1/e$-radius r_b much smaller than the cell radius b, the coefficients are given to a good approximation by:

$$G_i(\text{Gaussian}) = g_i$$

$$g_i = \frac{N_{01}}{ab^2[J_1(X_i)]^2} \exp(-\nu_i^2 r_b^2/4), \tag{21}$$

where N_{01} is the initial concentration of excited molecules at the center of the laser beam.

As the excited molecules are deactivated, they release their energy, which causes an increase in the gas temperature. The temperature increase drives the hydrodynamic response of the system. To derive an expression for the density fluctuation, the hydrodynamic continuity, momentum, and energy equations are combined, linearized (for small perturbations), and manipulated to produce a third-order partial differential equation for the density $\rho(r, t)$ as a function of position and time (Barker and Rothem, 1982).

To solve the partial differential equation and obtain $\rho(r, t)$, we followed Longaker and Litvak (1969) and Flynn and coworkers (Grabiner, Siebert, and Flynn, 1972; Siebert, 1973; Siebert, Grabiner, and Flynn, 1974), who expanded the density in terms of zero-order Bessel functions:

$$\rho(r, t) = \sum_{j=1}^{\infty} c_j(t) J_0(\nu_j r). \tag{22}$$

The proper boundary condition requires that the fluid velocity vanish at the cell wall, resulting in a similar expansion, but in terms of $J_0(\chi_j r/b)$, where the χ_j are the zeros of $J_1(\chi_j) = 0$. The latter expansion was used by previous workers, but the expansion of $N_1(r, t)$ in Eq. (18) is in terms of $J_0(\chi_j r)$. To make the derivation tractable, $\rho(r, t)$ was expressed in the same basis as $N_1(r, t)$. An alternative approach is to assume that the excited molecules are not deactivated at the walls and expand both functions in terms of $J_0(\chi_j r/b)$, which accurately describes the reflected acoustic waves, but less realistically describes the energy transfer. The expansion in Eq. (18) gives accurate results for the hydrodynamics except for values of r near the cell wall and for the reflected acoustic waves, which may exhibit a small anomalous phase shift due to the different boundary conditions. For the study of energy transfer and thermal lensing, the exact shape of the reflected acoustic waves is of little interest, but an improved version of the theory will accurately describe the reflected acoustic waves (Jacobs, 1988).

Substitution of the expansion for $\rho(r, t)$ in the third-order partial differential equation produces a third-order ordinary differential equation in the coefficients c_j. The ordinary differential equation can be solved for the appropriate initial conditions (Siebert, 1973) to give the following result, which was obtained by inverting a Laplace transform:

$$c_j = \left\{ C_1 \exp(\alpha_j t) + C_2 \exp(r_1 t) + \exp(-a_2 t)\left[C_3 \cos(b_2 t) + b_2^{-1}(C_4 + a_2 C_3)\sin(b_2 t)\right]\right\} \times \left(G_j \nu_j^2 / B\right)\left(U \nu_j^2 - k_e Q\right), \tag{23}$$

where r_1 is the real root and a_2 and b_2 are the real and imaginary parts of the complex conjugate roots ($r_2 = a_2 \pm ib_2$) of the cubic equation:

$$s^3 + s^2 E \nu_j^2 / B - s(W - F)\nu_j^2 / B - C \nu_j^4 / B = 0. \tag{24}$$

For a large number of test cases, the cubic Eq. (24) always had one real root and two complex conjugate roots; if the cubic equation has three real roots, the Laplace transform can also be inverted easily, but the resulting function will be somewhat different. The coefficients C_1 to C_4, are given by:

$$C_1 = \left\{(\alpha_j - r_1)\left[(\alpha_j - a_2)^2 + b_2^2\right]\right\}^{-1}, \tag{25a}$$

$$C_2 = -\left\{(\alpha_j - r_1)\left[(r_1 - a_2)^2 + b_2^2\right]\right\}^{-1}, \tag{25b}$$

$$C_3 = -(C_1 + C_2), \tag{25c}$$

$$C_4 = \left(\frac{a_2^2 + b_2^2 - \alpha_j r_1}{\alpha_j + r_1}\right)(C_1 + C_2) + \frac{2a_2(r_1 C_1 + \alpha_j C_2)}{(\alpha_j + r_1)} \tag{25d}$$

In these equations, the factors that have not yet been defined are easily calculated for ideal gases, as summarized in Table I. Note that these results differ from those in the original paper, where a sign error was committed and a term was omitted from the energy equation. The new results reported here were obtained from the full linearized energy equation (Jacobs, 1988) and then solved by Laplace transforms, just as before. The differences from the original paper are small and become apparent only at very low pressure (all figures presented here were calculated using the revised equations).

With the coefficients c_j in hand, we can now calculate the TDTL signal:

$$S = \frac{3A}{2M} Z^2 \sum_{j=1}^{\infty} c_j \left(\frac{\partial^2 J_0(\nu_j r)}{\partial r^2}\right), \tag{26}$$

$$S = -\frac{3A}{4M} Z^2 \sum_{j=1}^{\infty} Cj \nu_j^2. \tag{27}$$

When r_b and b are about 0.1 and 2 cm, respectively, about 75 terms are needed for convergence of the summation to within 10^{-6}.

In contrast to TDTL experiments, TROA measurements are made with small microphones positioned a short distance from the laser beam and

TABLE I

IDEAL GAS FACTORS FOR EQS. (23)–(25)

$B = \frac{C_v}{R}$

$C = -\frac{RT_0^2}{P_0 M}\lambda$

$E = \frac{T_0}{P_0}\lambda$

$F = \frac{RT_0}{M}$

$W = -\frac{C_v T_0}{M}$

$U = D_{12}\,\mathrm{kT}\,(1 - m_1/m_2)$

$Q = u_{int} + (\frac{3}{2})\,\mathrm{kT}\,(1 - m_1/m_2)$

C_v ≡ Molar heat capacity

M ≡ Molecular weight

m_1, m_2 ≡ Molecular masses for type 1 and type 2 molecules

k ≡ Boltzmann's constant

R ≡ Gas Law constant

k_T ≡ Coefficient of thermal diffusion (usually can be neglected)

u_{int} ≡ Excitation energy per molecule (excluding translations)

H_1 ≡ Enthalpy of excitation

T_0, P_0, N_0 ≡ Initial temperature, pressure, and total number density.

away from the cell wall. The microphones respond to the pressure fluctuations associated with the acoustic waves before the waves have been perturbed by the presence of the walls. Assuming isentropic expansion and compression (Herzfeld and Litovitz, 1959), the pressure disturbance is related to the density wave by the expression

$$P(r,t) = P_0\left\{1 - \left(\frac{\rho_0 + \rho(r,t)}{\rho_0}\right)^{\gamma}\right\}, \tag{28a}$$

$$P(r,t) \simeq P_0\gamma\frac{\rho(r,t)}{\rho_0}, \tag{28b}$$

where P_0 is the pressure in the absence of any disturbance, and $\gamma = C_p/C_v$. Because $P(r, t)$ is directly proportional to $\rho(r, t)$ in the limit of small fluctuations, and because the density is calculated directly from Eq. (22), only the density is presented in the figures discussed below.

VI. Experimental Methods

For TDTL experiments, the experimental setups are adaptations (Trevor, Rothem, and Barker, 1982; Bailey *et al.*, 1981) of the version used by Flynn and coworkers (Grabiner, Siebert, and Flynn, 1972; Siebert, Grabiner, and Flynn, 1974). The pump laser in Fig. 1 may be a pulsed tunable dye laser (possibly followed by a second harmonic generator), or it may be a CO_2 laser. Usually, the spatial mode quality of the pulsed dye lasers is relatively poor, whereas Q-switched CO_2 lasers can have excellent TEM_{00} output beams. We have found that the ultraviolet output of the second harmonic generator tends to have better mode quality than the fundamental (Toselli and Barker, 1987), but either may require spatial filtering for high-precision experiments.

The probe laser is usually a low-power helium–neon laser operating at 632.8 nm, but we have also used a He–Ne laser operating near 1.15 μm (Trevor, Rothem, and Barker, 1982). The pump- and probe-laser beams can be combined for coaxial propagation on a dielectric coated beam splitter, or on a quartz, or germanium flat, depending on the wavelength requirements of the pump laser. Tests using a pellicle beam splitter showed very large anomalous signals, apparently due to photothermal distortions of the plastic film membrane, but which obscure the gas-phase thermal lensing; this effect potentially could provide a method for testing optical components for absorption (Trevor, Rothem, and Barker, 1982).

The coaxial laser beams are steered through the cell containing the absorber gas, and then they must be separated to avoid stray light interference. Prior to entry into the cell, a lens may be used to control divergence of the beams. A diffraction grating can be used to separate the two beams, followed by laser line interference filters and polarizing filters. All of these components must be tested to ensure that they do not introduce spurious TDTL signals.

The iris C_2 is the "pinhole" that isolates the intensity of the laser beam center. The pinhole diameter must be chosen small enough relative to the pump-laser beam diameter so that the transmitted light is representative of the intensity at the beam center ($\leq 0.1 r_b$).

The iris C_1 and other irises in the optical train are used *only* for alignment of the laser beams; they must not be used for controlling beam diameters, because of the problems associated with Fresnel diffraction; a beam telescope is best used for controlling beam diameter and divergence (Bailey *et al.*, 1978). Optical alignment is the single most difficult aspect of TDTL experiments. Slight misalignments can easily lead to signals of the opposite polarity and widely different magnitudes.

For a visible probe laser, a photomultiplier is used for monitoring the signal. For infrared probe lasers, solid-state photodiode detectors must be used. The signal output is amplified and then co-added in an appropriate signal averager for noise reduction.

TROA experiments tend to be much simpler than TDTL experiments, because only one laser is used and alignment of the optical beam is not critical. Microphones are available from many sources, but we have found the Knowles model BT-1759 hearing-aid microphone to be very inexpensive and useful up to about 20 kHz (Presser, Barker, and Gordon, 1983; Barker *et al.*, 1985; Zellweger, Brown, and Barker, 1985). For higher frequencies, Gordon and coworkers (Beck, Ringwelski, and Gordon, 1985; Beck and Gordon, 1987) have used piezoelectric transducers, which have very fast response but relatively low sensitivity (see also Toselli and Barker, 1987). The microphone output is amplified and signal-averaged, as in the TDTL experiments.

VII. Observed and Calculated Results

A. Signals from Thermal Lensing

Typical experimental and calculated TDTL signals are presented in Fig. 2. The rising portion of the signal is governed mostly by the energy-deposition rate and the effects of the acoustic waves, whereas the falling portion of the signal is controlled mostly by thermal conductivity; diffusion plays a small role throughout the signal. The minor disturbances in Fig. 2a and the prominent oscillation at about 150 μs in Fig. 2b are due to returning acoustic waves, which have reflected from the cell walls.

Under the conditions described by Eq. (7), the decaying TDTL signal is approximately described by

$$S(t) = \gamma\left(\frac{\tau_t}{t + \tau_t}\right)^2, \tag{29}$$

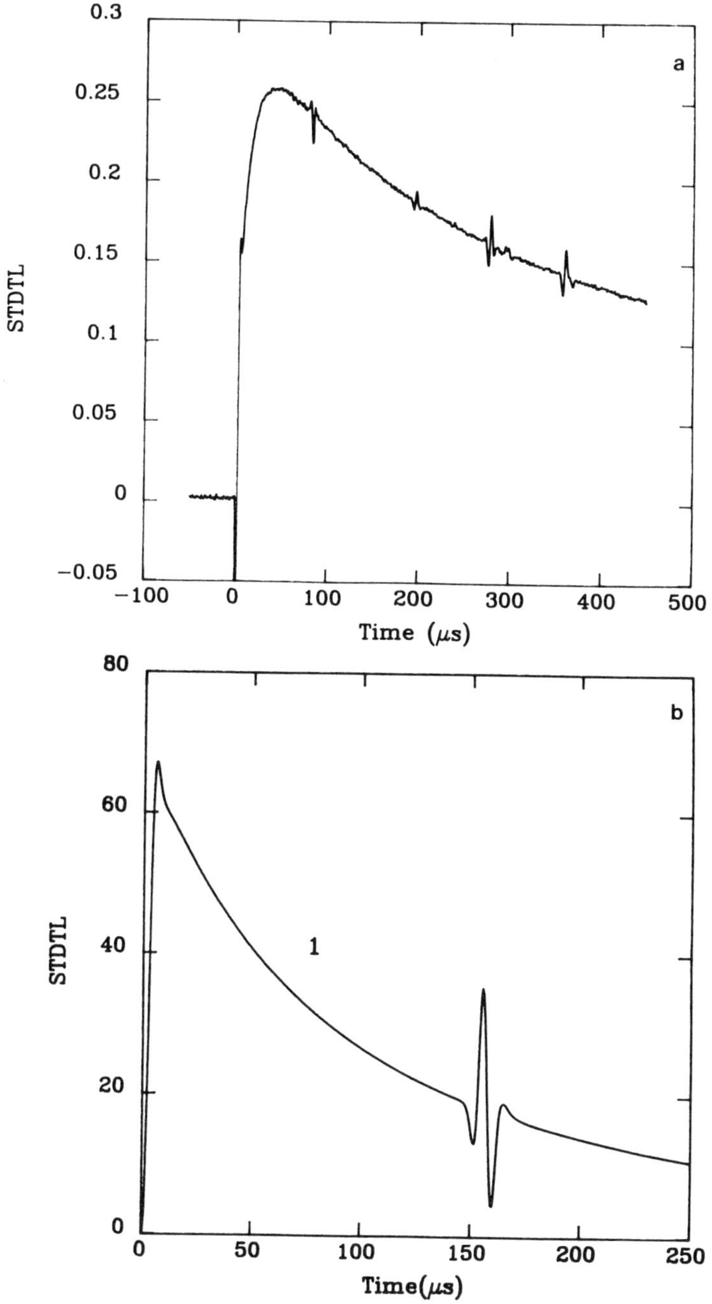

FIG. 2. (a) Experimental TDTL signal. Conditions: excitation wavelength: 540 nm; 298 K; 0.5 Torr NO_2 in 100 Torr argon. (b) Typical calculated TDTL signal. See Table II for the curve labels and for the physical conditions of these calculations.

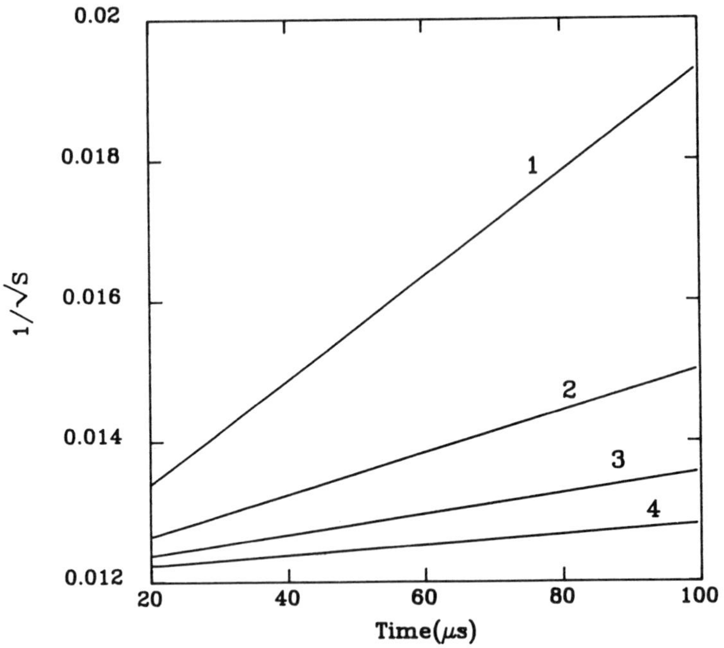

FIG. 3. Calculated $S^{-1/2}$ vs. t for Excited NO_2 in argon. See Table II for the curve labels and for the physical conditions of these calculations.

where the symbols have been defined earlier. A plot of $S^{-1/2}$ vs. t will produce a straight line with an intercept–slope ratio equal to τ_t. Plots for representative calculations are presented in Fig. 3, where it is clear that the approximate relation (29) is accurately obeyed. According to Eq. (5), τ_t is directly proportional to pressure, and the plot of τ_t vs. P presented in Fig. 4 demonstrates that proportionality. Bailey *et al.* (1981, 1982a, b, 1983b, 1984, 1987) have exploited the accuracy of this approximation to evaluate thermal conductivities of gases and vapors as functions of temperature and showed that the TDTL approach is superior to the hot-wire method. Some of their thermal conductivity data are presented in Fig. 5.

Although thermal conductivity depends on energy transfer, the relationship between thermal conductivity coefficients and elementary energy-transfer rate coefficients is not straightforward (Hirschfelder, Curtiss, and Bird, 1964). Under favorable circumstances, the TDTL technique can be used to measure energy-transfer rate coefficients directly. If the conditions satisfy Eq. (6), all effects except energy transfer can be neglected, and the signal is directly proportional to the heat deposited in the system. For a first-order

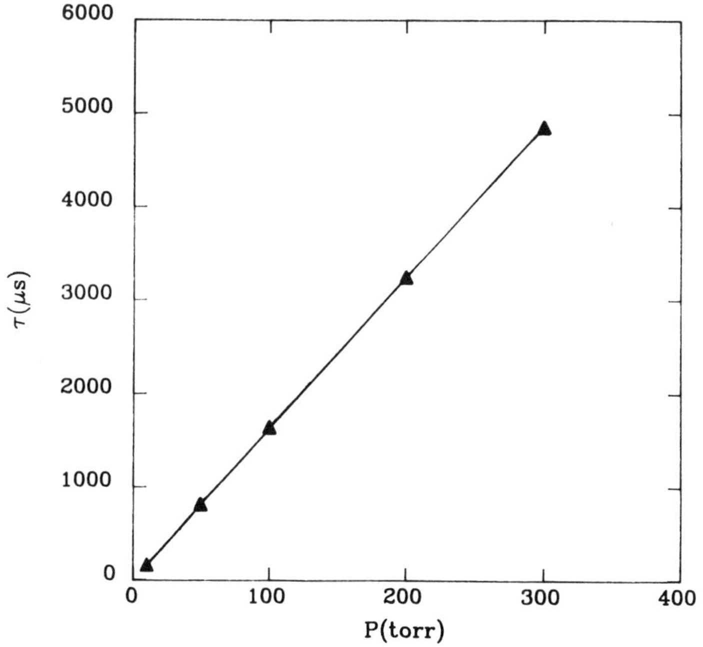

FIG. 4. Thermal-conductivity time constant vs. pressure. Conditions: same as Fig. 3.

energy-deposition process with a time constant given by Eq. (2), the TDTL signal is

$$S(t) \propto 1 - \exp(-t/\tau_e). \tag{30}$$

The conditions necessary for Eq. (30) to be valid are rarely attained in energy-transfer experiments. Usually, the rate coefficient for energy deposition is so large that a low pressure of collider gas must be used to increase τ_e to a value $\gg \tau_s$, the acoustic transit time. However, low pressure decreases both the diffusion and thermal conductivity time constants, tending to violate the conditions given in Eq. (6).

Remarkably, some energy-transfer systems can still fulfill the conditions in Eq. (6). For example, Flynn and coworkers (Grabiner, Siebert, and Flynn, 1972; Siebert, 1973; Siebert, Grabiner, and Flynn, 1974) observed slow energy transfer that satisfies the conditions necessary for Eq. (30) in the deactivation of several gases. Several of these gases also exhibited kinetic cooling in addition to the usual kinetic heating due to the deposition of heat from the pump laser. An example of "cooling" is presented in Fig. 6. This behavior is a clear indication that the temperature at the beam

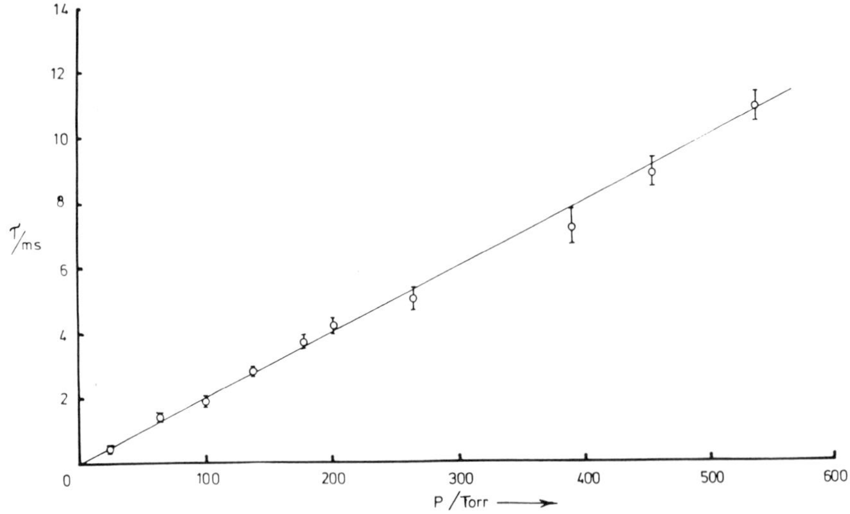

FIG. 5. TDTL thermal-conductivity time constant for methanol vapor (Bailey *et al.*, 1987). These high-precision data give a thermal diffusivity of 1.02 ± 0.10 N s^{-1} at 341 K.

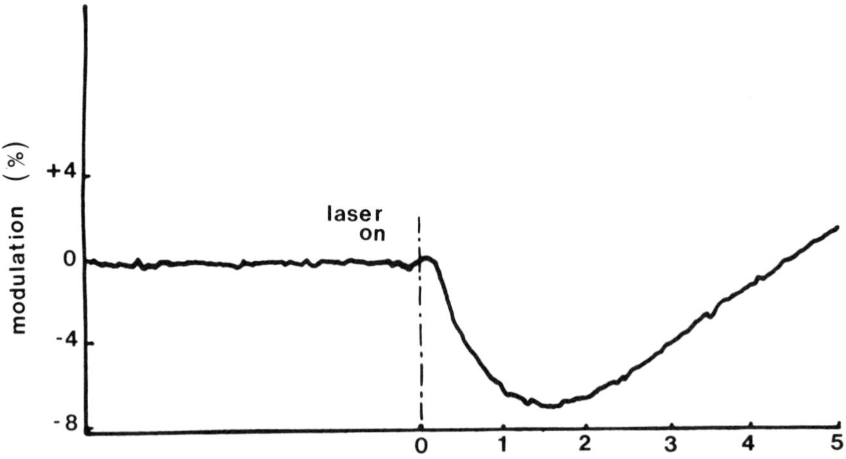

FIG. 6. TDTL trace for $CO_2(001)$ deactivation (Bailey *et al.*, 1985). CO_2 (144 Torr) + N_2 (435 Torr); time units: 10^{-5} s.

center initially decreases slightly during the energy-transfer process, resulting in the formation of a transient *converging* lens. As the energy transfer proceeds further, the energy deposited by the pump laser eventually appears in the translational degrees of freedom, and the initial cooling is superceded by the usual temperature rise, which results in a *diverging* lens.

In their experiments on CO_2, Bailey *et al.* (1985) used a TEA laser tuned to the 1046.58-cm^{-1} line to produce $CO_2(001)$. The first step of the deactivation by nitrogen is

$$CO_2(001) + N_2(0) \rightarrow CO_2(030) + N_2(1), \qquad (31)$$

which is exothermic by 18 cm^{-1}. The $CO_2(001)$ is produced by exciting $CO_2(002)$, which has an energy greater than the average thermal vibrational energy. The depleted population of $CO_2(002)$ is initially replenished through collisions that activate lower-energy CO_2 molecules, resulting in a net cooling of translational temperature. Figure 6, taken from Bailey *et al.* (1985), shows the characteristic cooling followed by warming, which is also observed in neat CO_2 and $CO_2(001)$ deactivation by carbon monoxide. For these conditions, the simplicity of the TDTL behavior permits straightforward interpretation of the results, but for faster energy transfer, more complicated behavior is observed, because the conditions in Eq. (6) are not satisfied.

In deactivation of many polyatomic molecules, energy transfer occurs on every collision, but not all of the energy is transferred in each step. The overall rate of energy deposition then depends on the collision frequency and on $\langle \Delta E \rangle$, the average amount of energy transferred in each collision. For a molecule such as azulene ($C_{10}H_8$) excited with 15 000–30 000 cm^{-1} of vibrational energy, < 50 collisions are needed for deactivation by unexcited azulene (Rossi, Pladziewicz, and Barker, 1983; Barker, 1984). Because the TDTL signals are limited in their time response by the acoustic transit time τ_s, very low pressures might be used to slow the energy-deposition rate. At low pressure, however, thermal conductivity and diffusion become relatively more important, and Eq. (30) cannot be used. Under these circumstances, the full unified theory must be used.

The influence of thermal conductivity, diffusion, and acoustic waves at relatively low pressure is shown in Fig. 7. In this model calculation, excited azulene is diluted in 10 Torr of argon at 370 K, the laser beam radius r_b is 0.1 cm, and the first-order rate constant for energy deposition is 2.6×10^4 s^{-1} (about two orders of magnitude slower than that actually observed for azulene). The acoustic disturbance persists for 6–8 μs, followed by behavior that qualitatively resembles that of Eq. (30). However,

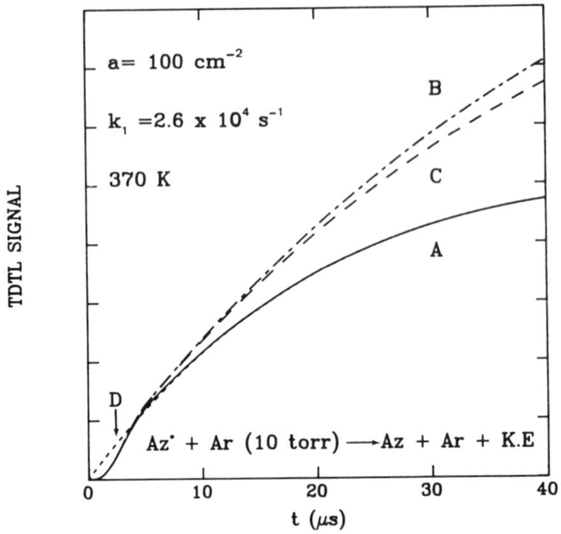

FIG. 7. Influence of various factors on TDTL signals (Barker and Rothem, 1982). Line A:5. full theory; line B: thermal conductivity neglected; line C: both thermal conductivity and diffusion neglected; line D: full theory, except that acoustic effects are neglected.

the actual functional form is more complicated, due to the influence of thermal conductivity and diffusion, as indicated in Fig. 7.

An experimental TDTL record for 35 mTorr of pure azulene is shown in Fig. 8 (Trevor, Rothem, and Barker, 1982). Nearly 70 000 laser shots were averaged to obtain the signal shown. The prominent spike at $t = 0$ due to electrical interference is followed by the acoustic disturbance, where the $r_b \simeq 0.07$ cm. Thermal conductivity and diffusion are very important at this low pressure and cause the signal to reach its maximum and decay after only ~ 4 μs. Under these conditions, the signal is very insensitive to the energy-deposition rate, as shown by the model calculation simulating these conditions (Fig. 8b). Clearly, simple reduction of pressure is not a good strategy for measuring fast energy-deposition rates, when thermal conductivity and diffusion can dominate.

The effects of the competition between energy-deposition rate and thermal conductivity (and diffusion) are illustrated in Fig. 9. Here, the pressure is held constant and only the energy-deposition rate is changed. In curve 7, the energy-deposition rate is substantially slower than thermal conductivity, whereas the reverse is true for curve 9. The behavior is complex, but the decaying portion of the curves depends on the energy-deposition rate as long as it is slower than thermal conductivity and diffusion.

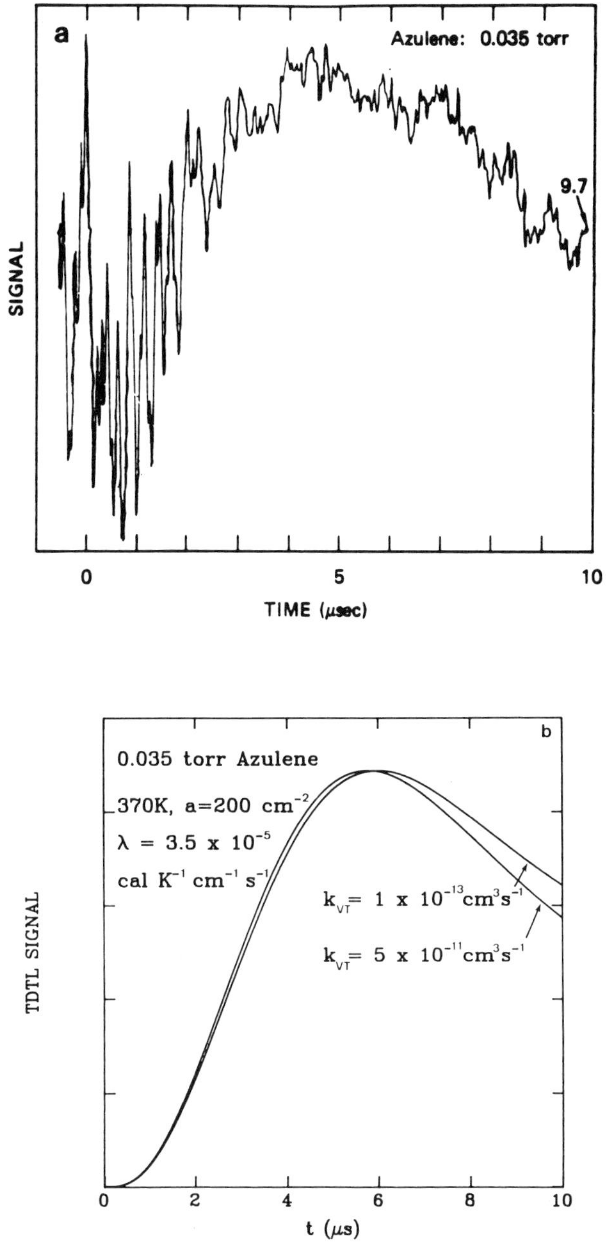

FIG. 8. (a) TDTL signal: 35 mTorr pure azulene (Trevor, Rothem, and Barker, 1982). (b) Model TDTL signal (Trevor, Rothem, and Barker, 1982), simulating panel (a). Note that the lens behavior is insensitive to the V–T rate constant when diffusion is the dominant process.

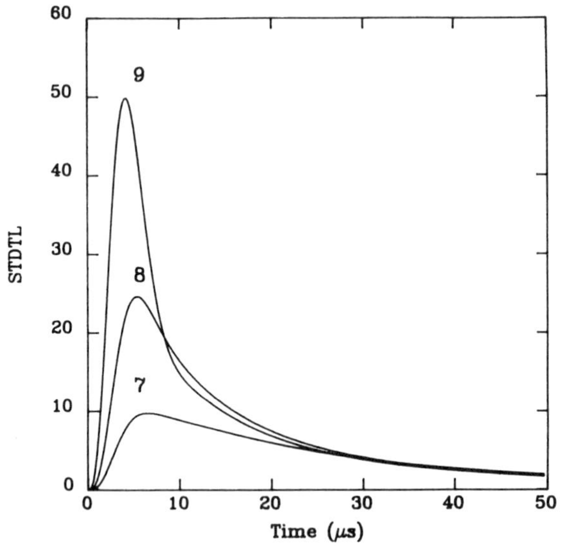

FIG. 9. Competition between energy deposition rate and thermal conductivity (and diffusion). See Table II for the curve labels and for the physical conditions of these calculations.

An experimental measurement using azulene in the regime $\tau_e < \tau_t \simeq \tau_D$ is shown in Fig. 10. Here, krypton was used as collider gas because its large mass slows thermal conductivity and diffusion. Although the signals are noisy even after averaging a large number of laser shots, the results indicate that the first-order energy-deposition rate is about $10^5\ s^{-1}$. The conclusion drawn from this experiment was that collisions with the unexcited azulene in the sample make a significant contribution to the energy-transfer process and that azulene–azulene collisions probably produce V–V, rather than V–T/R energy transfer (Trevor, Rothem, and Barker, 1982). Due to the rapid energy transfer, however, a stronger conclusion could not be reached on the basis of the TDTL experiments. (For recent data on azulene energy transfer, see Barker (1984), Shi *et al.* (1988), and Shi and Barker (1988).)

In all of the discussion so far, the acoustic transit time has been the shortest time constant in the TDTL system. However, when excited molecules, or chemical reactions, are rapid, the energy-deposition time can be substantially shorter than the acoustic transit time. Under these conditions, an initial spike is observed in the TDTL signal (Barker and Rothem, 1982; Bailey *et al.*, 1983a). This behavior is particularly apt to occur when pressures are high (i.e., thermal conductivity and diffusion are very slow)

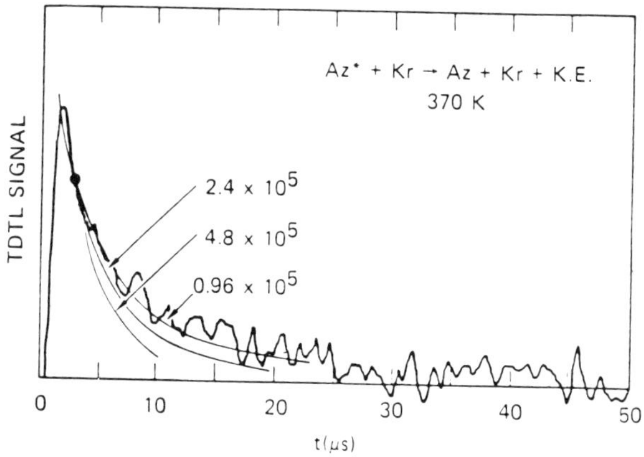

FIG. 10. Deactivation of excited azulene in krypton (Trevor, Rothem, and Barker, 1982). Excitation wavelength: 600 nm; 370 K, 87 mTorr azulene in 1.00 Torr krypton; the solid lines are calculated for three values of the first-order V–T rate constant.

and the pump-laser beam is of large diameter. Calculations predicting this behavior are presented in Fig. 11.

Experiments by Bailey *et al.* (1983a) have verified the existence of the spike in an energy-transfer system that consisted of SF_6 and cyclopropane excited by a CO_2 laser (Fig. 12). The signal is monotonic for $\tau_e \geq \tau_s$, but for faster energy deposition, a distinct spike is observed. Bailey *et al* (1983a) developed an approximate model for this effect by neglecting thermal conductivity and diffusion. The approximate model agrees well with the experimental results, except for the slow rise subsequent to the spike. They attribute this discrepancy to non-Gaussian laser beams and to the use of ray optics in deriving the expression for the TDTL signal. The ray-optics derivation neglects diffraction effects and assumes that the index of refraction is well represented by a second-order Taylor expansion at all radii. For Gaussian refractive-index profiles, these conditions are satisfied, but when the acoustic wave effects dominate, as in the spike, the profiles are not Gaussian, and it is not clear that the ray-optics approach is adequate (Bailey *et al.*, 1983a).

An extreme example of non-Gaussian refractive-index profiles is provided by the recent experiments of Guckert and Carr (1986). In these TDTL experiments, a high-power CO_2 TEA laser pumped cyclobutanone to high vibrational energies, and the resulting final temperatures were $\simeq$ 1000 K. The pump-laser beam was not Gaussian, but had a near-"top-hat" spatial profile. The lensing that resulted from these experiments was far from ideal, and refractive-index perturbations were very large, invalidating

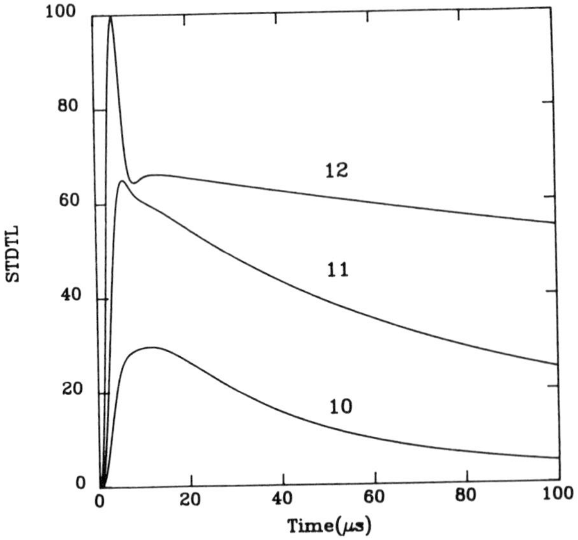

FIG. 11. Competition between energy deposition rate and acoustic transit rate. See Table II for the curve labels and for the physical conditions of these calculations.

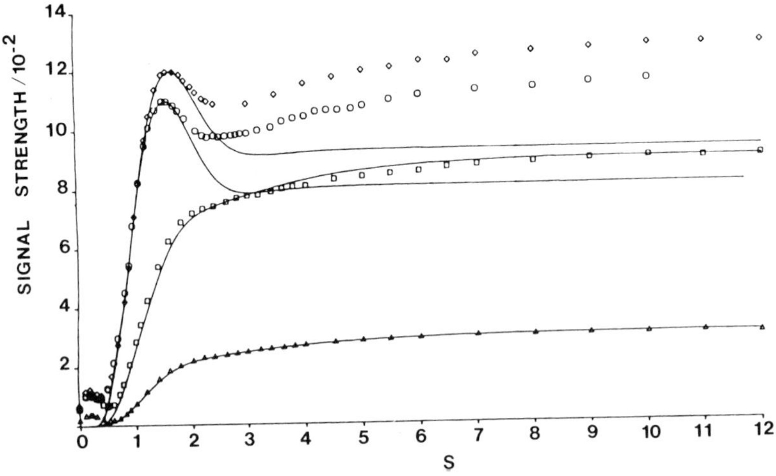

FIG. 12. SF_6/Cyclopropane energy transfer at 354 K (Bailey *et al.*, 1983a). The ordinate, S, is the reduced time t/τ_s, and the solid lines are from the approximate theory of Bailey *et al.* (1983a). From bottom to top, the pressures in Torr are 26.15, 42.1, 206.4, and 357.5.

the analysis leading to Eq. (9). However, Guckert and Carr observed that the TDTL signals could be interpreted in terms of acoustic transit times across the laser beam and were able to measure the temperature-dependent speed of sound in the heated region. From the speed of sound, they deduced the temperature in the heated region and compared it with that inferred from chemical reaction yields. Their conclusion was that a disequilibrium between vibrational and translational degrees of freedom persisted from a surprising period of time, under their conditions. This conclusion is controversial, but their ingenious interpretation of this nonideal TDTL experiment illustrates the versatility of photothermal methods.

B. Signals from Acoustic Waves

In all of the cases where thermal lensing can take place, acoustic waves are generated and can be used for measurements. Both techniques are limited in their time response by the acoustic transit time across the pump-laser beam, and both techniques depend on density perturbations; thus both techniques in principle have the same information content. Experimentally, TROA experiments are much simpler than TDTL experiments, which require wavelength separation of the two laser beams and very exacting coaxial alignment of the two beams. On the other hand, high-frequency-response microphones tend to be relatively insensitive, and the slower components in a system with multiple time constants tend to be lost in noise. Thus the choice of method is based on convenience, on whether the energy-deposition process is a single, or multiple exponential decay, and on the rate of the energy-transfer process.

The optoacoustic signal varies in both shape and amplitude as the time constant of energy deposition varies relative to the acoustic transit time. In Fig. 13 are presented the density fluctuations corresponding to the TDTL signals depicted in earlier figures and summarized in Table II. When $\tau_e \gg \tau_s$, the TROA signal consists of a narrow spike followed by a slow decay, which is a direct consequence of the slow energy deposition rate. As the energy deposition rate is increased, the slow decay becomes more rapid (Fig. 13b), until it merges with the major-density spike (Fig. 13c). As the energy-deposition rate increases still further, eventually dominating over thermal conductivity, a density minimum appears following the high-density spike (Fig. 13d, e).

The relative magnitude of the minimum and the density spike maximum in the acoustic wave is a measure of the energy-deposition time constant, relative to the acoustic transit time. Because the TROA signal is most

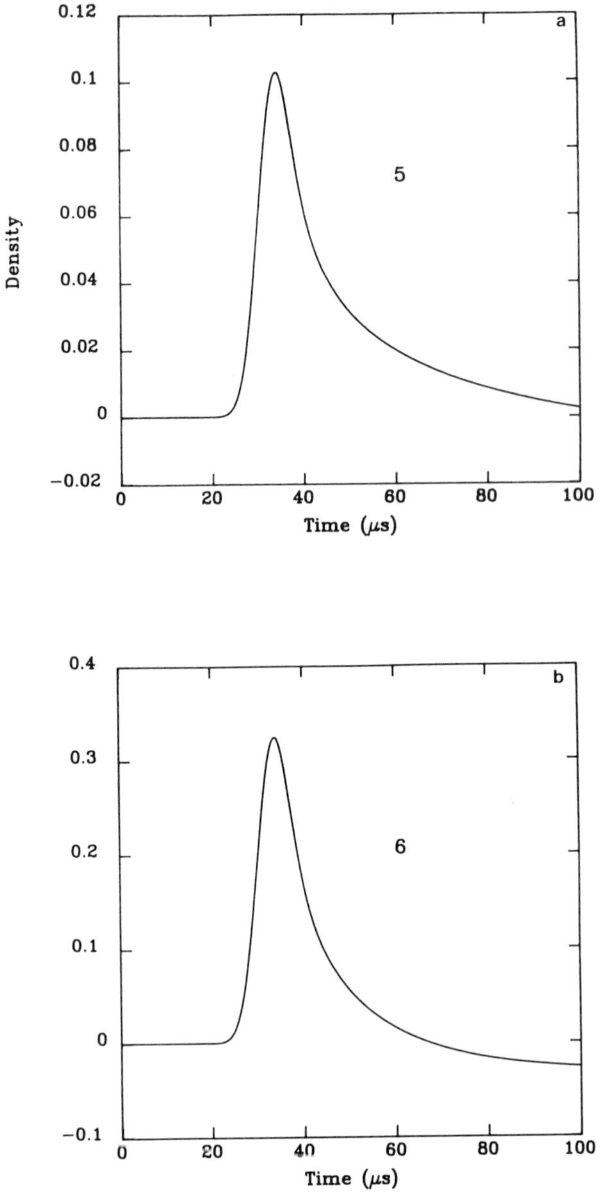

FIG. 13. Acoustic–wave density perturbations. The curve labeling is the same as for the TDTL curves. See Table II for the curve labels and for the physical conditions of these calculations.

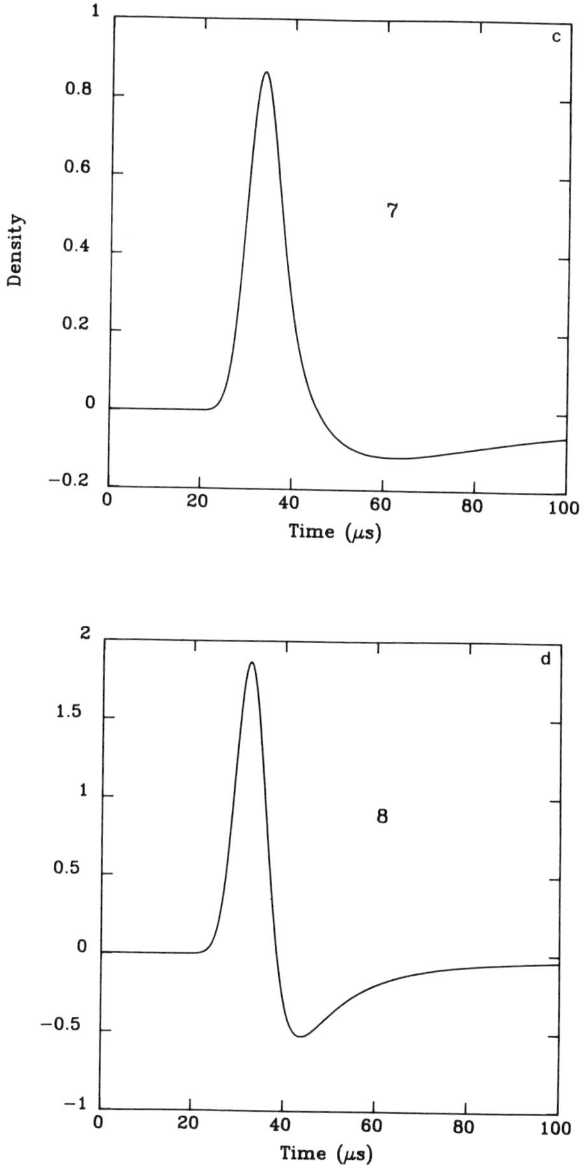

FIG. 13. *(continued)*

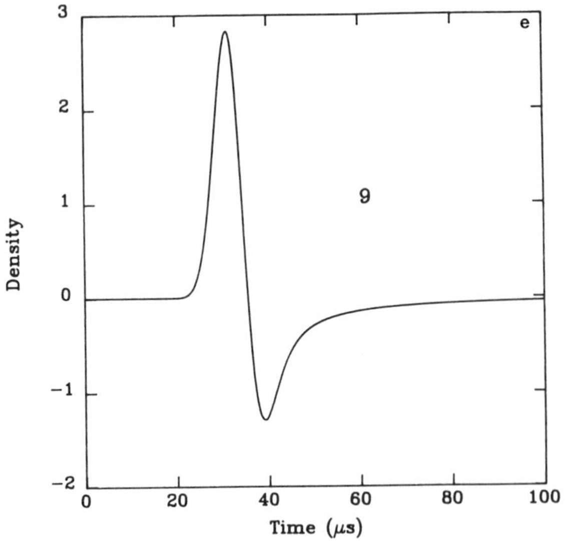

FIG. 13. (*continued*)

sensitive to events on the same time scale as the acoustic transit time (Fig. 14), the approximate theory of Bailey *et al.* (1983a) works well for the regime where $\tau_e \leq \tau_s$. Gordon and coworkers (Beck, Ringwelski, and Gordon, 1985; Beck and Gordon, 1987) have used the theory of Bailey *et al.* to calculate the ratio of maximum to minimum and thereby to determine the energy-transfer time constant for the deactivation of SF_6. The experimental results showed good agreement with the theory of Bailey *et al.*, as shown in Fig. 15.

TROA experiments involving chemical reactions have been performed without actual time resolution of the acoustic wave (Barker *et al.*, 1985). In this investigation, the microphone time response was not fast enough to resolve the wave, but the incident wave could be distinguished from subsequent reflected waves. K. S. Peters and coworkers have considered this situation for the analysis of heat release in rapid reactions and have termed the technique "photoacoustic calorimetry" (Rothberg *et al.*, 1983). Their interest was in the magnitude of the heat release, but rate information can also be extracted.

According to Rothberg *et al.* (1983), the acoustic wave is approximated as a delta function and is convoluted with the damped oscillator response function of the microphone. The microphone signal voltage per absorbed photon $v(t)$ has an amplitude controlled by microphone characteristics and

TABLE II

CONDITIONS FOR LABELED CURVES IN FIGURES. CALCULATED RESULTS FOR EXCITED NO_2 (0.05 TORR) IN ARGON. EXCITATION WAVELENGTH = 558 nm, T = 298 K, Z = 89 cm, b = 2.5 cm AND r_b = 0.1 cm. TIME CONSTANTS EXPRESSED IN μs.

No.	P(Torr)	τ_s	τ_e	τ_t	τ_d
1	10.	3.12	2.79	163.	266.
2	25.	3.12	1.21	405.	661.
3	50.	3.12	0.62	809.	1320.
4	100.	3.12	0.31	1616.	2639.
5	0.5	3.21	178.	9.55	14.5
6	0.5	3.21	52.9	9.55	14.5
7	0.5	3.21	16.7	9.55	14.5
8	0.5	3.21	4.98	9.55	14.5
9	0.5	3.21	1.17	9.55	14.5
10	1.0	3.12	9.36	33.8	54.1
11	8.8	3.12	3.12	143.	233.
12	50.	3.12	0.62	808.	1320.

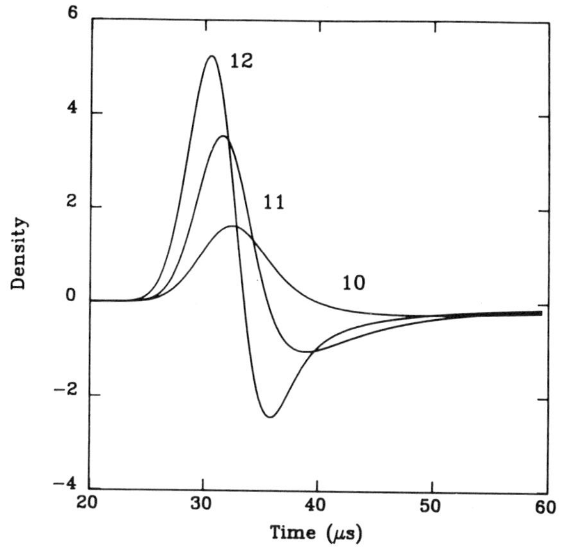

FIG. 14. Competition between energy-deposition rate and acoustic-transit rate in TROA experiments. See Table II for the curve labels and for the physical conditions of these calculations.

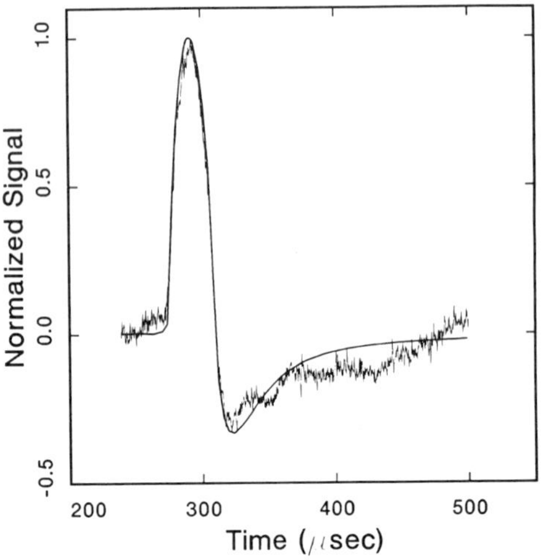

FIG. 15. TROA experiment on deactivation of laser-excited SF_6 by argon (Beck *et al.*, 1987). 0.26% SF_6 in argon at 18.1 Torr; an average of 1.5 laser photons absorbed at the P(20) line of the CO_2 9.6-μm transition.

by the amount of heat released:

$$v(t) = \frac{h_0\alpha}{4\pi r_m C_v} \frac{\nu\tau}{1+\nu^2\tau^2} \{\exp(-t/\tau_e) - \exp(-t/\tau_m) \times [\cos(\nu t) - (\nu\tau')^{-1}\sin(\nu t)]\}. \quad (32)$$

In this expression, the time origin is taken as the time of arrival of the acoustic wave at the microphone distance r_m, ν is the resonant frequency of the microphone, τ_m is the damping time constant, and $(\tau')^{-1} = (\tau_e)^{-1} + (\tau_m)^{-1}$. These parameters depend on the particular microphone used, and in our experiments they were $\nu \simeq 1.5 \times 10^5$ rad s^{-1} and $\tau_m = 155$ μs for a Knowles BT-1759 microphone (Barker *et al.*, 1985). The other parameters in Eq. (32) are the molar heat capacity at constant volume C_v, the amount of heat released per molecule h_0, and a proportionality factor α.

This expression is clearly an approximation, because it neglects the complicated structure exhibited by acoustic waves and simply assumes they can be represented by delta functions. Nevertheless, Eq. (32) gives a reasonable approximation to observations. For example, the reaction of O-atoms with NO_2 was investigated by photolyzing NO_2 with a pulsed laser

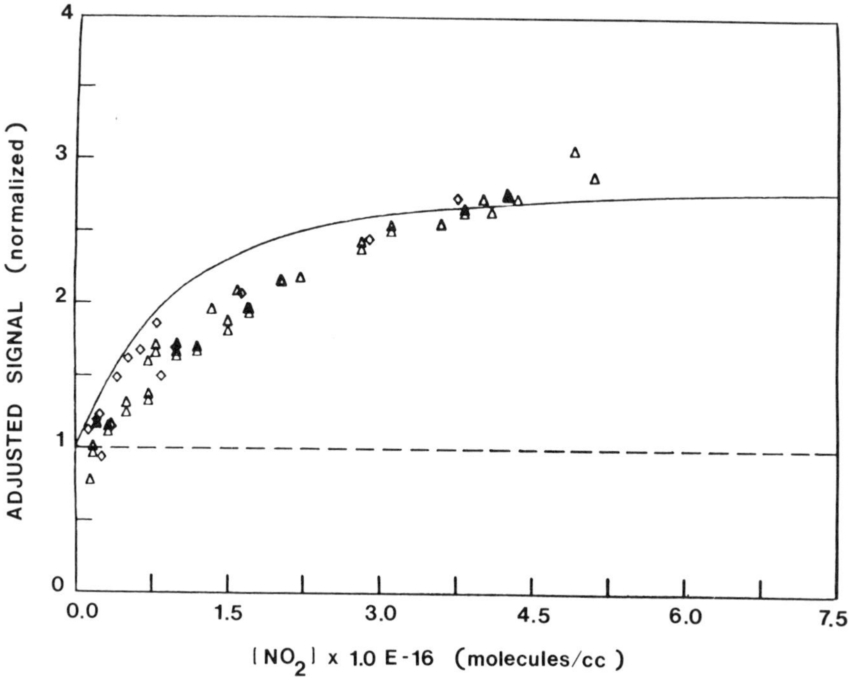

FIG. 16. TROA chemical reaction rate data (Barker *et al.*, 1985). The reaction is $O + NO_2 \rightarrow O_2 + NO$ in ~ 10 Torr of N_2; the solid line is the prediction based on thermochemistry and Eq. (32).

and observing the microphone signal at a fixed time following the laser pulse (the time of the first maximum in the oscillatory microphone output).

$$NO_2 + h\nu(302.5\text{ nm}) \rightarrow NO + O, \tag{33}$$

$$O + NO_2 \rightarrow O_2 + NO. \tag{34}$$

The heat release due the reaction (33) depends on the difference between the photon energy and the bond dissociation energy of NO_2, while that of reaction (34) is just the enthalpy of reaction; the bimolecular rate constant for reaction (34) also is well known. By varying the NO_2 concentration, the energy-deposition time constant can be varied: $\tau_e = \{k_{34}[NO_2]\}^{-1}$. The observed and predicted microphone signals are compared in Fig. 16. The agreement is very good, demonstrating that this method can be used to measure the rates of chemical reactions and energy-transfer processes.

VIII. Conclusions

In this chapter, we have presented the theory and practice of time-resolved photothermal energy-transfer and kinetics measurements. Although the TDTL and the TROA techniques superficially seem quite different, they both depend on the same fundamental processes and are described by the same theory. The two techniques are complementary in that the TDTL method is more appropriate for relatively slow energy-deposition rates, whereas the TROA method is better for rapid processes.

A major strength of the photothermal techniques is that they are quite general and that they require only that the molecule absorbs the pump-laser light and relaxes, at least partially, nonradiatively. In addition, calibration of the techniques depends on the major diluent (i.e., collider gas) and on the laser beam characteristics, not on the absorber identity. This feature simplifies intercomparisons among different absorber gases and constitutes another strength. Both techniques suffer from the same weakness: The energy deposited cannot be assigned specifically to a particular process, except on the basis of the overall behavior of the system. This is in contrast to techniques such as laser-induced fluorescence, which carry the spectroscopic signature of the emission process.

To date, relatively few studies have utilized photothermal techniques to investigate reaction rates and energy transfer, but the TDTL method has proven to be very useful for thermal conductivity measurements. In the future, both techniques will find more use in comparative energy-transfer studies. Since the photothermal techniques specifically measure the rate of energy deposition into the translational modes, they will continue to be used along with other techniques in investigations of energy-transfer mechanisms.

ACKNOWLEDGMENTS

We are grateful to our former colleagues, who were coauthors of several papers on which this review is based. JRB is also grateful to the Department of Energy, Office of Basic Energy Sciences for partial support, and BMT thanks CONICET of Argentina for a postdoctoral fellowship. Thanks also go to P. B. Hays and S. J. Jacobs for insightful discussions, and to F. R. Cruickshank, R. J. Gordon, and R. W. Carr for discussions and for copies of papers prior to publication. Special thanks go to F. R. Cruickshank and colleagues for a careful reading of the manuscript and several helpful suggestions, to an anonymous referee, and to S. J. Jacobs for exceptionally helpful discussions and other assistance with solutions of the hydrodynamics equations.

REFERENCES

Bailey, R. T., Cruickshank, F. R., Guthrie, R., Pugh, D., and Johnstone, W. (1978). *Chem. Phys. Lett.* **59**, 324.

Bailey, R. T., Cruickshank, F. R., Guthrie, R., Pugh, D., and Johnstone, W. (1980), *J. S. C. Faraday II* **76**, 633.

Bailey, R. T., Cruickshank, F. R., Guthrie, R., Pugh, D., and Johnstone, W. (1981). *J. C. S. Faraday II* **77**, 1387.

Bailey, R. T., Cruickshank, F. R., Pugh, D., McLeod, A., and Johnstone, W. (1982a). *Chem. Phys.* **68**, 351.

Bailey, R. T., Cruickshank, F. R., Guthrie, R., Pugh, D., and Weir, I. J. M. (1982b). *J. Mol. Structure* **80**, 433.

Bailey, R. T., Cruickshank, F. R., Pugh, D., F. R., Guthrie, Johnstone, W., Mayer, J., and Middleton, K. (1982c). *J. Chem. Phys.* **77**, 3453.

Bailey, R. T., Cruickshank, F. R., Guthrie, R., Pugh. D., and Weir, I. J. M. (1983a). *Mol. Phys.* **48**, 81.

Bailey, R. T., Cruickshank, F. R., Pugh, D., Guthrie, S., McLeod, A., Foulds, W. S., Lee, W. R., and Vankatesh, S. (1983b). *Chem. Phys.* **77**, 243.

Bailey, R. T., Cruickshank, F. R., Pugh, D., Weir, I. J. M. (1984). *Chem. Phys.* **87**, 125.

Bailey, R. T., Cruickshank, F. R., Guthrie, R., Pugh, D., and Middleton, K. M. (1985). *J. C. S. Faraday II* **81**, 255.

Bailey, R. T., Cruickshank, F. R., Guthrie, R., Pugh, D., and Weir, I. J. M. (1987). *Chem. Phys. Lett.* **134**, 311.

Barker, J. R. (1984). *J. Phys. Chem.* **88**, 11.

Barker, J. R., and Rothem, T. (1982). *Chem. Phys.* **68**, 331.

Barker, J. R., Brouwer, L., Patrick, R., Rossi, M. J., Trevor, P. L., and Golden, D. M. (1985). *Int. J. Chem. Kinetics* **17**, 991.

Beck, K. M., and Gordon, R. J. (1987). *J. Chem. Phys.* **87**, 5681.

Beck, K. M., Ringwelski, A., and Gordon, R. J. (1985). *Chem. Phys. Lett.* **121**, 529.

Born, M., and Wolf, E. (1975). *Principles of Optics*, 5th ed., p. 88, Pergamon Press, New York.

Braun, W., Scheer, M. D., and Kaufman, V. (1986). *J. Res. Nat. Bur. Standards* **91**, 313.

Friedrich, D. M., and Klem, S. A. (1979). *Chem. Phys.* **41**, 153.

Gordon, J. P., Leite, R. C. C., Moore, R. S., Porto, S. P. S., and Whinnery, J. R. (1964). *Bull. Am. Phys. Soc.* **9**, 501.

Gordon, J. P., Leite, R. C. C., Moore, R. S., Porto, S. P. S., and Whinnery, J. R. (1965). *J. Appl. Phys.* **36**, 3.

Grabiner, F. R., Siebert, D. R., and Flynn, G. W. (1972). *Chem. Phys. Lett.* **17**, 189.

Guckert, J. R., and Carr, R. W. (1986). *J. Phys. Chem.* **90**, 4286.

Herzfeld, K. F., and Litovitz, T. A. (1959). *Absorption and Dispersion of Ultrasonic Waves*. Academic Press, New York.

Hirschfelder, J. O., Curtiss, C. F., and Bird, R. B. (1964). *Molecular Theory of Gases and Liquids*. Wiley, New York.

Hu, C., and Whinnery, J. R. (1973). *Appl. Opt.* **12**, 72.

Jacobs, S. J., (1988). *Chem. Phys.*, submitted.

Jansen, K. L., and Harris, J. M. (1985). *Anal. Chem.* **57**, 1698.

Longaker, P. R., and Litvak, M. M. (1969). *J. Appl. Phys.* **40**, 4033.

Pao, Y.-H., Ed. (1977). *Optoacoustic Spectroscopy and Detection*. Academic Press, New York.

Presser, N., Barker, J. R., and Gordon, R. J. (1983). *J. Chem. Phys.* **78**, 2163.

Rohlfing, E. A., Gelfand, J., and Miles, R. B. (1982). *J. Appl. Phys.* **53**, 5426.
Rohlfing, E. A., Rabitz, H., Gelfand, J., and Miles, R. B. (1984). *J. Chem. Phys.* **81**, 820.
Rossi, M. J., Pladziewicz, J. R., and Barker, J. R. (1983). *J. Chem. Phys.* **78**, 6695.
Rothberg, L. J., Simon, J. D., Bernstein, M., and Peters, K. S. (1983). *J. Am. Chem. Soc.* **105**, 3464.
Shi, J., and Barker, J. R. (1988), *J. Chem. Phys.* **88**, 6219.
Shi, J., Bernfeld, D., and Barker, J. R. (1988). *J. Chem. Phys.* **88**, 6211.
Siebert, D. R. (1973). Ph.D. Dissertation, Columbia University, New York.
Siebert, D. R., Grabiner, F. R., and Flynn, G. W. (1974). *J. Chem. Phys.* **60**, 1564.
Smith, N. J. G., Davis, C. C., and Smith, I. W. M. (1984). *J. Chem. Phys.* **80**, 6122.
Tam, A. C., and Hess, P. (1987). *J. Chem. Phys.* **86**, 3950.
Toselli, B. M., and Barker, J. R. (1987). Unpublished work.
Tough, R. J. A., and Willetts, D. V. (1982). *J. Phys. D.* **15**, 2433.
Trevor, P. L., Rothem, T., and Barker, J. R. (1982). *Chem. Phys.* **68**, 341.
Twarowski, A. J., and Kliger, D. S. (1977a). *Chem. Phys.* **20**, 253.
Twarowski, A. J., and Kliger, D. S. (1977b). *Chem. Phys.* **20**, 259.
Wallington, T. J., Scheer, M. D., and Braun, W. (1987) *Chem. Phys. Lett.* **138**, 538.
Whinnery, J. R. (1974). *Acc. Chem. Res.* **7**, 225.
Wrobel, J., and Vala, M. (1978). *Chem. Phys.* **33**, 93.
Zellweger, J.-M., Brown, T. C., and Barker, J. R. (1985). *J. Chem. Phys.* **83**, 6251.

CHAPTER 6

MIRAGE DETECTION OF THERMAL WAVES

P. K. KUO, L. D. FAVRO, AND R. L. THOMAS
Department of Physics and Astronomy and
Institute for Manufacturing Research
Wayne State University
Detroit, MI

I. Imaging of Surface and Subsurface Defects in Solids 191
A. Introduction 191
B. Application to Detection of Planar Cracks 193
C. Application to a Difficult Geometry: A Bolt Hole 197
II. Characterization of Material Properties 198
A. Determination of Thermal Diffusivity of a Solid 198
B. Thicknesses and Thermal Property Determinations for Layered Structures . 204
Appendix 209
References 211

I. Imaging of Surface and Subsurface Defects in Solids

A. INTRODUCTION

The general experimental methods for mirage detection of thermal waves have been described in Chapter 2. A detailed block diagram of our experimental arrangement (Favro, Kuo, and Thomas, 1987) for applying these methods to the imaging of surface and subsurface defects in solids is shown in Fig. 1. The argon-ion laser beam is chopped by an acousto-optic modulator and focused on the sample in order to provide a periodic localized surface heat source. The range of modulation frequencies is from one Hz to several hundred kHz, providing thermal wave penetration depths ranging from about a centimeter to roughly a micron in typical materials. A He–Ne laser beam, skimming the surface of the sample with a beam waist of the order of 100 μm, is used to probe the thermal waves in the air (or other gas) in the vicinity of the surface heat source by means of the mirage effect in the air. The resulting vector deflection of this probe beam is monitored in both magnitude and phase by means of a position-sensitive optical detector. The vector nature of the deflection implies that the

ISBN 0-12-636345-5

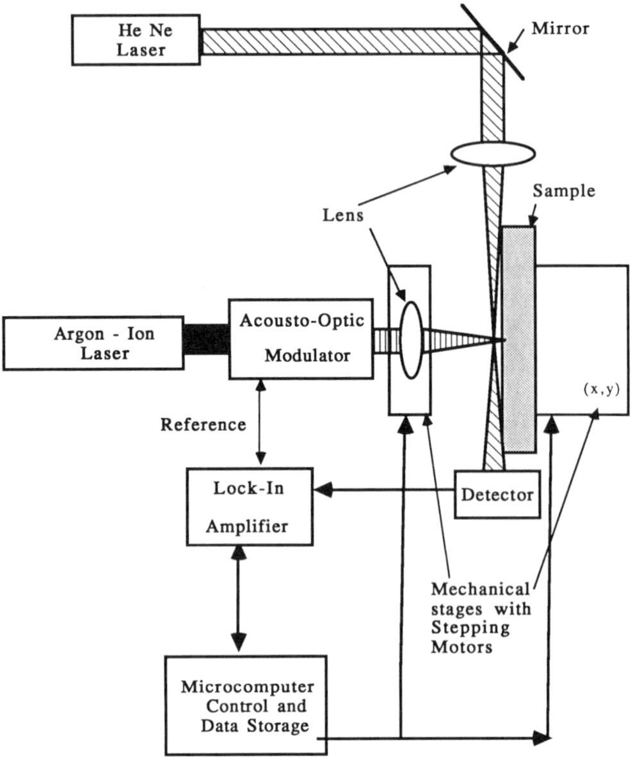

FIG. 1. Block diagram of our experimental arrangement (Favro, Kuo, and Thomas, 1987) for mirage thermal wave imaging of surface and subsurface defects in solids.

magnitude of the signal has two spatial components. The components are referred to as the normal deflection (φ_n, normal to the sample surface) and the tangential deflection (φ_t, tangential to the sample surface). They are measured separately because of the distinctly different information each carries in regard to thermal wave scattering in the solid. Thermal wave images of the subsurface structure are formed by scanning the position of the sample relative to the two (fixed) laser beams and storing the magnitude and phase data from a lock-in amplifier in the memory of a microcomputer, which later is used to display the image on a monitor or an oscilloscope screen.

A recent comparison of the various thermal wave imaging techniques, together with an extensive bibliography, can be found elsewhere (Thomas *et al.*, 1986). Mirage detection has both advantages and disadvantages in comparison with other thermal wave techniques. For example, in contrast to the gas-cell technique, which in principle is incapable of imaging closed

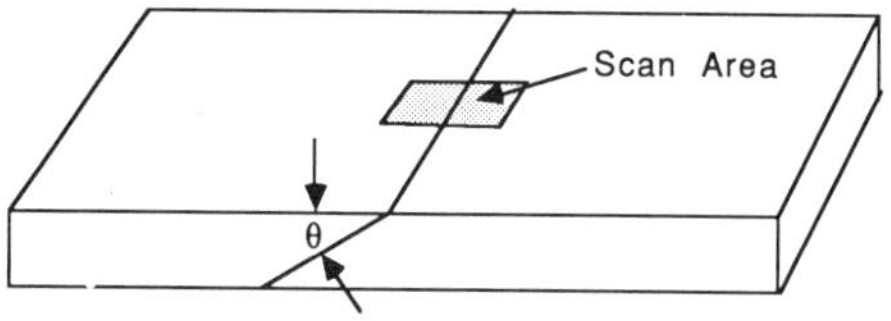

FIG. 2. Geometry for a tightly closed, model slanted crack. The shaded region represents the area on the top surface, intersected by the crack, which is scanned by the heating beam in thermal wave imaging.

vertical cracks, the mirage scheme produces images of such cracks with excellent contrast. Another advantage over gas-cell and piezoelectric transducer detection techniques is that it is totally contactless. It also has a signal-to-noise advantage over infrared (IR) thermal wave imaging techniques for samples of low emissivity. A disadvantage of the method is its restriction to samples that are either flat or convex (one exception is the case of a cylindrical hole, to be described below). A practical disadvantage is the difficulty of maintaining the relative alignment of *two* laser beams, as well as the height of the probe beam above the sample while scanning.

B. Application to Detection of Planar Cracks

1. Planar Cracks in Model Samples

Figure 2 shows the geometry for a tightly closed, model slanted crack. Such slanted "cracks" were fabricated in metal samples by cutting the slab sample, polishing the cut faces, and gluing the sample back together. When the angle θ is appreciably smaller than 90°, such cracks are readily imaged but, for $\theta = 90°$, are unobservable using gas-cell detection (Grice *et al.*, 1983a). However, by using mirage detection, either with the normal deflection, φ_n, with the heating and probe beams displaced relative to one another (to destroy the left–right symmetry), or with the tangential deflection, φ_t, (which has a left–right antisymmetry, even with the beams aligned), such a crack is readily imaged. Figures 3 and 4 (Grice *et al.*, 1983a) demonstrate this fact. For example, in Fig. 3b (when the beams are aligned), a line scan of the magnitude of φ_n is symmetrical about the crack, making it just barely observable. In the φ_n images of the crack with the two beams displaced, similar line scans are antisymmetrical about the crack, and the crack is readily imaged (Figs. 3a and 3c). Changing the displacement of the heating beam from the left to the right of the probe beam reverses the black–white contrast. In the φ_t images of the crack with the

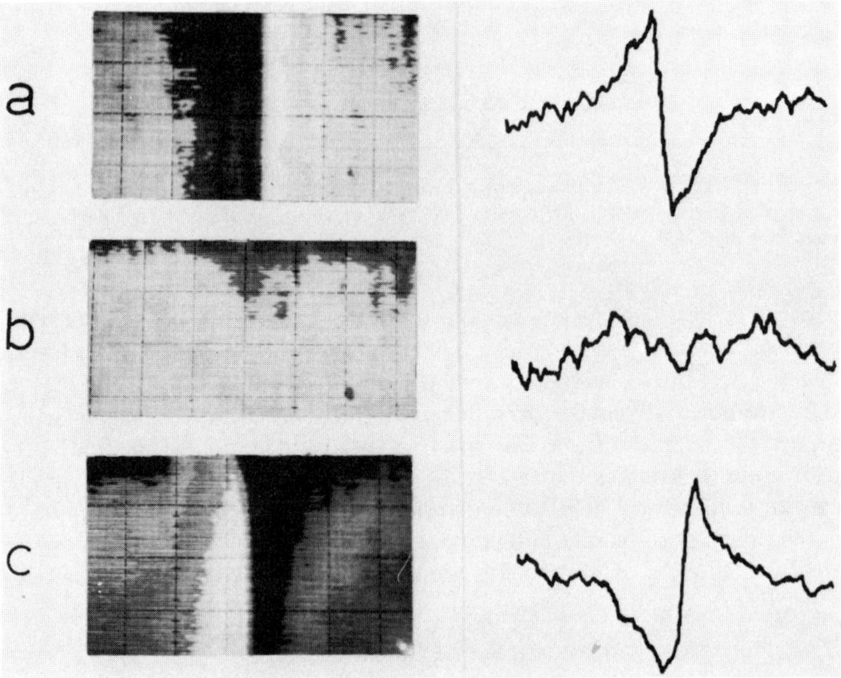

FIG. 3. Mirage-effect thermal wave images and corresponding line scans for a fabricated vertical crack ($\theta = 90°$ in Fig. 2), using the variations in magnitude of φ_n. (a) The heating beam is to the left of the probe beam; (b) the heating and probe beams intersect, and (c) the heating beam is to the right of the probe beam. (Grice *et al.*, 1983a).

two beams aligned (Fig. 4), the magnitude image is symmetrical about the crack, but with zero magnitude at the position of the crack, leading to good contrast and resolution. The phase image is antisymmetrical about the crack, and the 180° phase change across the crack gives rise to even better contrast. It should be noted that in acquiring these images, the probe beam was aligned to be parallel to the crack. If it should happen that the probe beam is exactly perpendicular to the crack, the crack would again be unobservable.

2. Fatigue Cracks in Aluminum Alloy Samples

An important practical example of a nearly vertical crack is that of a fatigue crack. To illustrate the excellent sensitivity of the mirage technique for imaging such cracks, as well as the utility of destroying the left–right

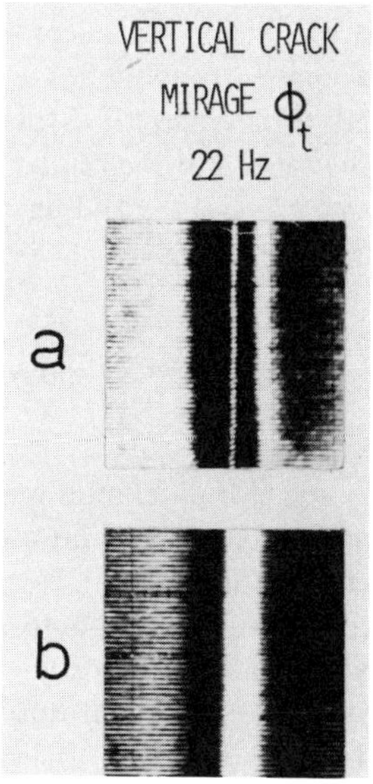

FIG. 4. Mirage-effect thermal wave images of the same fabricated vertical crack as in Fig. 3, using the variations in magnitude (a) and phase (b) of φ_t (Grice *et al.*, 1983a).

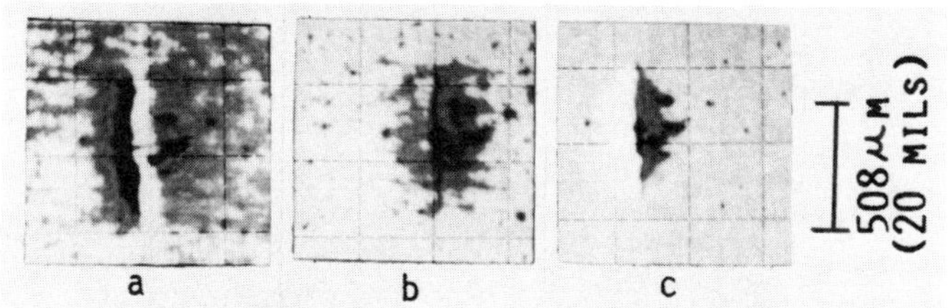

FIG. 5. Mirage-effect magnitude images of a fatigue crack at 212 Hz, using φ_n. Images correspond to the probe positioned to the left (a), through the center (b), and to the right (c) of the heating beam (Thomas *et al.*, 1982).

symmetry (which produces the black–white contrast in the images), in Fig. 5 we present φ_n images taken of a roughly 500-μm-long fatigue crack in an aluminum alloy sample with three different displacements of the two beams (Thomas *et al.*, 1982). Such cracks are sometimes observed by using gas-cell thermal wave imaging (Grice *et al.*, 1983b), but only because the cracks are not quite vertical or not tightly closed.

3. Brittle Fracture Cracks in Structural Ceramic Materials

Brittle fracture cracks in structural ceramic materials are very difficult to detect by conventional nondestructive evaluation techniques. For example, the eddy-current techniques that are sometimes applied for the detection of fatigue cracks cannot be used on these insulating materials. Furthermore, ultrasonic techniques must operate at very high frequencies if they are to detect the small cracks that are of practical importance in these materials. This, in turn, restricts ultrasonic methods to highly polished samples. To illustrate the ease with which mirage thermal wave imaging can detect such cracks, in Fig. 6 we show φ_t images of two such cracks that were created on

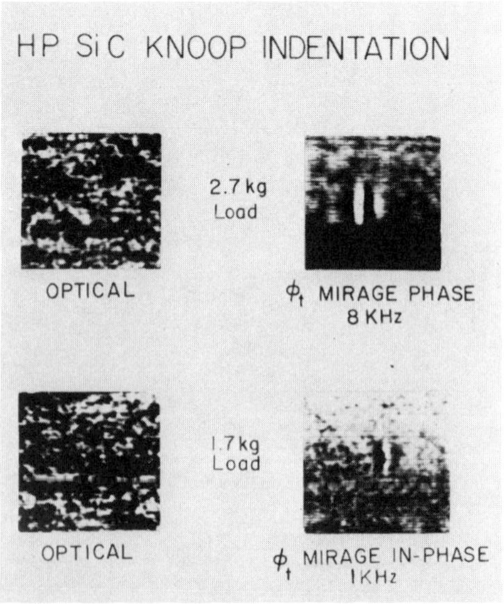

FIG. 6. Comparison of scanned optical and corresponding transverse deflection mirage (φ_t) thermal wave images for SiC surfaces containing half-penny-shaped cracks caused by Knoop indentations of 2.7-kG and 1.7-kG loads (Lin *et al.*, 1985). The surface damage was removed by machining prior to the scans.

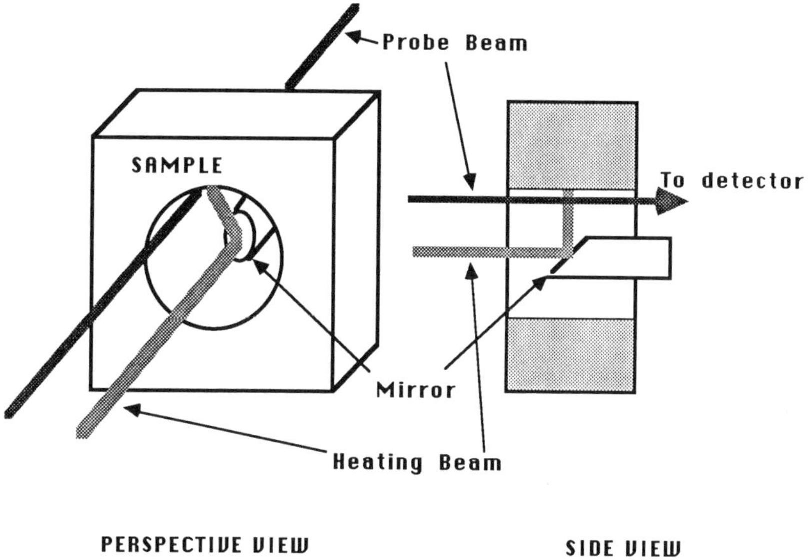

FIG. 7. Schematic diagram of an arrangement for mirage-effect thermal wave imaging of the inner surface of a cylindrical bolt hole.

a SiC specimen by means of a diamond tip indentor (Lin *et al.*, 1985). In this sample, the surface indentation was ground away prior to imaging, leaving no visible trace of the indentation. The images of Fig. 6 were made by using a single channel of the lock-in amplifier, with the signal in phase with the heating beam. It should be noted in Fig. 6 that in the larger of the two cracks (the one made with a 2.7-kg load on the diamond indenter), the black–white contrast of the image results from three phase reversals. In the smaller of the two cracks, only one phase reversal is observable. The central phase reversal locates the position of the crack. The outer phase reversals result from oscillations in the thermal waves reflected from the crack.

C. Application to a Difficult Geometry: A Bolt Hole

Mirage thermal wave imaging has also been used to study cracks in a geometry that is quite difficult for most nondestructive evaluation techniques: nearly vertical fatigue cracks on the inside surface of a cylindrical hole in a metal alloy (Grice *et al.*, 1984). The geometry of the experimental arrangement is given in Fig. 7. Except for the fact that a rotary, rather than a linear translational stage was used to scan the inner surface of the bolt

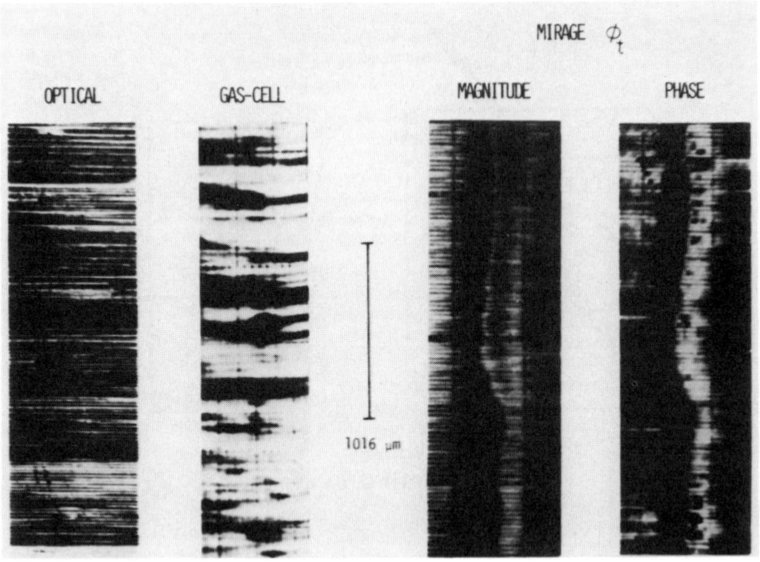

FIG. 8. Comparison of optical, gas-cell thermal wave, and mirage-effect (φ_t) thermal wave images of a fatigue crack on the inner surface of a cylindrical bolt hole (Grice *et al.*, 1984).

hole (with the two laser beams fixed and the phase variations of φ_t being recorded), the experiment is identical to that described above for mirage detection of vertical cracks in flat samples. The resulting mirage thermal wave images are given in Fig. 8, along with optical and gas-cell thermal wave images of the same region for comparison. It may be noted that although the surface was quite rough (boring-tool marks are evident in the optical image of Fig. 8), the fatigue crack is easily distinguished from the surface features in the mirage thermal wave image by its characteristic 180° phase reversal.

II. Characterization of Material Properties

A. Determination of Thermal Diffusivity of a Solid

1. Experimental Technique

The thermal diffusivity of a uniform material can be determined by means of mirage-effect thermal wave measurements. The experimental arrangement is similar to that shown in Fig. 1, except that the procedure is

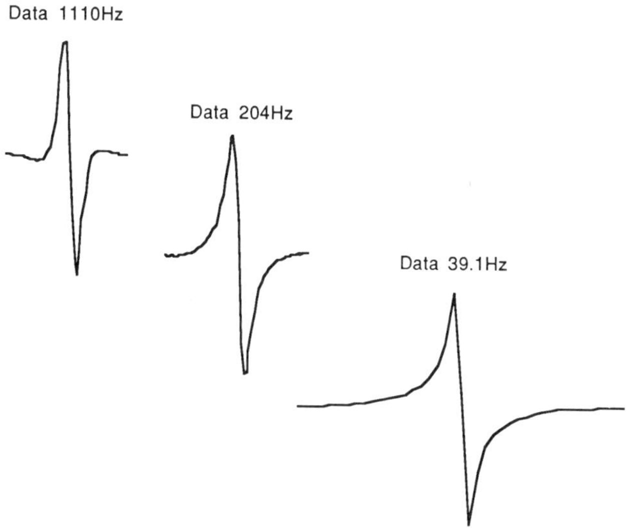

FIG. 9. Experimental plot of the in-phase component of φ_t versus the offset distance, x, between the heating and probe beams for a pure single crystal of chromium.

to keep the position of the sample fixed and to use another stepping-motor stage (not shown) to scan the heating beam at right angles to the probe beam across the sample surface. A single channel of the lock-in amplifier is used to monitor the in-phase component of φ_t as a function of the offset distance between the two beams. An example of such a plot for chromium is shown in Fig. 9. The antisymmetry of the plot results from the fact that the probe-beam deflection switches from left to right as one passes from one side of the heated region to the other. The zero crossings on either side of the central crossing correspond to points that are shifted in phase by $\pm 90°$ relative to the central position. Thus they can be used to determine the value of the thermal wavelength as a function of frequency in the *air* just above the sample. However, since the heat capacity of the air is so small compared to that of the solid, the temperature distribution in the air very close to the surface of the sample is determined by the temperature distribution in the solid. Therefore, if the probe beam is sufficiently close to the sample as it skims the surface (or equivalently, if the limiting frequency range of the measurement is sufficiently low), the beam is effectively sampling the temperature distribution in the solid, and the $\pm 90°$ phase points can be used to determine the thermal wavelength in the *solid*. The slope of a plot of that wavelength versus the reciprocal of the square root of the frequency can be used to determine the thermal diffusivity of the *solid*,

$\alpha = \kappa/\rho c$. Alternatively, one can solve the three-dimensional thermal diffusion problem for the measured quantity in the air as a function of the diffusivity in the solid and make a detailed comparison with experiment to determine α.

2. Theory for a Pure Material

We have described such a theory of the mirage-effect theory elsewhere (Favro, Kuo, and Thomas, in press). The resulting expressions for the normal (z-direction) and tangential (x-direction), deflections of the probe beam are given by

$$M_{\text{norm}} = \frac{-1}{\pi} \frac{1}{n} \frac{dn}{dT} e^{-q_g^2 R_2^2/4} \int_0^\infty dk \, \frac{k_g \cos(kx) e^{ik_g h} e^{-k^2 R_1^2/4}}{(\kappa_s k_s + \kappa_g k_g)} \tag{1}$$

and

$$M_{\tan} = \frac{-i}{\pi} \frac{1}{n} \frac{dn}{dT} e^{-q_g^2 R_2^2/4} \int_0^\infty dk \, \frac{k \sin(kx) e^{ik_g h} e^{-k^2 R_1^2/4}}{(\kappa_s k_s + \kappa_g k_g)}. \tag{2}$$

In these expressions, n is the index of refraction of the air, T is the temperature, q and κ (with the appropriate subscripts) are the thermal wavenumber and conductivity in the two media, R_1 and R_2 are the heating- and probe-beam radii, respectively, x is the separation between the two laser beams, and k_g and k_s are given by

$$k_s = (q_s^2 - k^2)^{1/2} \tag{3}$$

and

$$k_g = (q_g^2 - k^2)^{1/2}, \tag{4}$$

respectively.

The technique (Thomas *et al.*, 1985; Thomas *et al.*, 1986; Kuo *et al.*, 1986a; Kuo *et al.*, 1986b; Favro, Kuo, and Thomas, 1987) for measuring the thermal diffusivity involves measuring $M_{\tan}$ as a function of the offset distance, x, in the limit of very small probe height, h. The dependence of $M_{\tan}$ on the thermal diffusivity of the sample, α_s, is through the quantity k_s, which appears in the denominator of the integrand. The dependence of k_s on α_s is, in turn, through q_s, as was defined in Eq. (3). Except for the overall phase factor involving the probe-beam radius, R_2 (henceforth to be ignored),

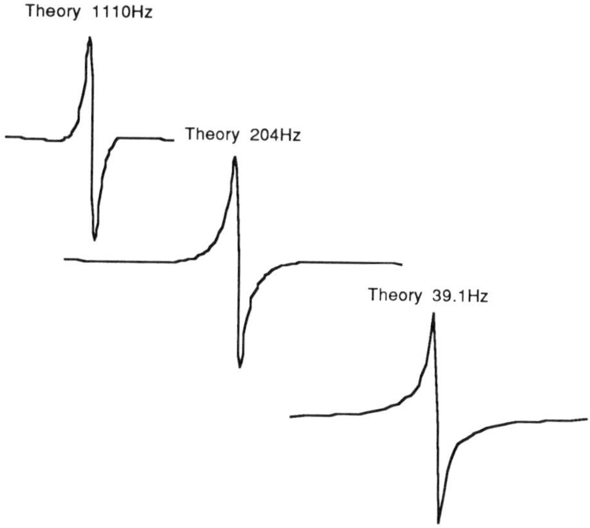

FIG. 10. Plots of the real part of Eq. (2), describing M_{tan} ($\propto \varphi_t$) versus the transverse offset, x, between the probe beam and the heating beam.

its dependence on k_g, at least for small values of h, is quite weak. This is true because, while k_s and k_g are ordinarily of about the same order of magnitude, κ_s is typically orders of magnitude larger than κ_g, thus making the term involving κ_g in the denominator negligible. Another consequence of this fact is that it makes the dependence on κ_s very much like an over-all factor, and thus it has very little influence on the shape of M_{tan}. It is the shape, rather than the normalization, of this function that provides most of the physically measurable information, due to the fact that too many factors contribute to the over-all signal strength. The function M_{tan} is related to, but not quite the same as, the special variety of Bessel functions known as Kelvin functions. Plots of the real part of the function versus the offset, x, for a small value of R_1 are given in Fig. 10. The antisymmetry of the function is quite apparent, as is the fact that it represents a very heavily damped wave. The two points at which this plot of the real part of M_{tan} first goes to zero on either side of the central zero are of particular interest. Since the real part of the function vanishes there, the function is purely imaginary, corresponding to values of the phase of $\pm\pi/2$. Inasmuch as we are dealing with wave propagation, one might naively think that these ninety-degree points would correspond to a quarter-wavelength distance on each side of the center. This is not so for two reasons. First, the effect of the finite heating-beam size, R_1, is such as to effectively add a constant to this distance. Second, cylindrical waves are not exactly periodic, and the first zeros do not occur precisely $\pm\pi/2$ radians from the origin. A numerical

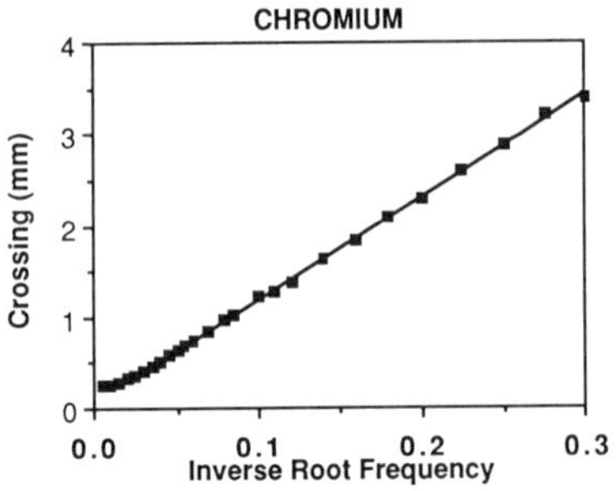

FIG. 11. Experimental (squares) and theoretical (line) plots of x_0 (see text) versus the reciprocal of the square root of the frequency for a pure single crystal of chromium.

analysis shows that the distance x_0 between the two ninety-degree phase points on either side of the origin is given by

$$x_0 = d + \sqrt{1.4}\,\frac{\lambda}{2} = d + \sqrt{1.4\pi\alpha/f}, \tag{5}$$

where d is a distance on the order of the heating-beam diameter and where f is the frequency ($= \omega/2\pi$). Thus, a plot of x_0 versus the reciprocal of the square root of the frequency should have a slope given by $(1.4\pi\alpha)^{1/2}$ and an intercept that is dependent on the size of the heating beam. A theoretical and an experimental plot of this type for chromium is shown in Fig. 11. It is important, when making experimental measurements, that the *low-frequency* portion of this plot be given more weight in determining the slope. This is because we have assumed that the probe-beam height, h, is small. "Small" in this case means small compared to the thermal wavelength in air. The thermal wavelength increases with decreasing frequency, decreasing the effective height of the beam and making our assumption more valid. The high-frequency portions of both the theoretical and experimental curves show deviations from linearity associated with the finite beam height and the properties of the gas.

3. Experimental Results

a. *Elements*

Experimentally, one can achieve a smaller value of h by utilizing the "bouncing" optical probe technique (Reyes *et al.*, 1987), shown schematically in Fig. 12. In this technique, the probe beam is incident at an angle of

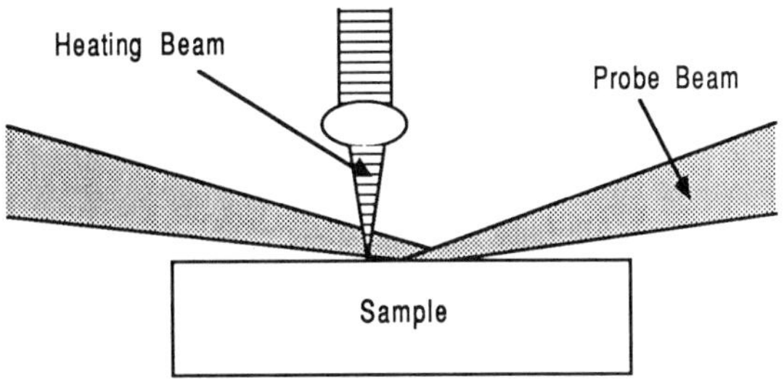

FIG. 12. Schematic diagram of the "bouncing" optical probe technique (Reyes *et al.*, 1987).

a few degrees from the sample surface and reflects from the surface into the position sensor. A comparison between measured diffusivities of pure elements using the "bouncing technique" and the nominal values computed from the handbook values of κ, ρ, and c is given in Fig. 13 (Reyes *et al.*, 1987).

b. *Compound Semiconductors*

In addition to the results described above for pure elements, mirage measurements of thermal diffusivity have been made for a number of compound semiconductors (Kuo *et al.*, 1986b). Representative plots of x_0 vs. $f^{-1/2}$ are given in Fig. 14. The resulting diffusivities are in reasonable agreement with handbook values.

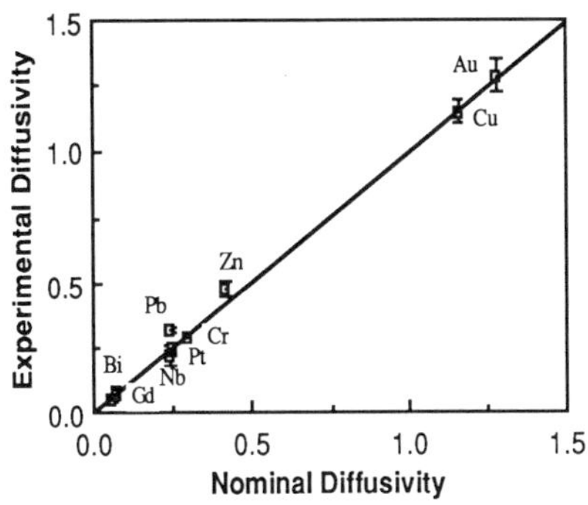

FIG. 13. Comparison between "bouncing" mirage-effect determinations of α, using the relation $(1.4\ \pi\alpha)^{1/2}$ and nominal values of α computed from the handbook values of κ, ρ, and c (Reyes *et al.*, 1987).

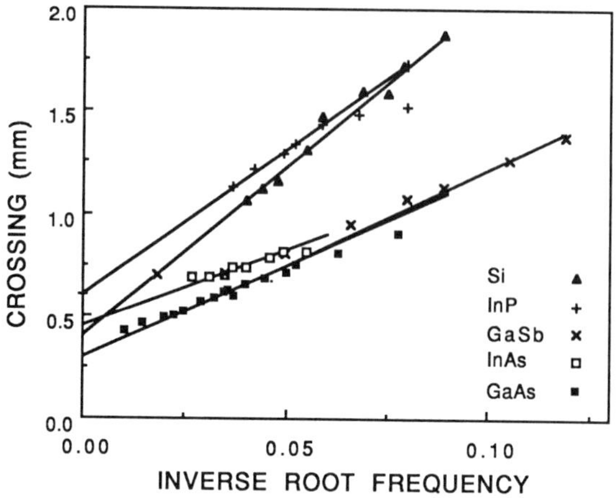

FIG. 14. Experimental plots of x_0 (see text) versus the reciprocal of the square root of the frequency for compound semiconductors (Kuo *et al.*, 1986b).

B. Thicknesses and Thermal Property Determinations for Layered Structures

The method used above to measure thermal diffusivity has also been applied to layered structures (Kuo *et al.*, 1986c; Favro, Kuo, and Thomas, in press). Our theoretical description of the mirage effect for a coated material is very similar to that of the uncoated material. In the following, we extend the theoretical treatment to a more general class of situations including periodically repeated multilayered (superlattice) structures. We begin with the differential equation satisfied by the temperature distribution $\overline{T}(x, z)$, where the bar indicates an average over the y-coordinate (along the probe-beam direction),

$$\nabla \cdot \left[\kappa_i \nabla \overline{T}_i(x, z)\right] + \kappa_i q_i^2 \overline{T}_i(x, z) = -\delta(x)\,\delta(z) \tag{6}$$

This equation is to be satisfied for each homogeneous region (layer) with a different set of material constants for that layer. A subscript "i" has been added to remind one of that fact. As in Favro *et al.* (in press), we define the Fourier transform of $\overline{T}_i$,

$$\overline{T}_i(x, z) = \int_{-\infty}^{\infty} dk\, e^{ikx} t_i(k, z), \tag{7}$$

which satisfies the differential equation

$$\frac{\partial}{\partial z}\left[\kappa_i \frac{\partial}{\partial z} t_i(k, z)\right] + \kappa_i k_i^2 t_i(k, z) = \frac{-1}{2\pi}\,\delta(z); \qquad k_i^2 = q_i^2 - k^2. \quad (8)$$

We have chosen the coordinate system such that the surface of the sample is at $z = 0$ and the region occupied by the gas is $z < 0$, and that by the sample, $z > 0$. The solutions to (8) take the form,

$$t_g(k, z) = C_g \exp(-ik_g z); \qquad z < 0 \text{ (gas)}, \quad (9)$$

$$t_1(k, z) = C_1 \sinh(\theta_1 + ik_1 z); \qquad 0 < z < a_1, \quad (10)$$

$$t_i(k, z) = C_i \sinh[\theta_i + ik_i(z - a_{i-1})]; \qquad a_{i-1} < z < a_i, \quad i = 2, 3, \ldots . \quad (11)$$

The form of Eqs. (10) and (11) was chosen to include waves propagating in both the positive and negative z-directions. This is because the wave propagating away from the source (at $z = 0$) will now scatter at each of the boundaries between two adjacent layers at $z = a_i$. The constant C_g describes the amplitude and phase of the thermal waves propagating into the gas. This wave is the one that is responsible for the deflection of the probe beam that gives rise to the mirage signal. The (complex) constant θ_1 describes the relative amplitude and phase between the forward-going and backward-going waves in layer 1. With the only source at $z = 0$, this quantity describes the collective response to a forward-going thermal wave of all layers in the half-space $z > 0$. Indeed, when we apply the boundary conditions that both the temperature and the heat flux be continuous at $z = 0$, we obtain,

$$C_g = C_1 \sinh(\theta_1) \quad (12)$$

and

$$\kappa_g k_g C_g + \kappa_1 k_1 C_1 \cosh(\theta_1) = i/2\pi. \quad (13)$$

The value of the constant C_g is all that is necessary to describe the temperature distribution in the gas, and hence to calculate the probe-beam deflection. This value can be determined if the value of θ_1 is known:

$$C_g = \frac{i}{2\pi(\kappa_1 k_1 \coth\theta_1 + \kappa_g k_g)}. \quad (14)$$

The value of θ_1 can be determined if the value of θ_2 is known, the value of θ_2 can be determined if the value of θ_3 is known, etc. This follows from the fact that the constants C_i can be eliminated from the boundary conditions at the interface between two successive layers.

$$\tanh[\theta_i + ik_i(a_i - a_{i-1})] = \frac{\kappa_i k_i}{\kappa_{i+1} k_{i+1}} \tanh \theta_{i+1}. \tag{15}$$

The iterative process will continue until one of three situations arises:

(i) The $(n + 1)$th layer is thermally thick.

If the $(n + 1)$th layer is thermally thick, there is no wave reflected from the next layer. This requires that the constant θ_{n+1} be infinite, which in turn leads to the condition

$$\tanh \theta_{n+1} = 1, \tag{16}$$

which determines all of the previous θ_i.

(ii) The $(n + 1)$th layer is made of an insulating material.

Since the thermal conductivity, κ_{n+1}, vanishes, there can be no heat flux into the $(n + 1)$th layer, and the boundary conditions yield

$$\cosh[\theta_n + ik_n(a_n - a_{n-1})] = 0, \tag{17}$$

which determines the value of θ_n, and hence all of the previous θ's.

(iii) The $(n + 1)$th layer behaves like a heat sink.

This case will occur whenever the $(n + 1)$th layer has an extremely high thermal conductance and has sufficient thermal capacitance. Under these conditions, the $(n + 1)$th layer cannot sustain any appreciable ac temperature amplitude, so we have instead,

$$\sinh[\theta_n + ik_n(a_n - a_{n-1})] = 0, \tag{18}$$

which again determines the θ's.

Another interesting class of structures is that of superlattice materials, in which two or more layers of film are repeatedly deposited on a substrate. If the periodicity of the structure is n, we can take advantage of the periodicity by assigning

$$\theta_{n+1} = \theta_1. \tag{19}$$

Then we need only to solve for the remaining n-independent θ_i's. A rigorous justification of this heuristic procedure is provided in the Appendix. It should be noted that the frequency-dependent parameter θ_1 determined in this manner contains a complete description of all the thermal properties of the composite structure that are necessary to calculate the mirage deflection.

We now apply the general treatment to two practical situations: (a) a substrate material coated with a single layer of thin film of a different material, and (b) a superlattice material in which alternate layers of two different materials are repeated indefinitely.

Case (a): Since the subscript "2" now refers to the substrate, which we shall consider to be thermally thick, we set $\tanh\theta_2 = 1$. Consequently,

$$\theta_1 = -ik_1a_1 + \frac{\kappa_1 k_1}{\kappa_b k_b}, \tag{20}$$

where the subscript "b" stands for the bulk substrate.

Case (b): We let subscripts "1" and "2" refer to the two films that make up the basic cell of the superlattice material. Since we have $\theta_3 = \theta_1$, the conditions for determining θ_1 and θ_2 are

$$\tanh[\theta_1 + ik_1a_1] = R\tanh\theta_2, \tag{21}$$

$$\tanh[\theta_2 + ik_2(a_2 - a_1)] = \frac{1}{R}\tanh\theta_1, \tag{22}$$

where we have used the abbreviation

$$R = \frac{\kappa_1 k_1}{\kappa_2 k_2}. \tag{23}$$

By solving these equations for θ_1, we have

$$\theta_1 = \operatorname{arctanh}\left[\frac{(R^2-1)b_1b_2 + \sqrt{(R^2-1)^2 b_1^2 b_2^2 + 4R(b_1R + b_2)(b_1 + b_2R)}}{2(b_1R + b_2)}\right], \tag{24}$$

where

$$b_1 = i\tan(k_1a_1); \qquad b_2 = i\tan[k_2(a_2 - a_1)]. \tag{25}$$

The mirage deflections for these two cases differ from the expressions for the mirage deflections for a bulk sample, Eqs. (1) and (2), simply by the insertion of the factor $\coth\theta_1$ in the first term in the denominator of each.

$$M_{\text{norm}} = \frac{-1}{\pi}\frac{1}{n}\frac{dn}{dT}e^{-q_g^2R_2^2/4}\int_0^\infty dk\,\frac{k_g\cos(kx)e^{ik_gh}e^{-k^2R_1^2/4}}{\kappa_1k_1\coth\theta_1+\kappa_gk_g}, \quad (26)$$

$$M_{\tan} = \frac{-i}{\pi}\frac{1}{n}\frac{dn}{dT}e^{-q_g^2R_2^2/4}\int_0^\infty dk\,\frac{k\sin(kx)e^{ik_gh}e^{-k^2R_1^2/4}}{\kappa_1k_1\coth\theta_1+\kappa_gk_g}. \quad (27)$$

With Eqs. (26) and (27) it is possible to extract information about the substrate as well as the coating from experimental plots of $M_{\tan}$. In this case, however, one needs more information than just the slope of x_0 versus $f^{-1/2}$. The full procedure is to make a multiparameter fit of the formulae to the entire curves of both the real and imaginary parts of $M_{\tan}$ versus the offset distance, x. Depending on the magnitudes of the parameters involved, one can find any or all of the coating thicknesses, the diffusivities of the substrate, coating, and gas, and such experimental parameters as the probe-beam height, etc. If some of these parameters are known from independent measurements, the fitting process is of course much quicker. Also, in some cases it is possible to do the fitting, not on the entire signal, but on an x_0 versus $f^{-1/2}$ plot and obtain a particular piece of information. For instance, if the diffusivities of the gas, the substrate and the coating, as well as the beam height, are known, the coating thickness can be obtained by fitting such curves. An example of this is shown in Fig. 15

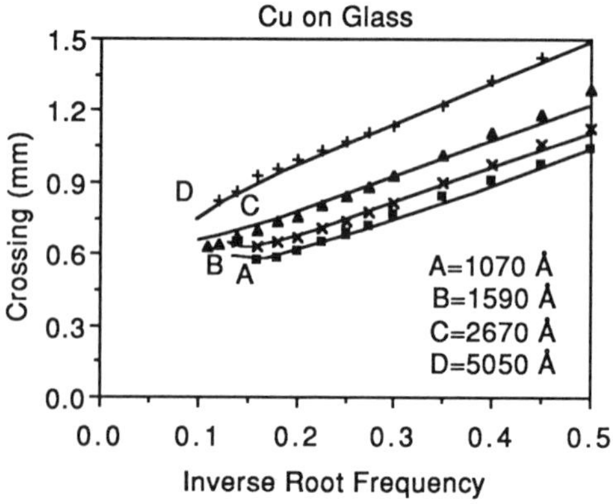

FIG. 15. Experimental (symbols) and theoretical (lines) plots of x_0 (see text) versus the reciprocal of the square root of the frequency for thin (1000 Å to 5000 Å) Cu films on glass substrates (Jaarinen *et al.*, in press).

(Jaarinen *et al.*, 1987), where the thicknesses of independently measured 1000-Å to 5000-Å films of copper on glass were determined by this kind of curve fitting. In this regard it should be pointed out that these measurements were made in the low-frequency regime, with thermal wavelengths that were orders of magnitude larger than the thicknesses of the films, thus demonstrating that it is not necessary to use ultra-high frequencies to measure thin films.

Appendix

Here we give a more detailed justification of the heuristic procedure stated in Sec. II.B. Given the boundary conditions of the first n layers,

$$\tanh[\theta_i + ik_i(a_i - a_{i-1})] = \frac{\kappa_i k_i}{\kappa_{i+1} k_{i+1}} \tanh \theta_{i+1} \tag{A1}$$

one can solve for θ_1 in terms of θ_{n+1},

$$\theta_1 = F(\theta_{n+1}). \tag{A2}$$

The function F will include as parameters all the physical constants such as thickness, thermal diffusivity, thermal conductance, etc., of the n layers of film. If the n-layered structure is repeated m times, the resulting relation is an mth order iteration of the F function,

$$\theta_1 = F^{(m)}(\theta_{mn+1}) = F^{(m-1)}\big(F(\theta_{(m-1)n+1})\big). \tag{A3}$$

The claim made in Sec. II.B is that as $m \to \infty$, the resulting θ_1 approaches a solution to the equation

$$\theta_1 = F(\theta_1). \tag{A4}$$

Implicit in the above statement is the fact that as $m \to \infty$, θ_1 also becomes independent of θ_{mn+1}. To prove this fact, we resort to a more conventional representation of the thermal waves in the first and the $(n+1)$th layer,

$$t_1(k, z) = A_1 e^{ik_1 z} + B_1 e^{-ik_1 z} \tag{A5}$$

$$t_{n+1}(k, z) = A_{n+1} e^{ik_{n+1}(z-a_n)} + B_{n+1} e^{-ik_{n+1}(z-a_n)}. \tag{A6}$$

The relations between these new constants and the old ones used in Sec.

II.B are,

$$A_i = C_i e^{\theta_i}, \qquad B_i = C_i e^{-\theta_i}, \tag{A7}$$

etc. The boundary conditions at $z = a_i$ then give rise to linear relations among the constants. We cast these relations in a matrix form,

$$\begin{bmatrix} A_1 \\ B_1 \end{bmatrix} = \mathbf{M} \begin{bmatrix} A_{n+1} \\ B_{n+1} \end{bmatrix}, \qquad \text{where} \quad \mathbf{M} = \begin{bmatrix} \alpha & \beta \\ \gamma & \delta \end{bmatrix}. \tag{A8}$$

This matrix contains all the thermal wave propagation information through the n-layers. In terms of the elements of the matrix M the relation (A3), for instance, takes the form,

$$\theta_1 = \frac{1}{2} \ln \left(\frac{\alpha e^{2\theta_{n+1}} + \beta}{\gamma e^{2\theta_{n+1}} + \delta} \right). \tag{A9}$$

Certain facts are known about this matrix. The first is that its determinant is equal to unity, due to the fact that the matrix corresponding to each individual layer has the same property. This implies that the product of the two eigenvalues of M is also unity. We shall call the eigenvalues λ and λ^{-1}, with the stipulation that $|\lambda| \geq 1$. The second is that the possibility that $|\lambda| = 0$ can be ruled out, since the thermal waves are always attenuated (the wave number k_i always has a positive imaginary part). Let (x_1, y_1) and (x_2, y_2) be the eigenvectors of M with eigenvalues λ and λ^{-1}, respectively. We can define the similarity transform which diagonalizes M,

$$S = \begin{bmatrix} x_1 & x_2 \\ y_1 & y_2 \end{bmatrix}. \tag{A10}$$

It then follows that

$$M = S \begin{bmatrix} \lambda & 0 \\ 0 & \lambda^{-1} \end{bmatrix} S^{-1} \tag{A11}$$

and,

$$M^m = S \begin{bmatrix} \lambda^m & 0 \\ 0 & \lambda^{-m} \end{bmatrix} S^{-1}. \tag{A13}$$

This last relation, together with the fact $|\lambda| > 1$, permits us to compute the asymptotic form for M^m, for large m,

$$M^m \cong \begin{bmatrix} x_1 y_2 & -x_1 x_2 \\ y_1 y_2 & -x_2 y_1 \end{bmatrix} \frac{\lambda^m}{x_1 y_2 - x_2 y_1}. \tag{A14}$$

In terms of these constants the relation (A4) now takes the form, for large m,

$$\theta_1 = \frac{1}{2}\ln\left(\frac{x_1 y_2 X - x_1 x_2}{y_1 y_2 X - x_2 y_1}\right) = \frac{1}{2}\ln(x_1/y_1), \qquad \text{where } X = \exp(2\theta_{mn+1}), \tag{A15}$$

a result indeed independent of θ_{mn+1}. We recall that the ratio x_1/y_1 can be computed from the matrix elements of M, thus (A15) can be expressed as,

$$e^{2\theta_1} = \frac{x_1}{y_1} = \frac{\lambda - \delta}{\gamma}. \tag{A16}$$

On the other hand, if we set $\theta_1 = \theta_{n+1}$ in (A10), a quadratic equation for the quantity $\exp(2\theta_1)$ results,

$$\gamma Y^2 + (\delta - \alpha)Y - \beta = 0, \qquad \text{where} \quad Y = \exp(2\theta_1). \tag{A17}$$

By substituting (A16) into (A17), we obtain,

$$(\alpha - \lambda)(\delta - \lambda) - \beta\gamma = 0, \tag{A18}$$

which is just the equation for the eigenvalues of the matrix M. This proves that the result of (A3) approaches the solution to (A4) as $m \to \infty$.

ACKNOWLEDGEMENTS

We would like to thank C. B. Reyes for her assistance in preparing the figures. This work was sponsored by the Army Research Office under Contract No. DAAG-29-K-0152 and by the Institute for Manufacturing Research, Wayne State University.

REFERENCES

Favro, L. D., Kuo, P. K., and Thomas, R. L. (1987). In *Review of Progress in Quantitative Nondestructive Evaluation*, (D. O. Thompson and D. E. Chimenti, eds.), Vol. 6A, pp. 293–299. Plenum, New York.

Favro, L. D., Kuo, P. K., and Thomas, R. L. (1987). In *Photoacoustic and Thermal Wave Phenomena in Semiconductors* (A. Mandelis, ed.), pp. 69–96. Elsevier, New York.

Grice, K. R., Inglehart, L. J., Favro, L. D., Kuo, P. K., and Thomas, R. L. (1983a). *J. Appl. Phys.* **54**, 6245–6255.

Grice, K. R., Favro, L. D., Kuo, P. K., and Thomas, R. L. (1983b). In *Review of Progress in Quantitative Nondestructive Evaluation*, (D. O. Thompson and D. E. Chimenti, eds.), Vol. 2B, pp. 1019–1028. Plenum, New York.

Grice, K. R., Inglehart, L. J., Favro, L. D., Kuo, P. K., and Thomas, R. L. (1984). In *Review of Progress in Quantitative Nondestructive Evaluation*, (D. O. Thompson and D. E. Chimenti, eds.), Vol. 3B, pp. 763–768. Plenum, New York.

Jaarinen, J., Reyes, C. B., Oppenheim, I. C., Favro, L. D., Kuo, P. K., and Thomas, R. L. (1987). In *Review of Progress in Quantitative Nondestructive Evaluation*, (D. O. Thompson and D. E. Chimenti, eds.), Vol. 6B, pp. 1347–1352. Plenum, New York.

Kuo, P. K., Reyes, C. B., Favro, L. D., Thomas, R. L., Kim, D. S., and Zhang, Shu-yi. (1986a). In *Review of Progress in Quantitative Nondestructive Evaluation*, (D. O. Thompson and D. E. Chimenti, eds.), Vol. 5B, pp. 1519–1523. Plenum, New York.

Kuo, P. K., Lin, M. J., Reyes, C. B., Favro, L. D., Thomas, R. L., Kim, D. S., Zhang, Shy-yi, Inglehart, L. J., Fournier, D., Boccara, A. C., and Yacoubi, N. (1986b). *Can. J. Phys.* **64**, 1165–1167.

Kuo, P. K., Sendler, E. D., Favro, L. D., and Thomas, R. L. (1986c). *Can. J. Phys.* **64**, 1168–1171.

Lin, M. J., Inglehart, L. J., Favro, L. D., Kuo, P. K., and Thomas, R. L. (1985). In *Review of Progress in Quantitative Nondestructive Evaluation*, Vol. 4B, pp. 739–744. Plenum, New York.

Reyes, C. B., Jaarinen, J., Favro, L. D., Kuo, P. K., and Thomas, R. L. (1987). In *Review of Progress in Quantitative Nondestructive Evaluation*, Vol. 6A, pp. 271–275. Plenum, New York.

Thomas, R. L., Favro, L. D., Grice, K. R., Inglehart, L. J., Kuo, P. K., Lhota, J., and Busse, G. (1982). *Proc. 1982 IEEE Ultrasonics Symposium* (B. R. McAvoy, ed.), pp. 586–590.

Thomas, R. L., Inglehart, L. J., Lin, M. J., Favro, L. D., and Kuo, P. K. (1985). In *Review of Progress in Quantitative Nondestructive Evaluation*, (D. O. Thompson and D. E. Chimenti, eds.), Vol. 4B, pp. 859–866. Plenum, New York.

Thomas, R. L., Favro, L. D., Kim, D. S., L. D., Kuo, P. K., Reyes, C. B., and Zhang, Shy-yi. (1986). In *Review of Progress in Quantitative Nondestructive Evaluation*, (D. O. Thompson and D. E. Chimenti, eds.), Vol. 5B, pp. 1379–1382. Plenum, New York.

Thomas, R. L., Favro, and Kuo, P. K. (1986). *Can. J. Phys.* **64**, 1234–1237.

CHAPTER 7

FLUID VELOCIMETRY USING THE PHOTOTHERMAL DEFLECTION EFFECT

JEFFREY A. SELL
Physics Department
General Motors Research Laboratories
Warren, MI

I. Introduction 213
II. Photothermal Deflection Effect 214
A. Static Fluids or Solids 216
B. Moving Fluids 217
III. Related Research 220
IV. Laminar Flow (Low Velocity) 221
V. Laminar Flow (High Velocity) 228
A. Multipoint-Velocity Measurements 232
B. Transient Thermal Grating 235
VI. Turbulent Flows 240
VII. Two-Dimensional Velocity Measurements 242
VIII. Velocities for Flowing Liquids 243
IX. Temperature and Pressure Dependence of PD Signals 245
X. Conclusions 246
XI. Appendix 246
A. Additional Comments on PDV Equipment 246
References 248

I. Introduction

The subject of laser velocimetry of fluids has become an active and broad-based field of research; measurements of gas velocity are important for studies of fluid mechanics, aerodynamics, and combustion. By far the most widely used laser-based technique is Doppler anemometry or velocimetry (LDA or LDV) (Durst, Melling, and Whitelaw, 1981; Drain, 1980). Here, particles seeded into the flow move through stationary fringes set up by interference of two laser beams. The velocity is determined from the Doppler shift of the scattered laser radiation. (Many different variants have developed, but they will not be reviewed here.) These techniques have so many advantages that it is likely they will remain the best methods for many more years. However, uncertainty over particle lag and tracking, and reluctance to seed with particles in some circumstances due to interfer-

ISBN 0-12-636345-5

ence with other laser diagnostics has prompted research in alternative techniques. Among these are phosphorescence marking (Hiller *et al.*, 1985), Doppler-shifted fluorescence (Hiller and Hanson, 1985; Cheng, Zimmermann, and Miles, 1983), speckle velocimetry (Meynart, 1983), particle tracking (Chang, Wilcox and Tatterson, 1984), and Raman spectroscopies (Moosmuller, Herring, and She, 1984; Gustafson, McDaniel, and Byer, 1981). The Doppler fluorescence technique requires a high velocity (5 m/s) to achieve a measurable Doppler shift. It also requires seeding by a species that absorbs and emits in the desired spectral range; this is also true of the phosphoresence method, which has the additional requirement of a long emitting-state lifetime. The speckle- and particle-tracking methods require seeding by particles, which is not always desirable. Raman-based methods are applicable only for high velocities.

Several different researchers have recently shown that photothermal deflection (PD), photoacoustic deflection, and thermal lensing (TL) methods are viable techniques for velocity measurements of fluids. Although these experiments have only recently been performed, it appears that there is a sufficient body of literature already and that the technique has some very promising advantages to warrant a review. This report reviews the literature and current state of the art of fluid velocimetry by the photothermal deflection method. This review will concentrate on the PD methods since much more work has been published on them, but it will also briefly review the thermal lensing and photoacoustic deflection studies.

II. Photothermal Deflection Effect

The PD effect, also called PDS or PTDS for photothermal deflection spectroscopy, relies on the creation of an index-of-refraction gradient in a medium due to the absorption of radiation, typically from a modulated pump-laser beam. The index gradient is directly related to the absorption of this radiation and subsequent relaxation to thermal energy, forming a gradient in density, and therefore in index. Typically, a beam from another laser, usually a low-power continuous-wave (CW) laser such as a He–Ne laser, is directed through the medium in an orientation so that it is deflected by the index gradient. Figure 1 gives a diagram of a typical PDS experiment. Usually, the pump and probe beams cross or nearly cross in the medium (as opposed to thermal lensing spectroscopy, where the beams are often collinear). The deflection of the probe is monitored with a position-sensitive detector or a photodiode with a razor blade or other aperture in front of it. Normally, the detector is connected to a lock-in amplifier, boxcar gate, or computer, so that the deflection is detected synchronously

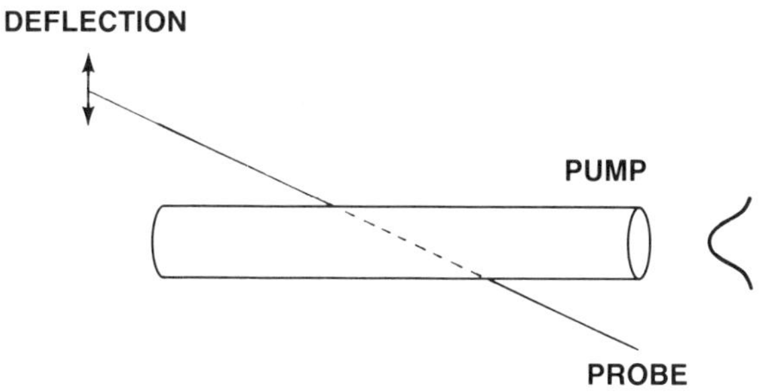

FIG. 1. Schematic diagram of a typical crossed-beam photothermal deflection experiment. The probe-beam waist is small compared to the pump-beam waist.

with the modulation of the pump beam. In this way, signal averaging can be accomplished that averages out any other beam deflection due to index gradients that may be present but unrelated to the pump-beam absorption. In many studies, there is a sufficiently high signal-to-noise ratio (SNR) to permit signal shot detection. It will be shown below that the probe-beam deflection is related to the absorber concentration, pump-beam energy, beam parameters, and, most importantly for this work, to the velocity of the medium.

A simple calculation of the temperature elevation of the sample is useful for determining how significant the perturbation is. Assuming C_2H_4 is the absorber at 100 ppm in N_2, a pump pulse energy of 6 mJ, a pump-beam radius of 0.02 cm, and a velocity of 1 m/s, gives a temperature elevation of 18 K maximum (using the conservation-of-energy approach (Sell 1985a) and the absorption coefficient for C_2H_4 of 3.0×10^{-5} (ppm cm)$^{-1}$ (Brewer, Bruce, and Mater, 1982)). This can be considered a negligible temperature rise for many applications (since physical parameters such as thermal conductivity and chemical parameters such as reaction rate constants will not change appreciably). There will, however, be situations where an 18-K temperature rise is not negligible; reducing the absorbed pump-beam pulse energy may be necessary in these cases, if the signal amplitude is sufficient. For the example outlined, the deflection angle will be roughly 10^{-4} radian; thus, if the detector is located 1 m away, then the deflection amplitude will be 0.01 cm. For a typical detector sensitivity of 50 V/cm, a 0.5-V signal will result.

We also note here the complementarity of the PD effect to laser fluorescence spectroscopy. For maximum signal strength, fluorescence needs minimal quenching of the enhanced population of the excited state created by the absorbed pump radiation. The PD effect requires a large degree of quenching. Note that for absolute quantitative concentration measure-

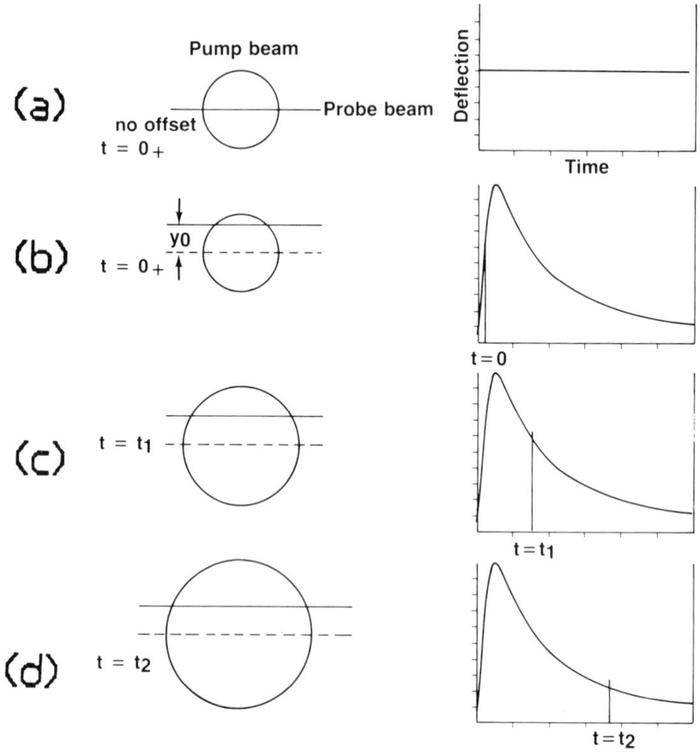

FIG. 2. Temporal development of the PD signal for static fluids or solids.

ments, they both require that the degree of quenching be known. For velocity measurements by the PD methods, the degree of quenching need not be known.

A. Static Fluids or Solids

Prior to discussing velocimetry using the PD effect, it is advantageous to first discuss the PD effect for static fluids or solids. Figure 2 shows a schematic representation of a crossed-beam (also called transverse) PD experiment for such situations. The figure shows, in the left side, the probe beam indicated by a single ray passing through the heated medium. (Imagine the pump beam emerging from the plane of the paper.) The radius of the pump beam is shown as a circle; at a time just after the pump beam is pulsed, the temperature of the medium has the same profile as the pump beam (here we assume rapid relaxation of the absorbed radiation to thermal energy). As time passes, the area of elevated temperature will grow and is shown by increasingly larger radius circles as one goes down the figure. In the right-hand side, we show the PD signal shape that would be observed on an oscilloscope. In part (a), the probe ray is shown passing through the center of the heated region; in this case, no signal is observed, since the temperature gradient is zero at the center. Part (b) shows the probe ray displaced by y_0 from the center of the pump beam; here, a signal will be observed that will rapidly rise (determined by the rate of energy relaxation to thermal energy, the displacement y_0, and the response time of the detector). Part (c) indicates the signal at a later time t_1 for the same pump–probe offset. The signal amplitude is decreasing due to thermal conduction of the heat from the center of the heated region. Finally, at time t_2, the signal has decreased further as the heated region grows and the gradient at the probe-ray position decreases.

Additional information on the static fluid PD signal dependence on parameters such as the absorbed pump-beam energy, pump-beam radius, and pump–probe offset was recently gained through optical ray tracing (Sell, 1987).

B. Moving Fluids

A schematic of the crossed-beam PD effect for a moving fluid is indicated in Fig. 3; this is similar to Fig. 2, except the heated volume is shown moving upward with a velocity v. Note that at a short time $t = 0_+$ after the pump-laser pulse, there will be a PD signal observed even for zero offset in pump–probe beams (see part (a) of the figure). This is because the symmetry is destroyed by the flow such that there is a nonzero gradient. If the probe ray is offset at a distance y_0 downstream of the pump beam, then the PD signal will show two lobes. The first corresponds to an upward deflection of the probe beam when the heated volume is below it, and the second to a downward deflection of the probe when the heated volume has moved above the probe (see part (b) and (c) of the figure). At later times, the PD signal will sweep through the other lobe and eventually decrease to zero.

The theory of the PD effect for moving fluids has been covered in Chapter 3 of this volume and only the key results are stated here. Sontag and Tam (1985) derived an expression for the deflection angle ϕ as a

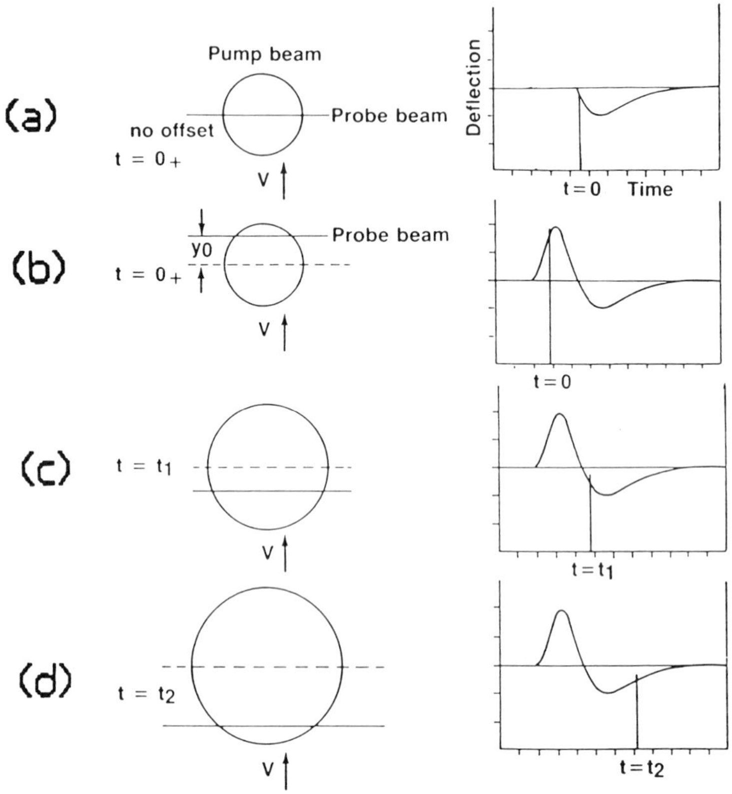

FIG. 3. Temporal development of the PD signal for fluids moving upward with velocity v.

function of time t, where

$$\phi = -\frac{1}{n}\frac{dn}{dT}\frac{4\alpha DE_0(vt - y_0)}{\sqrt{2\pi}\lambda(a_0^2 + 8Dt)^{3/2}}\exp\left[-\frac{2(vt - y_0)^2}{(a_0^2 + 8Dt)}\right] \tag{1}$$

for a fluid moving with velocity v where y_0 is the pump–probe offset. The above equation assumes a weakly absorbing medium and a Gaussian profile pump beam that is temporally rectangular, and a pump-pulse duration t_0 shorter than the diffusion time $\tau_d = a_0^2/4D$, where D is the diffusion coefficient and a_0 is the Gaussian pump beam $(1/e^2)$-intensity radius. Also, λ is the thermal conductivity, α is the absorption coefficient, and E_0 is the pump pulse energy.

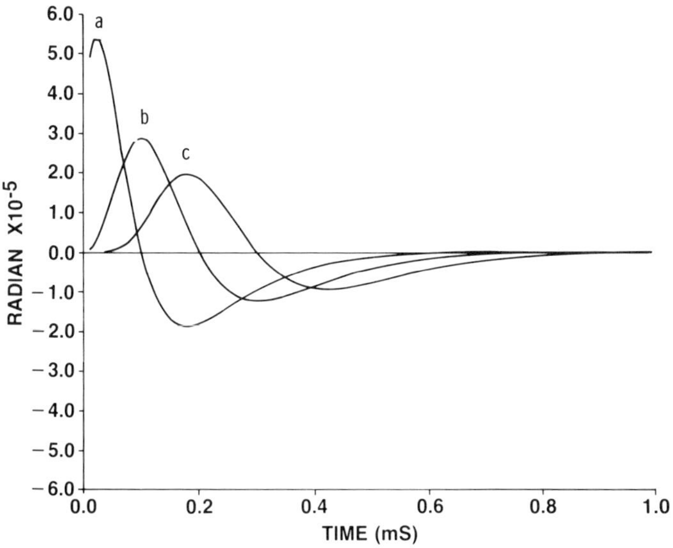

FIG. 4. PD angle in radians computed with Eq. (2) as a function of time for three pump–probe offsets: (a) 1.0 mm, (b) 2.0 mm, (c) 3.0 mm (see Rose, Vyas, and Gupta, 1986).

A more general equation was derived by Rose, Vyas, and Gupta (1986)

$$\phi = -\frac{1}{n}\frac{dn}{dT}\frac{8\alpha E_0}{\sqrt{2\pi}\,\rho C_p t_0}\int_0^{t_0}\frac{[x - v_x(t-\tau)]}{[a^2 + 8D(t-\tau)]^{3/2}} \times \exp\Big[-2[x - v_x(t-\tau)]^2 \big/ [a^2 + 8D(t-\tau)]\Big]\, d\tau, \tag{2}$$

which is valid for a pump pulse duration of arbitrary length t_0.

Figure 4 shows the PD signal as a function of time for three different pump–probe offsets: 1, 2, and 3 mm, computed using Eq. 2. This figure illustrates the simplest method of velocimetry using the PD effect, namely time-of-flight, also called transit-time velocimetry. If one knows the separation of the pump and probe beams, then the velocity can be determined directly from the time delay between the (short-duration) pump pulse and the zero crossing of the PD signal. This is the most straightforward method of determining velocity, but we shall postpone further discussion of it until Section V.

III. Related Research

Before further discussing the PD effect, we shall mention some indirectly related work. In the area of thermal lensing spectroscopy for flowing samples, Dovichi and Harris (1981), Nickolaisen and Bialkowski (1985, 1986), and Weimer and Dovichi (1985) have discussed the effect of fluid flow in thermal lensing experiments. These studies were not intended as measurements of the fluid flow, but rather considered the effects of flow on the TL signal amplitude when, for example, TL was used as a detector for gas or liquid chromatography. The PD and TL experiments are very similar; the PD signal is basically due to the first spatial derivative of the thermal profile of the fluid, whereas the TL signal is essentially the second derivative. Bialkowski (1986) recently compared thermal lensing to photothermal deflection spectroscopy. The conclusions of these TL studies regarding the dependence of the TL signal amplitude are somewhat dependent on the residence time of the fluid in the TL cell and on whether CW or pulsed pump excitation was used.

Nickolaisen and Bialkowski (1986) have raised an issue that is relevant to this study and that is the question of the rate of thermalization of the absorbed laser energy. In their study, they found that the risetime of the signal was determined by the rate at which the absorbed energy relaxed into translational modes. The relevance to the present work is that if the thermalization rate is very slow, then this can affect the velocity measurements to be discussed below. However, this will rarely be the case, since vibrational-and rotational-to-translational energy transfer rates are typically 10^9 to 10^{12} sec^{-1} for gases at atmospheric pressure. These rates correspond to decay times much shorter than the pump pulse duration and time-of-flight in the present experiments. For further information on this point, see the chapter in this volume by Barker and Toselli and the paper by Sontag, Tam, and Hess (1987).

Zapka and Tam (1982a, b) and Tam *et al.* (1983) have utilized photoacoustic deflection of probe beams to measure ultrasonic velocities, fluid velocities, and temperatures. These studies demonstrated noncontact, all-optical acoustic methods that are useful for hostile environments and in situations where good temporal resolution is needed. The technique for measuring fluid velocity was to use three probe beams (one upstream from the focused "pump" beam and two downstream). Ultrasonic velocity in both directions was measured by the time-of-flight, and the difference of the two was calculated to measure the flow velocity. Good agreement with a mechanical flow measurement was found over the velocity range studied, 30 to 130 cm/s. The spatial resolution was not as good as for photothermal deflection, however, due to the separation of the probe beams.

IV. Laminar Flow (Low Velocity)

The theory of the PD effect for flowing samples has been covered previously (Sell, 1985a; Rose, Vyas, and Gupta, 1986); we rely on these results for measurement of low velocities by the relaxation-rate method, also called the signal-shape method (Sell, 1985a). This technique uses the fact that the rate at which the PD signal decays (and, to some extent, rises) is dependent on the gas velocity. Physically, both conduction and convection contribute to the dissipation of thermal energy from the volume of space sampled by the probe beam. By fitting the experimental PD signal to the theory that uses the velocity as a parameter, the velocity can be determined because conduction rates are constant as long as the temperature elevation of the medium is small.

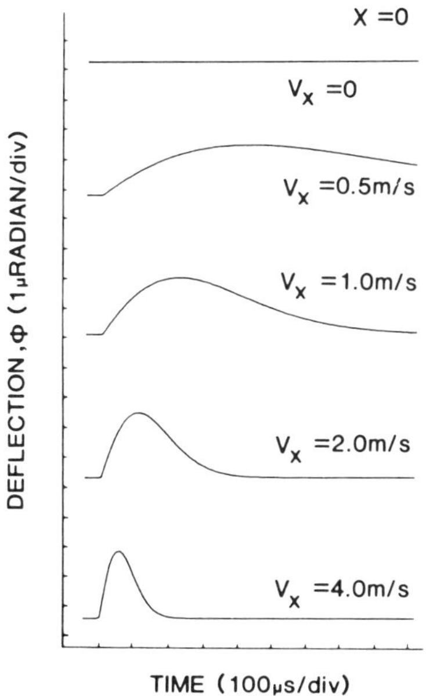

FIG. 5. PD angle as a function of time computed with Eq. (2) for several different values of fluid velocity v_x (see Nie, Hane, and Gupta, 1986).

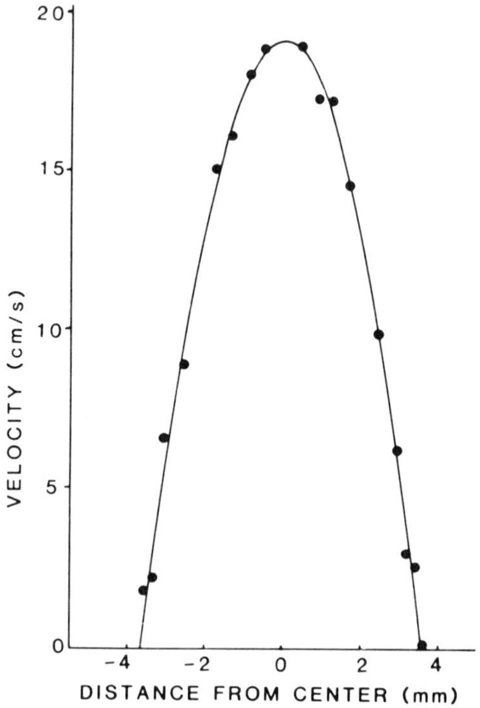

FIG. 6. Velocity profile of a laminar open jet as measured by the relaxation-rate PD method. Circles are the experimental data points (averaged over 100 pump-laser pulses), and the solid line is a best fit using Eq. (2) and assuming a parabolic profile (see Nie, Hane, and Gupta, 1986).

Figure 5 shows PD signals computed as a function of velocity for overlapping pump and probe beams (Nie, Hane, and Gupta, 1986). Clearly, the signal shape is velocity dependent. Nie, Hane, and Gupta (1986) have used a 1-μs pump pulse from a dye laser to record the PD signals above a NO_2-seeded N_2 jet and measured the velocity profile shown in Fig. 6. Velocities were obtained by fitting the data to Eq. (2) above; the profile has the expected parabolic shape. Nie, Hane, Gupta (1986) have also plotted the peak deflection signal as a function of flow rate for various pump pulse durations, shown in Fig. 7. The result shows the $(1/v_x)$-dependence of the deflection angle over a restricted velocity range discussed previously by the present author (Sell, 1984). This dependence can be used to determine the velocity, assuming that the concentration of the absorbing species does not vary during the experiment.

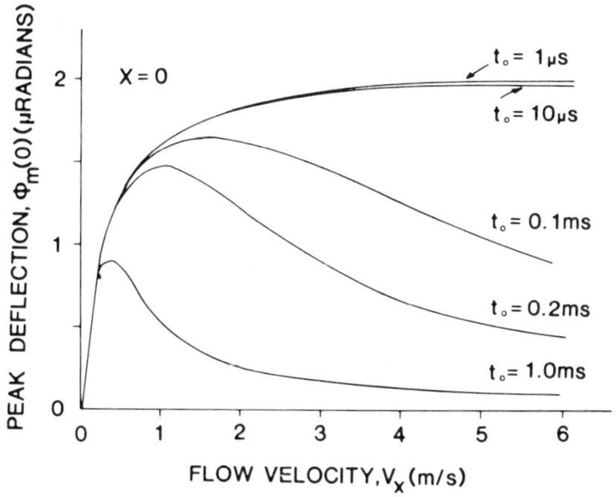

FIG. 7. Peak value of the probe-beam deflection at $x = 0$, $t = 0$, plotted as a function of v_x for several different pump-laser pulse lengths (see Nie, Hane, and Gupta, 1986).

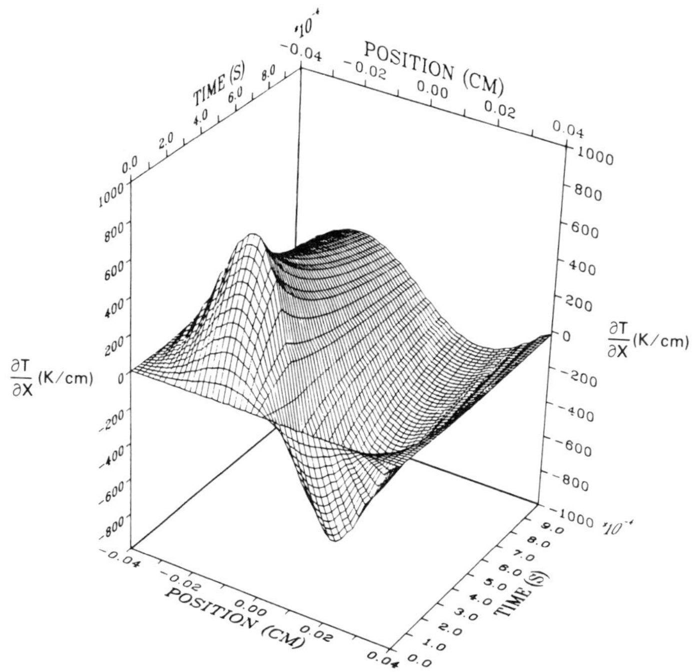

FIG. 8. Temperature gradient versus time (up to 1 ms) versus position for a static gas computed by using a finite-difference solution to the energy Eq. (3).

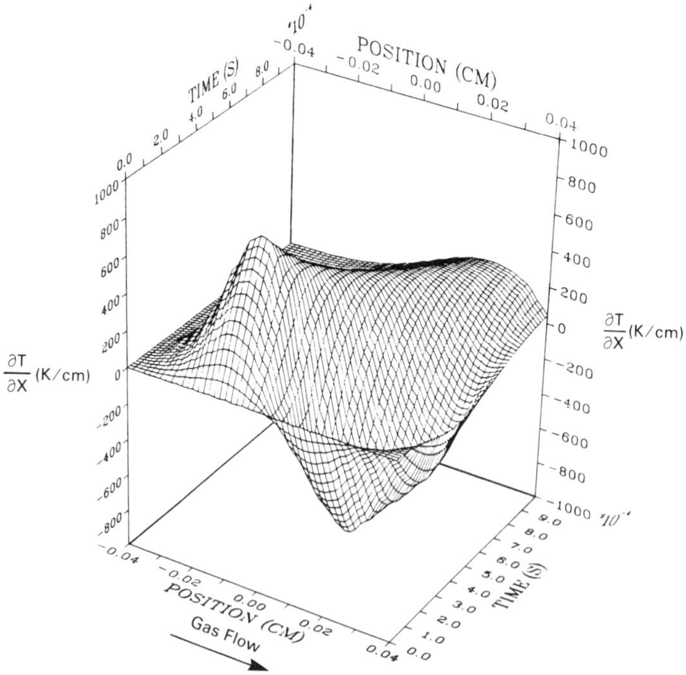

FIG. 9. Temperature gradient versus time (up to 1 ms) versus position to a gas moving with velocity $v = 50$ cm/s computed by using a finite-difference solution to the energy Eq. (3).

We turn to Fig. 8 and Fig. 9, which show the results of calculations of the PD signal shape on gas velocity from the present author's work (Sell, 1985a). Plots of the temperature gradient versus time versus x-position of the gas after absorption of the pump beam (centered about $x = 0$, and lasting for 0.2 ms) are shown. The temperature gradient is linearly related to the deflection angle ϕ and therefore to the PD signal measured. These gradients were computed by numerically solving the energy equation

$$\rho C\left(\frac{\partial T(x,t)}{\partial t} + v_x \frac{\partial T(x,t)}{\partial x}\right) = \lambda \nabla^2 T(x,t) + Q(r,t), \tag{3}$$

where

$$Q(r,t) = \begin{cases} \dfrac{2\alpha E_0}{\pi a_0^2 t_0} \exp(-2r^2/a_0^2)\exp(-\alpha z), & \text{if} \quad 0 \le t \le t_0, \\ 0, & \qquad t > t_0. \end{cases} \tag{4}$$

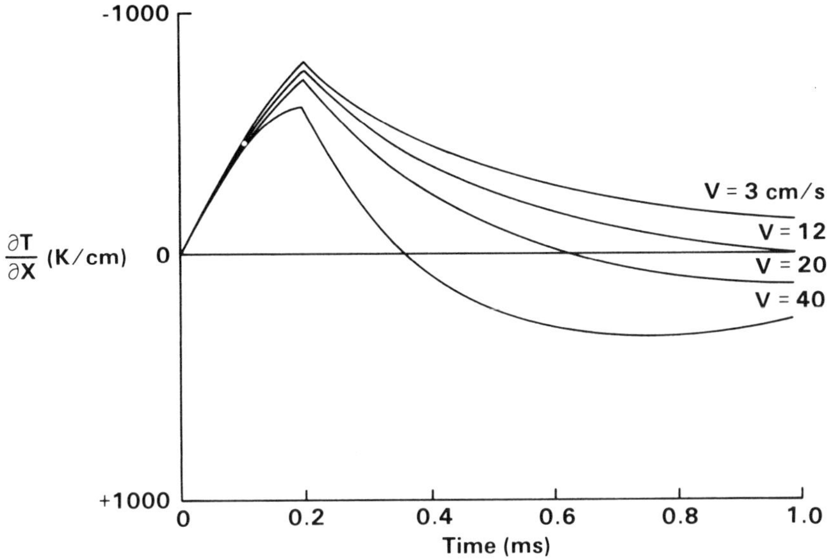

FIG. 10. PD signal as a function of time computed with Eq. (3) for several different velocities. Note that that pump pulse duration was 0.2 ms, giving rise to the long signal onset.

by using a finite-difference approach. Figure 8 gives results of a calculation for a static gas ($v = 0$), and Fig. 9 for $v = 50$ cm/s. For the static gas, the temperature gradient is symmetric about $x = 0$ (except for a sign change); the symmetry is destroyed in the flowing-sample case. This shows that a PD signal will be observed in the latter case even for perfectly overlapped pump and probe beams as discussed above.

Figure 10 shows the signal shape at a fixed pump–probe separation ($x = 0.010$ cm) for several different velocities. The signal shape was somewhat different than that shown in Fig. 3 because the pump-pulse duration in this case was 0.2 ms, leading to the long risetime in the signals shown in Fig. 10. Nevertheless, the dependence of the PD signal on velocity is illustrated and was used in the measurement of velocity. Figure 11 shows a comparison of the experimental PD signals and the computed ones for two different flow rates. The experimental data was obtained above a N_2 laminar flow seeded with C_2H_4; Sell (1985a) discussed in more detail the procedure used to fit the experimental data with the velocity-dependent theory to determine the velocity. Figure 12 shows the velocity at the center of the jet measured by this method compared to the velocity computed from the mass flow rate and the cross-sectional area of the jet, assuming a parabolic distribution. The agreement was good; the nonzero y-intercept is believed to be due to a calibration error in the mass flow measurement.

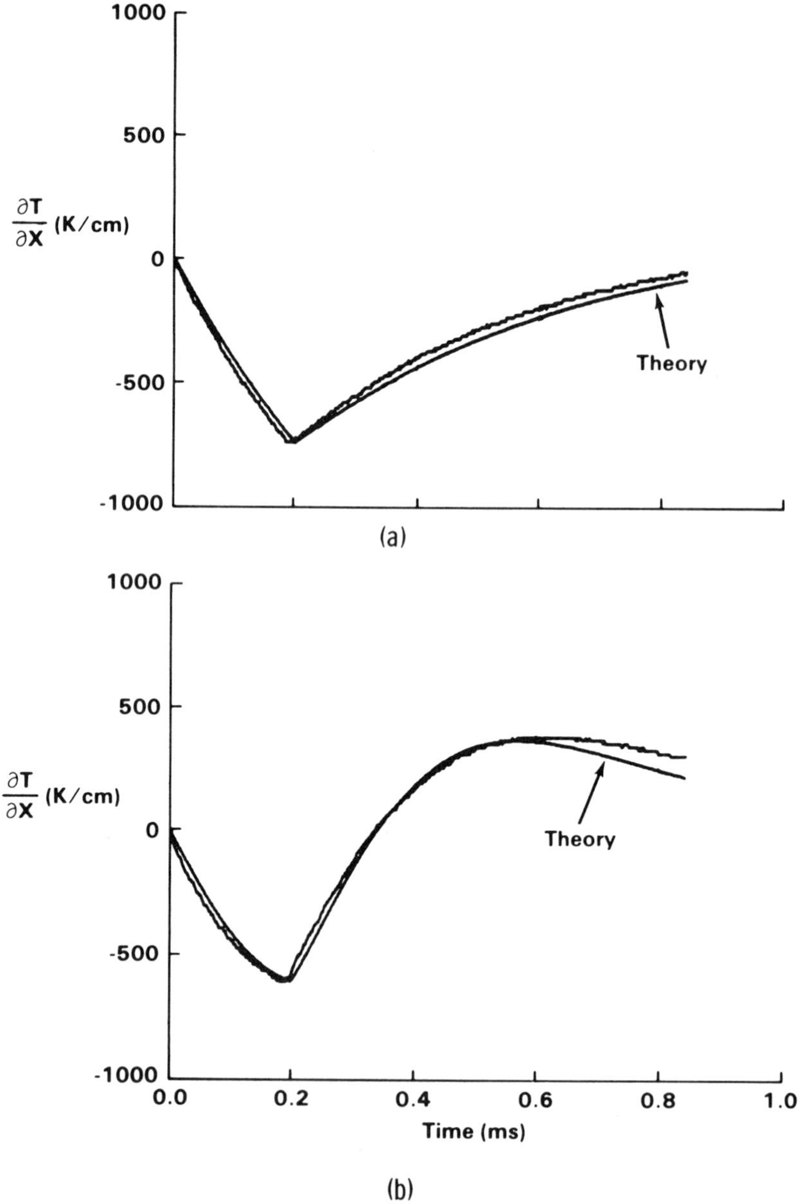

FIG. 11. Comparison of experimental PD traces averaged 1024 times and best-fit theoretical ones for two different velocities. The velocity derived by this approach was (a) 14 cm/s, and (b) 65 cm/s.

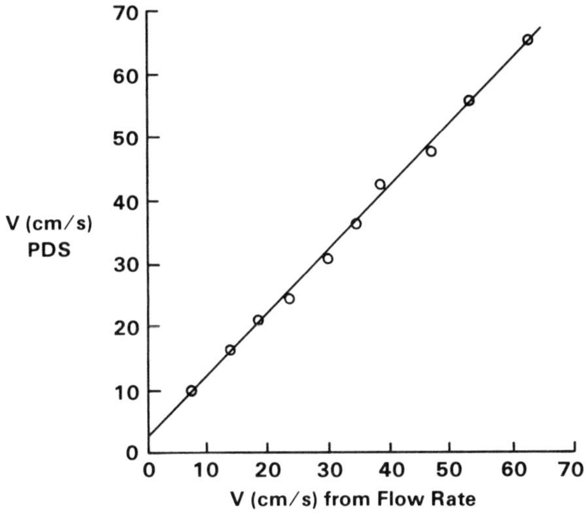

FIG. 12. Plot of velocities measured by the PD relaxation-rate method versus those calculated by the mass flow rate. The PD data was signal averaged over 1024 shots.

In addition, velocities can be measured simultaneously with a relative measurement of absorber concentration. This is possible because the peak deflection signal is linearly dependent on the absorber concentration (Sell, 1985a), whereas the peak deflection signal is only weakly dependent, and in some cases (Nie, Hane, and Gupta, 1986) independent of the velocity over a limited range of velocity. It was possible to separate the two and to use the signal relaxation rate to determine the velocity, and the peak height to determine the relative absorber concentration. This is shown in Fig. 13, where the velocity exhibits the expected parabolic shape, and the concentration profile was somewhat flatter near the center, but falls off near the jet edges due to diffusional mixing.

The above results were obtained by averaging over several pump-laser shots to improve the signal-to-noise ratio. However, single-shot measurements have been made (Sell, 1985a; Nie, Hane, and Gupta, 1986). The temporal resolution was roughly 1 ms, since that was the duration of the PD signal. As will be shown below, this can be improved using the transient grating approach. Note that the spatial resolution was also very good—roughly the volume of overlap of the focused pump and probe beams, which in this case was 10^{-5} cm^3. The spatial resolution of this relaxation-rate method is somewhat better than that of the time-of-flight technique to be discussed below, because the beams can be perfectly overlapped (Sell, 1985a; Nie, Hane, and Gupta, 1986).

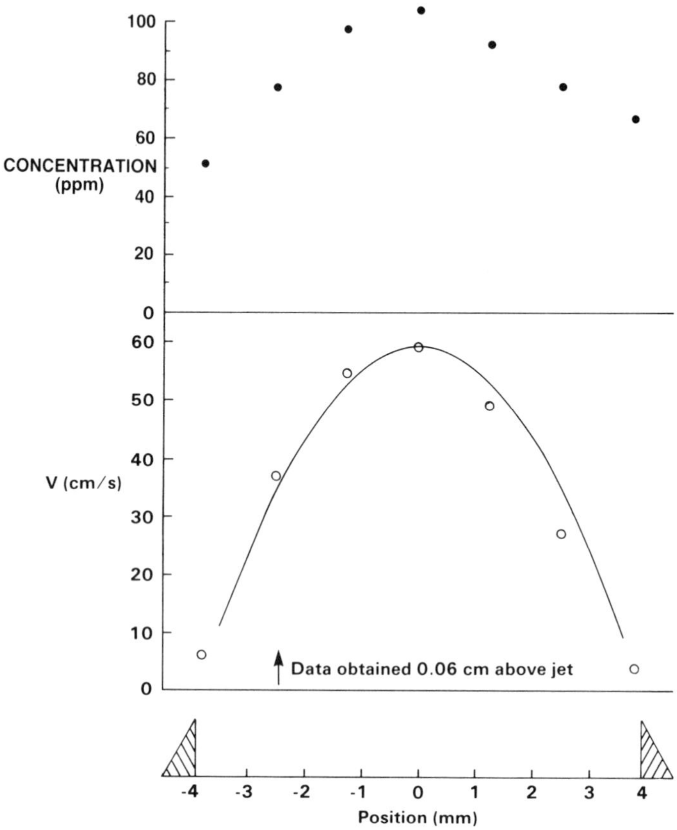

FIG. 13. Velocity and concentration profiles above the jet (measured simultaneously). The concentration measurements were relative, assuming a centerline concentration of 100 ppm, which was the jet feed concentration.

V. Laminar Flow (High Velocity)

For measurements of higher velocities (but still within the laminar regime), we rely on the time-of-flight technique. Physically, this uses a "labelling" or "tagging" of a flowing packet of fluid by heating it with the pulsed pump beam. Ideally, the temperature elevation of the fluid is only a few degrees, so that method can be considered nonintrusive for many applications. The focused probe beam crosses the fluid stream at a small distance (usually less than 1 mm) downstream from the pump beam. When the heated fluid passes through the probe, it is deflected; the time delay

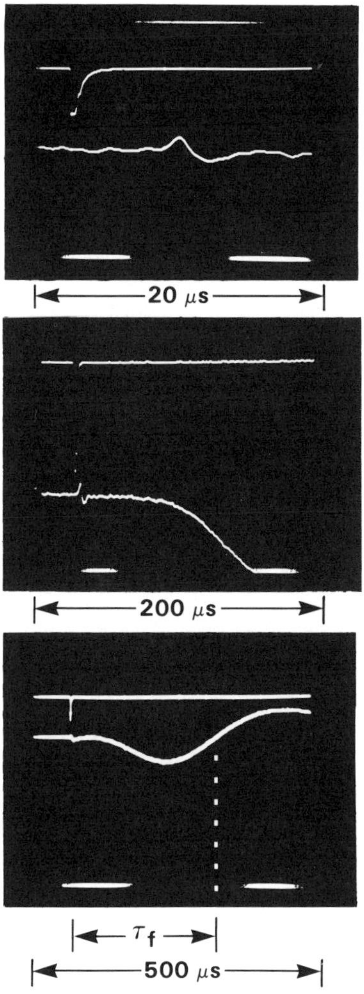

FIG. 14. Oscilloscope traces of the pump-laser intensity (top trace) and PD signals (bottom traces). Three different time scales are shown: 20, 200, and 500 μs (full scale). Both the photoacoustic and photothermal deflection signals are shown. The photoacoustic deflection signal is used to measure the pump–probe offset as discussed in the text.

between the short-duration pump pulse and the zero crossing (see Fig. 3) of the PD signal gives the velocity if the pump–probe offset is known.

Figure 14 shows an example where a Q-switched CO_2 laser was used as the pump and a laminar ethylene–nitrogen jet provides the moving fluid whose velocity is to be measured. Here, the oscilloscope traces show a photoacoustic deflection signal that occurs much sooner than the photother-

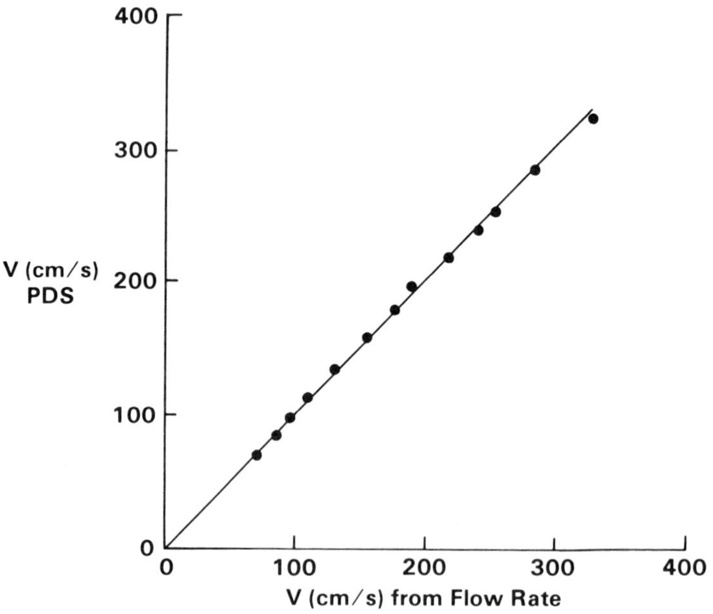

FIG. 15. Plot of gas velocities from the PD time-of-flight method versus those from the mass flow rate.

mal deflection signal. Assuming that the acoustic pulse travels at the speed of sound for the ambient fluid temperature and pressure, the time delay for the photoacoustic deflection signal can be used to compute the separation of the pump and probe beams (Sell, 1985a). Once the pump–probe separation is known, the velocity can be derived by dividing the separation by the time delay from the pump pulse to the zero crossing of the deflection signal. The velocities measured by this method agreed very well with those computed from the mass flow rate (see Fig. 15).

This same approach has also been used by Sontag and Tam (1985), but was called the traveling thermal lens approach. Figure 16 shows some of their PD signals (averaged over 32 shots) for various pump–probe separations. They have also measured the velocity profile above their jet that was parabolic; demonstrated simultaneous absorber concentration and velocity measurements; and obtained very good SNR for single-shot data. The latter was fit very well by an analytical expression for the deflection angle (Eq. (1) above) that is valid for a short pump pulse duration. For the conditions used in their study, the time-of-flight measurements should be most accurate for velocities above 0.6 m/s. Below this limit, the diffusion of the fluid complicates the data interpretation. The upper velocity limit for

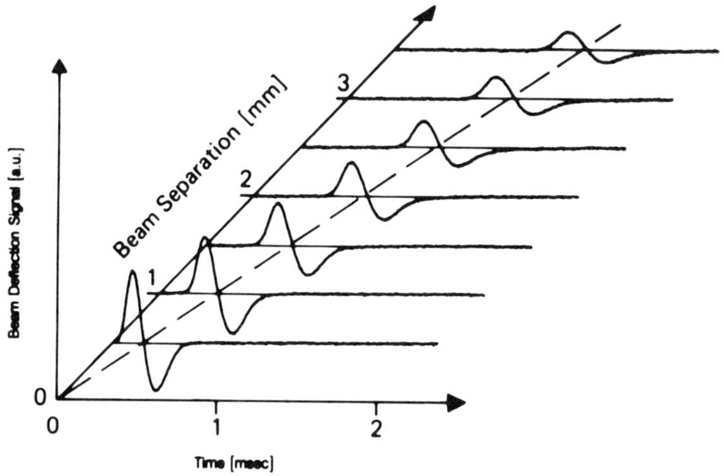

FIG. 16. PD signals averaged 32 shots for different pump–probe separations. The velocity of 4.24 m/s was derived from the time-of-flight (see Sontag and Tam, 1985).

this method was determined by the pump pulse duration; when the velocity is high enough that the fluid moves appreciably while the pump pulse is on, then the fluid tagging is smeared, and the signal shape will be distorted. For a 1-μs pulse whose waist is 100 μm, this upper limit would be roughly a velocity of 100 m/s. However, in practice, turbulence will probably limit the upper velocity measurable by this approach, since the signal will be considerably reduced and complicated by turbulence.

Velocimetry by the PD effect is not limited to noncombusting flows. Rose and Gupta (1985) have used the time-of-flight approach to measure gas velocities in H_2/O_2 flames (equivalence ratio of 1). Here, a 1-μs pump pulse from a dye laser was tuned to an absorption line of OH. Figure 11 of Chapter 8 shows the velocity profile, which has a maximum velocity of over 50 m/s; the profile deviated slightly from a parabolic one due to turbulence near the center of the jet. In addition to demonstrating the technique for high velocities, the importance of this work is that no seeding was necessary —OH is a radical present in the flame. Note, however, that the deflection amplitude was considerably reduced at elevated temperatures such as in post-flame combustion gases, so that it is more difficult to obtain sufficient signal amplitude to make velocity measurements (Rose and Gupta, 1986).

Rose and Gupta (1986) have also measured velocities in sooting flames. Here a rich ($\phi = 8.5$) C_2H_2/O_2 flame was studied, again using a pulsed dye laser as a pump. In addition to the velocity (see Fig. 17), the soot concentration profile was also recorded by using the amplitude of the PD

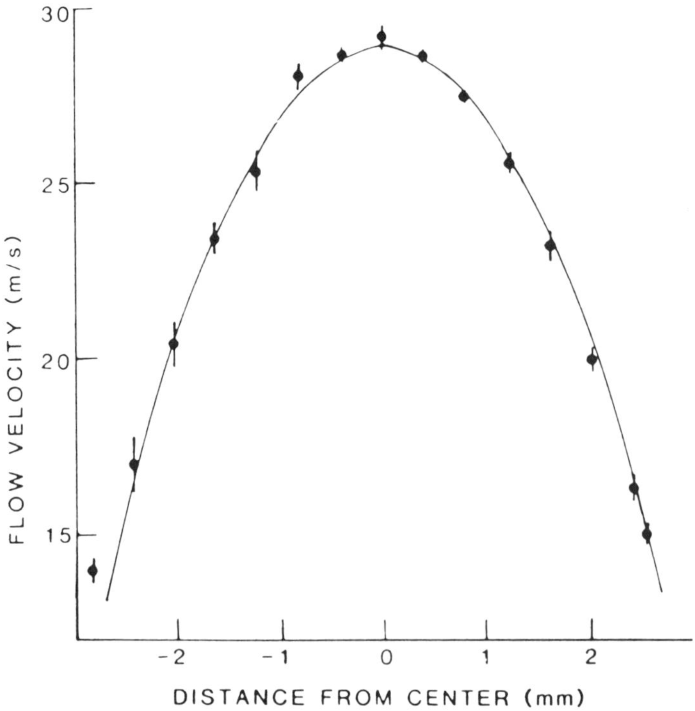

FIG. 17. Velocity profile of a laminar C_2H_2/O_2 flame 40 mm above the torch tip (see Rose and Gupta, 1986).

signal. Furthermore, the temperature in the postflame region was measured from the photoacoustic deflection velocity. The PD signal in a sooting flame was somewhat more complicated than that due to gases because of contributions from vaporizing soot to the PD signal. This point deserves further research, since velocity measurements in real combustors such as engines will most likely be complicated by soot.

A. Multipoint-Velocity Measurements

Sell and Cattolica (1986) and Cattolica and Sell (1986) recently demonstrated the measurement of gas velocity at many points simultaneously by using the PD effect. In this work, a planar probe beam was employed that crossed the focused CO_2 pump beam above a laminar jet (see Fig. 18). Instead of a position-sensing detector, a high-speed video camera was used

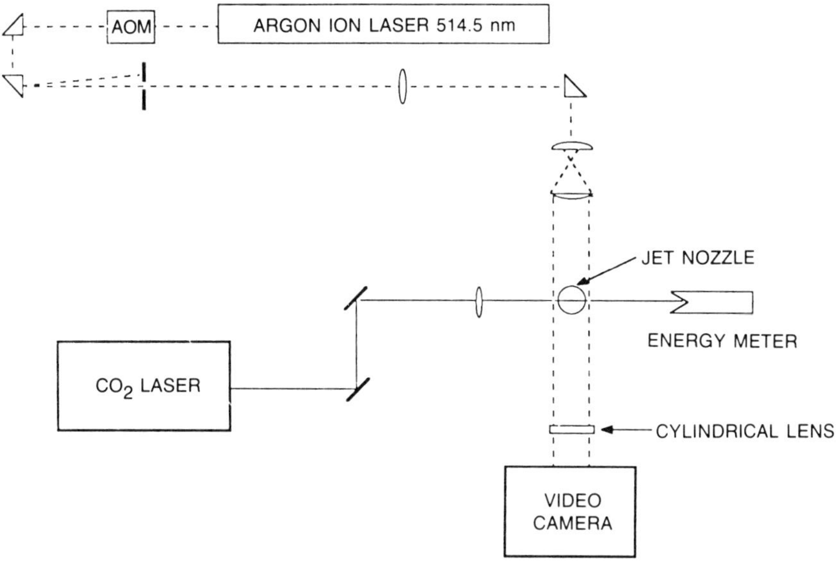

FIG. 18. Block diagram of the linear imaging velocity experiment. A planar probe beam was formed from the argon laser beam using cylindrical lenses.

to record the deflection of the plane as a function of time after the pump pulse was fired. Figure 19 shows a schematic diagram of the planar deflection. Note that the heated gas takes on the parabolic profile of the flow and will deflect the planar probe beam first near the center of the jet where it will reach the plane first. The initial deflection of the plane will therefore be upward in the center. Later, the slower-moving edges of the heated gas packet will approach the plane and deflect it upward further from the center of the jet. Actually, for high enough centerline velocity, the plane directly above the center of the jet will deflect downward at about the same time when the plane nearer the edges of the jet was still deflecting upward. This corresponds to the two lobes in the cylindrical beam signals shown in Fig. 3. In order to obtain quantitative velocities from these images, the y-centroid Y_c of the images was computed

$$Y_c(m) = \frac{\sum_n I(m, n) \times n}{\sum_n I(m, n)}, \tag{5}$$

where $I(m, n)$ is the intensity at each pixel, and m and n are the horizontal and vertical pixel numbers, respectively. The transit time approach was then used to compute the velocity across the jet. This procedure yielded the

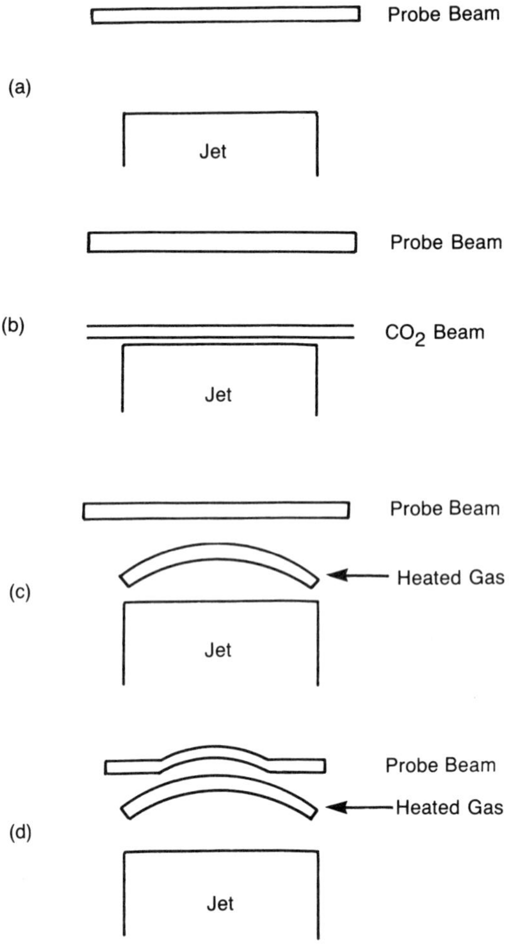

FIG. 19. Idealized drawing of planar deflection experiment. (a) End view of probe beam and side of jet before pump laser is fired, (b) pump-beam position with respect to (a), (c) parabolic distribution of heated gas above jet, and (d) initial upward deflection of probe beam as hot gas flows into it from underneath.

gas-velocity profiles shown in Fig. 20 for four different runs corresponding to four different flow rates. It should be emphasized that this data was obtained with single shots from the pump laser; the velocities were obtained simultaneously for 238 points separated by 30 μm. Although the demonstration of the method used an expensive high-speed video camera, it is probably possible to achieve similar results with a horizontal razor blade positioned in front of a linear diode array oriented horizontally. For small

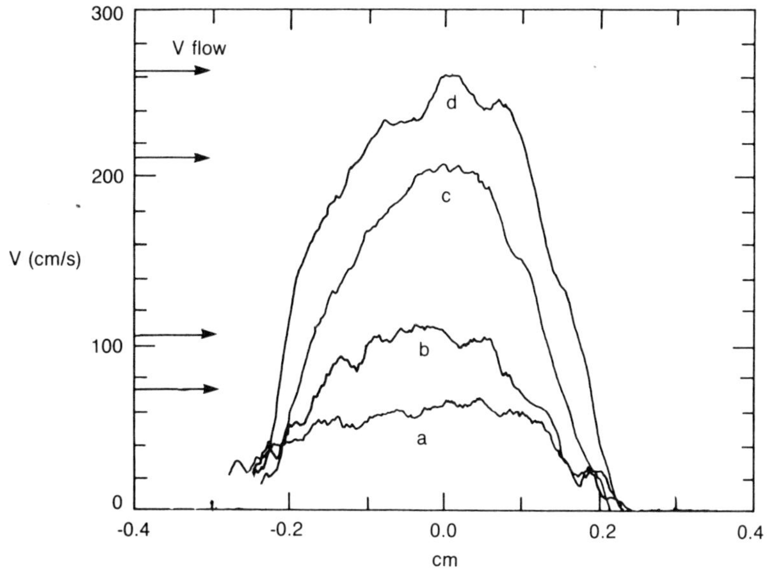

FIG. 20. Gas velocity (cm/s) by the transit time technique calculated from the *y*-centroids of the planar deflection as discussed in text. Four different runs are shown corresponding to four flow rates: (a) 8.2, (b) 16.3, (c) 32.6, and (d) 40.8 cc/s.

beam displacements, it can be shown that the intensity of each diode is related to the position of the centroid of the probe beam. In fact, this approach may be a more sensitive detection of the planar beam deflection. One of the disadvantages with the technique as demonstrated was the necessity of using a high pump pulse energy to get sufficient beam deflection, resulting in a large temperature rise of the fluid.

B. Transient Thermal Grating

A new technique for making higher-precision velocity measurements was recently discussed by Dasch and Sell (1986). Here, the pump beam from a CO_2 laser was split into two beams that were then crossed in a vertical plane at a small crossing angle (see Fig. 21). Due to the long coherence length of the CO_2 laser, the two beams were still in phase at the crossing point. Constructive and destructive interference occurred at the volume of intersection creating a volume grating in laser intensity and therefore in the temperature of the gas that absorbs the pump beam. The spacing of the

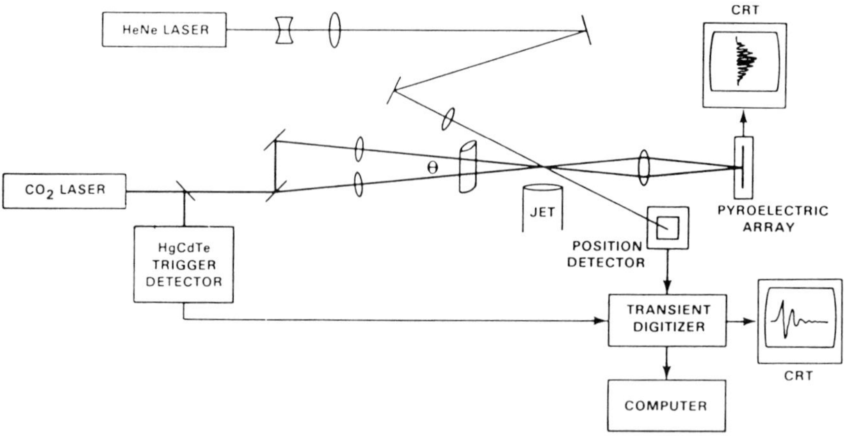

FIG. 21. Block diagram of the transient thermal grating PDV experiment.

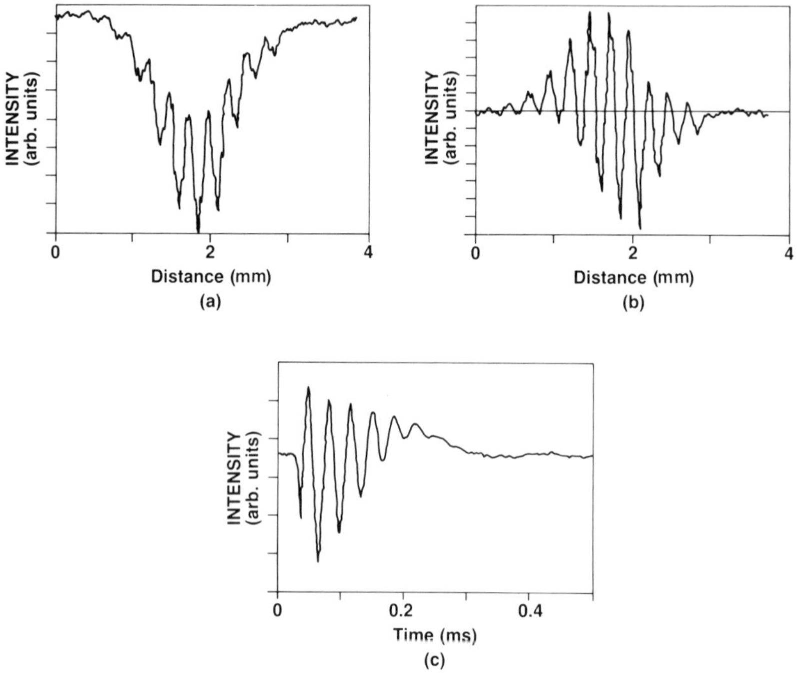

FIG. 22. (a) Laser-beam intensity grating recorded by the array detector, (b) first derivative of (a), and (c) a PDV signal obtained with the probe beam 0.5 mm downstream from the pump beam.

grating fringes l is given by

$$l = \frac{\lambda_l}{\theta}, \tag{6}$$

where θ is the crossing angle (see Fig. 21) and λ_l is the wavelength. The grating was imaged onto a self-scanned pyroelectric array so that the fringes could be observed directly. The volume grating was 1.1-mm radius ($(1/e^2)$-intensity) in the vertical direction and 300-μm horizontal radius (accomplished with a cylindrical lens, see Fig. 21). A He–Ne beam was focussed to a 60-μm waist and crossed the vertical plane formed by the two pump beams at an angle of 55°, a somewhat arbitrary value. The fringe separation was 245 μm, so that seven or eight fringes were formed in the grating; the modulation depth was $\approx 60\%$ due to imperfectly matched pump beams. The duration of the pump beam was 5 μs; the grating was therefore imprinted on the fluid and traveled downstream with it. Diffusion broadened the fringes after roughly 100 μs, but the peak-to-peak fringe separation remained at 245 μm. The position of the probe beam with

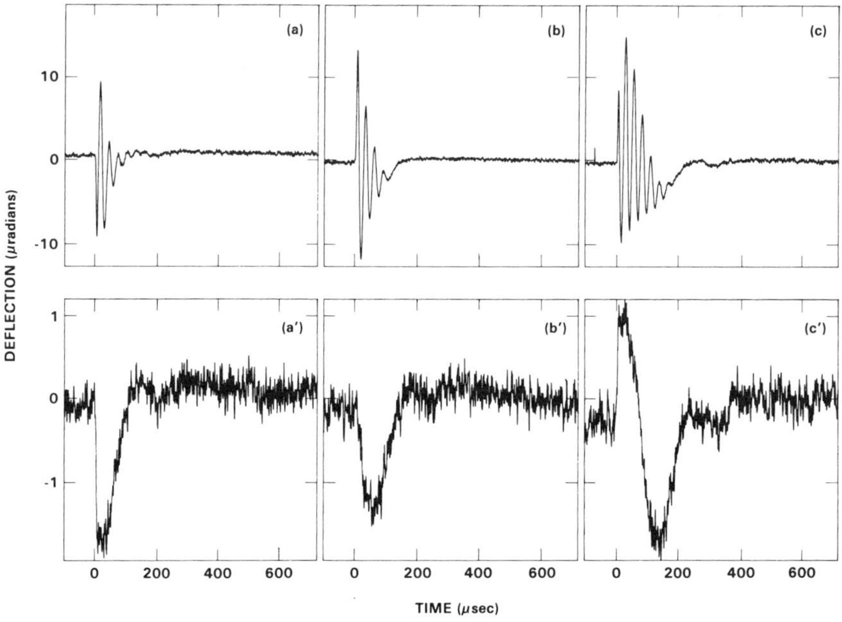

FIG. 23. Transient grating (upper panels) and ordinary PD signals (lower panels) for three different pump–probe offsets: (a) probe beam offset 0.5 mm upstream from center of pump beam, (b) probe beam coincident with pump, and (c) probe beam offset 0.5 mm downstream from pump. Data averaged 64 shots.

respect to the volume grating in the vertical direction was somewhat arbitrary. The probe beam was deflected due to the index gradients formed by the grating; the deflection was monitored with a position-sensitive detector located 0.6 m away. To determine the velocity, the signal was fast-Fourier-transformed (FFT), and the velocity v computed from

$$v = \nu l, \tag{7}$$

where ν is the frequency corresponding to the fringe spacing in the FFT, and l is the fringe spacing.

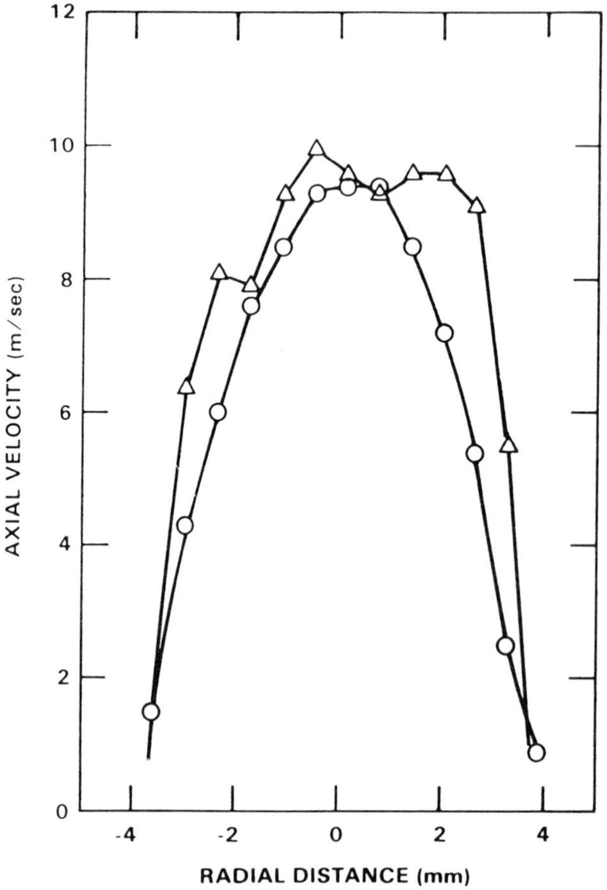

FIG. 24. Velocity profiles recorded 7 mm from jet with the transient grating PD method for two different flow rates: open circle, 4.8 m/s; open triangle, 7.1 m/s plug flow.

Figure 22 shows an example of the intensity grating as recorded by the pyroelectric array detector as well as the first derivative of it compared to an actual grating PD signal. Ideally, the PD signal should look identical to the first derivative of the intensity grating. However, the differences observed in Fig. 22 are due to the fact that the two pump beams are not perfectly matched, the probe beam is not infinitesimally small, and diffusion has occurred. Also, the probe beam was not located far enough downstream to monitor the whole grating.

Figure 23 illustrates some of the advantages of the transient grating approach to the conventional PD signal method of measuring velocity. In this figure, we compare the conventional PD signal obtained by blocking one of the two pump beams to the transient grating signal for three different vertical offsets of the pump–probe beams. An enhancement of roughly a factor of 10 was observed for the transient grating signal over the conventional PD signal due to the small grating spacing (therefore large gradients). Furthermore, each pair of gratings is essentially an independent velocity measurement leading to enhanced precision. Also, the precise spacing of the fringes leads to improved accuracy in the velocity measurements.

Figure 24 shows velocity profiles measured above the 7-mm diameter jet for two different mass flow rates corresponding to 4.8 m/s and 7.1 m/s plug flow. The higher flow run shows a profile somewhat flatter near the

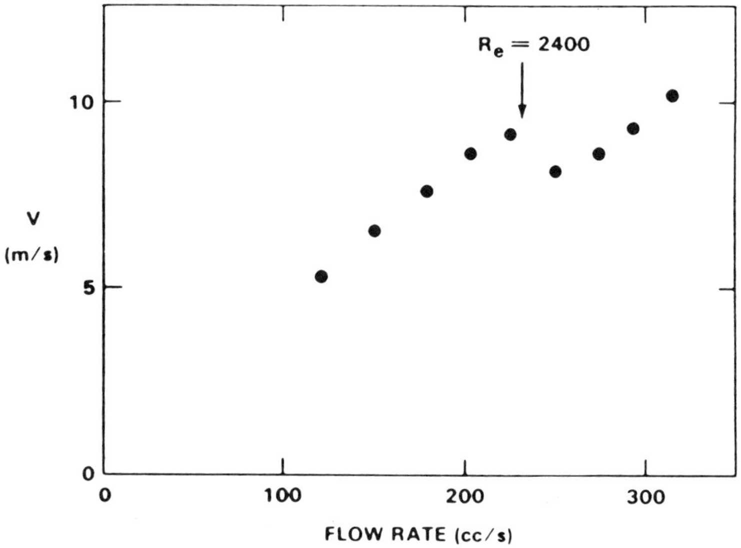

FIG. 25. Plot of centerline velocity by the transient grating versus mass flow rate.

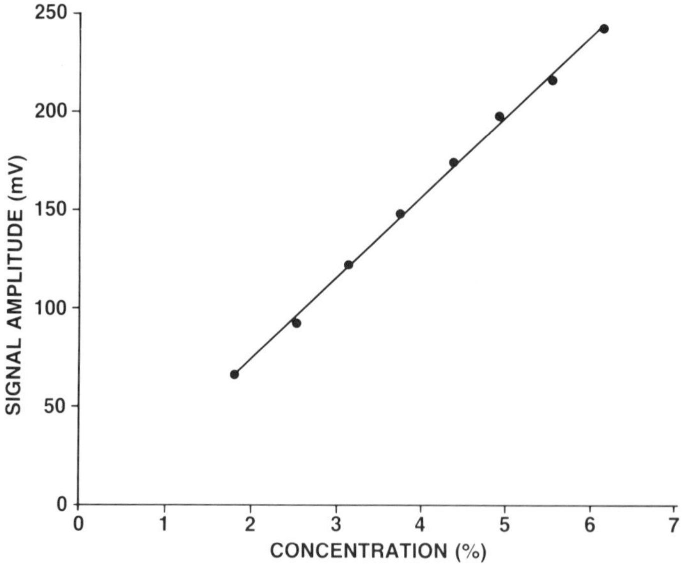

FIG. 26. Peak PDV signal amplitude versus ethylene concentration.

center than the parabolic profile measured for the 4.8 m/s run. In both cases, the integrated flow was within 5% of the actual flow (assuming cylindrical symmetry). Figure 25 shows the velocities measured at the jet centerline, 0.1 mm above the jet, as a function of mass flow rate. The onset of turbulence is observed at a Reynolds number of 2400, where the centerline velocity drops, corresponding to a flattening of the profile.

In addition to making flow-velocity measurements, it may be possible to make simultaneous measurements of absorber concentration from the peak signal from one of the fringes. Figure 26 shows the linear response of the signal obtained by this method with ethylene concentration. Due to this linearity, simultaneous absorber-concentration measurements should be relatively straightforward.

VI. Turbulent Flows

The only photothermal deflection velocimetry (PDV) for turbulent flows that this author is aware of was done with the transient grating approach discussed above (Dasch and Sell, 1986). Turbulent flows were induced by a wire mesh or grid held above the jet. For turbulent flows, the noise was

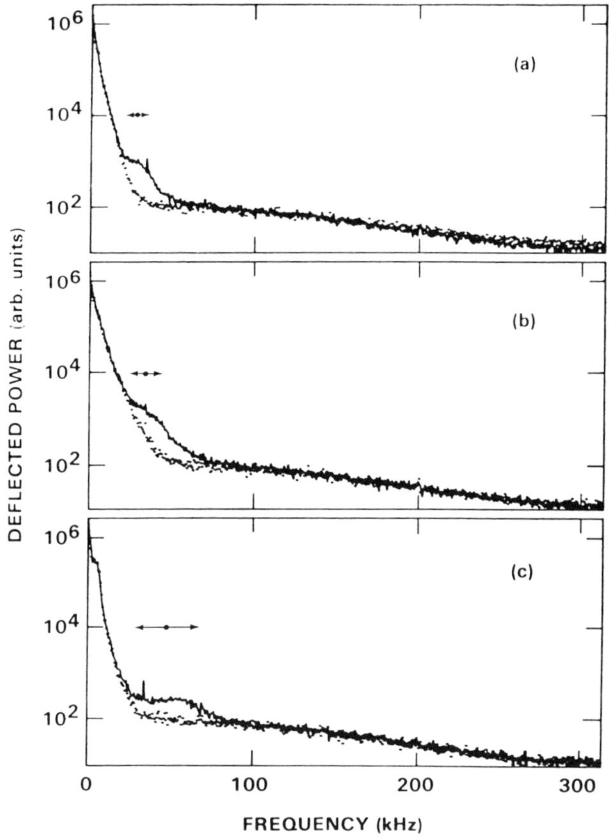

FIG. 27. Power spectrum of PD signals for turbulent flows obtained (a) 28 mm, (b) 15 mm, and (c) 2 mm from grid. The deduced FWHM of the velocity is indicated by (↔). The data were averaged over 64 shots; data obtained with pump laser on (—) and off (· · ·).

much larger than the signal. The noise was probably due to large-scale air entrainment by the jet but may also be due to density, and hence index, fluctuations in the fluid. For these studies, signal averaging was required; the FFT power spectra summed over many shots was taken as the velocity probability distribution function. Figure 27 shows results obtained 28 mm, 15 mm, and 2 mm above the grid. Close to the grid, the length scale was short, and this washed out the fringes considerably. At distances further from the grid, the turbulence as measured by the width of the velocity distribution, and the mean velocity, decreased as expected.

The velocity range accessible by this method can be estimated from the consideration that the pump pulse duration Δt_1 must be shorter than half the time Δt_c for convection across one fringe; also, the convection time should be shorter than the time Δt_d for diffusional blurring of the fringes:

$$2\,\Delta t_l < \Delta t_c < \Delta t_d. \tag{8}$$

Since $\Delta t_c = l/v$ and $\Delta t_d = \dfrac{l^2}{4\pi^2\lambda}$, where λ is the thermal conductivity, then

$$4\pi^2\frac{\lambda}{l} < v < \frac{l}{2\,\Delta t_l}. \tag{9}$$

For the conditions used in this work, the accessible velocity range was roughly 1 to 25 m/s. Shorter pump pulses would be required to increase the maximum measurable velocity.

Note that the theoretical treatment of the transient thermal grating approach for turbulent flows has not yet been developed. One might expect the fringes to be rapidly washed out by the turbulence, but that would depend on the turbulent length scale, the fringe spacing, and other factors. A quantitative treatment of the technique would be very helpful.

VII. Two-Dimensional Velocity Measurements

So far as this author is aware, measurements of two or three spatial components of velocity by PD techniques have not yet been demonstrated. However, in principle, extension of the techniques demonstrated for one spatial component to two should be straightforward. For example, one could imagine setting up the time-of-flight experiment with two probe beams oriented perpendicular to each other, and the use of two position-sensitive detectors to measure two components of the velocity vector. (Extension to three components would be considerably more difficult, but feasible.) If low velocities are being studied, then extension of the relaxation-rate approach would be the technique of choice. Many commercial position-sensing detectors are sensitive to beam displacements in two directions. Thus, a single probe beam that is aligned center-to-center with the pump beam could be used. Each of the two signals from the detector would be processed independently, giving two components. The transient grating approach would be more difficult to extend to two dimensions, requiring four pump beams. However, given the enhanced SNR and precision of the velocities, it may be worth the effort required to set up.

VIII. Velocities for Flowing Liquids

Considerably less work has been performed in PDV of liquids than of gases. However, all of the above techniques should work. The transient grating approach may, in fact, be easier in liquids than in gases due to reduced diffusional blurring of the fringes. Note that the deflection amplitude for liquids may be larger or smaller than that for gases, depending on the value of dn/dT (see Eq. (2)). The higher density for liquids will tend to reduce the amplitude compared to gases.

Zharov and Amer (1985) have demonstrated the time-of-flight technique for measuring laminar flows of liquid water through a square cross-sectional tube. In this work, a Nd:YAG laser with 10-ns, 50-mJ pulses was used as the pump, and a He–Ne was used as the probe beam. Figure 28 shows the

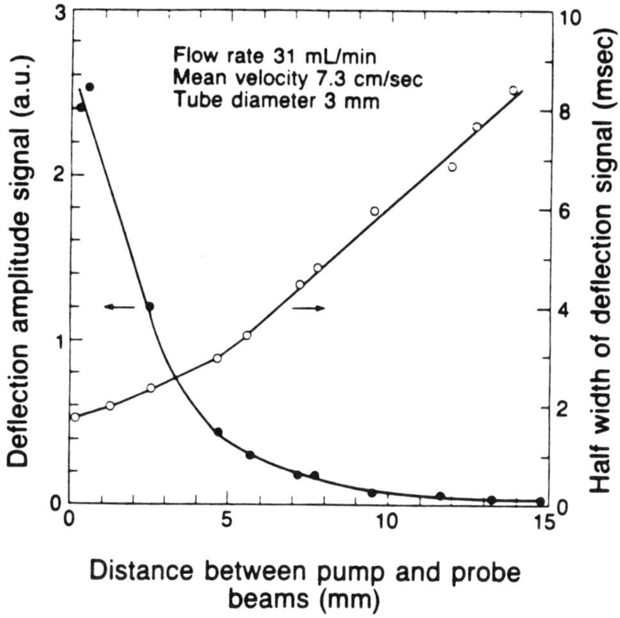

FIG. 28. Dependence of PD signal on pump–probe separation for liquid water samples (see Zharov and Amer, 1985).

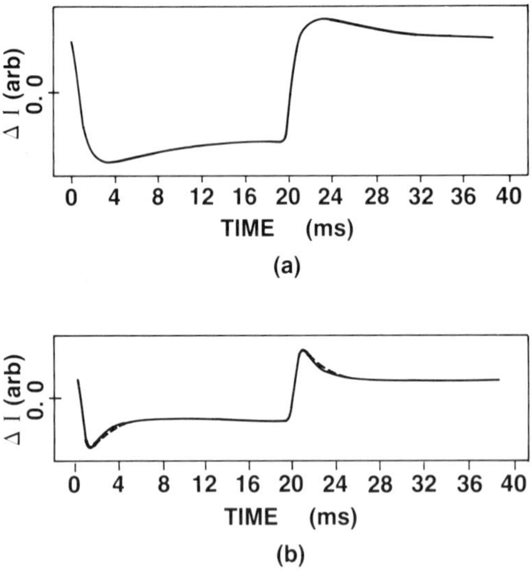

FIG. 29. Time-resolved thermal lensing data for a flowing liquid sample. A mechanically chopped pump beam is used as the excitation source; data was averaged 100 times. The least-squares fit of the theory is also shown, which gave flow velocities of (a) 7.6 mm/s and (b) 14 mm/s (see Weimer and Dovichi, 1985a).

dependence of the signal amplitude on the pump–probe separation. Note that the separation could be as large as 1 cm and still have a measurable signal; this is a much larger separation than could be used for gases due to the reduced diffusion of liquids. Zharov and Amer (1985) mention the possibility of making simultaneous concentration, flow velocity, and temperature measurements.

Weimer and Dovichi (1985a, b) have employed the thermal lensing approach (which they also call photothermal refraction) for measurements of flow velocity of liquids. They showed the theory for the time-resolved cross-beam thermal lens signal for flowing samples and for impulse excitation. They then computed the signal for trapezoidal excitation, since their experiments were conducted with a mechanically chopped pump beam that was best approximated with a trapezoidal function. The velocity measurements were conducted with a methanol–water mixture, which was doped with a compound that absorbs the He–Cd laser used as a pump beam. Figure 29 shows an example of their signal response and the theory fit to the data to obtain the velocity. This technique is therefore equivalent to the relaxation-rate method for the PD technique. Note that these very slow velocities were measurable due to the low diffusivity of the liquids.

IX. Temperature and Pressure Dependence of PD Signals

Rose, Vyas, and Gupta (1986) have studied the temperature dependence of the PD signal amplitude. Unfortunately, it depends on the gas velocity, pump–probe offset, and pump-beam waist in a complex way. However, examining the expression for the deflection amplitude ϕ (Eq. (1) above) term by term in their P and T dependence in the limit $v = 0$ is informative. The diffusion coefficient D goes like $T^{1.7}/P$, the term dn/dT goes like P/T^2, the thermal conductivity is roughly independent of pressure and goes like $T^{0.8}$. Thus, ignoring for the moment the absorption coefficient α, and the exponential term, the deflection angle ϕ should be independent of P and depend on T as $T^{-1.1}$. Rose, Vyas, and Gupta (1986) state that they have found the latter to be true (but did not examine the pressure dependence). It is expected that the exponential term would contribute somewhat to the temperature and pressure dependence of the diffusion coefficient, however. The dependence of the absorption coefficient has to be examined in each specific case; this can be done analytically for molecules small enough that their partition functions and spectral overlap with the pump laser can be evaluated. If this is not possible, then it may be necessary to measure these quantities separately.

This analysis becomes more complicated for the situations that we are most concerned with here, namely $v \neq 0$. However, in some cases, one is not concerned with the absolute magnitude of the PD signal, for example in the time-of-flight approach, or in the transient thermal grating method. One needs to know, first of all, if a signal can be observed, and then if the velocity derived by the method is accurate. For all of the PDV techniques discussed here, it is likely that at elevated temperatures, the signal amplitude will be reduced due to the $(1/T^2)$-dependence of the dn/dT-term. Probably the largest effect of T and P is on the diffusion coefficient. At elevated temperatures, for example, the larger diffusion coefficient will tend to wash out the fringes in the grating method more rapidly, thus increasing the lower velocity limit. At ambient temperature, but elevated pressure, the converse would be true. In addition, at reduced pressure, one must consider the relaxation rate of electronic/vibrational/rotational excitation to thermal energy. At 1 atm, this is usually fast compared to convection, conduction, and even most pump-laser pulse durations. However, it may be significantly slower at reduced pressure.

The pressure and temperature dependence of the PD signal should be studied further, since it will need to be well understood prior to the application of PDV in practical combustors.

X. Conclusions

PDV is especially useful for velocimetry of laminar gaseous flows. Advantages include excellent spatial resolution (10^{-5} cm^3), very good temporal resolution (ms), and lack of need for particle seeding. The velocity range measurable depends on the setup chosen and ranges from a few cm/s for the relaxation-rate method, to a few m/s for the time-of-flight approach, and up to 30 m/s for the transient grating technique. Simultaneous velocity, temperature, and absorber-concentration measurements appear feasible. Measurements of two spatial components of the velocity vector have not yet been demonstrated but would seem to be a straightforward extension of the one-dimensional studies.

XI. Appendix

A. Additional Comments on PDV Equipment

1. Pump Lasers

Much of the work discussed above was performed with a CO_2 laser which could be either electrically modulated, operated CW and mechanically chopped, or *Q*-switched. (Mechanical chopping is useful when one desires a temporally rectangular pulse (Sell, 1984)). In many ways, this is an ideal pump laser. The long coherence length contributed to the success of the transient grating technique, and the near-Gaussian spatial profile made data interpretation and modeling more straightforward than would be the case of a TEA laser. We have previously (Sell, 1984) listed many molecules of interest in energy and environmental studies, including propane (Sell, 1985b), that absorb the 10.6-μm wavelength. However, others have found a dye laser useful as a pump source, and even a N_2 laser. To this author's knowledge, an excimer laser has not been used for PD velocimetry, but certainly the high pulse energy makes it an attractive candidate. Probably the biggest disadvantage of multimode lasers is the complex and variable spatial profile, which complicates the modeling. Another possible complication would be the short pulse duration (10–20 ns), which leads to an increased possibility of laser-induced pressure waves, multiphoton effects, and laser-induced chemistry. We have previously shown (Sell, 1987) that reducing the pump pulse duration to be much shorter than the thermal diffusion time does not increase the PD signal (except perhaps for very high velocities)—it merely increases the power density. One must be wary of such complications at high power densities.

2. Probe Lasers

The bulk of the research conducted to date has been accomplished with a He–Ne laser as a probe-beam source. The requirements on the probe beam are not as stringent as the pump beam. One would prefer good intensity and pointing stability, as well as the ability to focus the beam tightly. He–Ne lasers do well on all these requirements for most applications. However, Tam *et al.* (1983) have used semiconductor diode lasers due to their superior pointing and intensity stability. One difficulty with diode lasers is their long wavelength (780 nm); an invisible probe beam is more frustrating to set up and align.

3. Detectors

Position-sensing detectors are very convenient for PDS and PDV. They are commercially available, inexpensive, and sensitive, but yet have a large linear range. Some are available that have two-dimensional sensitivity, permitting two-component measurements of the velocity vector (in principle). The two common types are the lateral cell detector and the quadrant detector. As far as this author can tell, there are no great advantages to one of these over the other. Both can be electrically wired so that intensity fluctuations are effectively ignored.

Another common technique used to measure changes in beam position is to set a razor blade so that it intersects part of the probe beam, and then to use a photodiode to measure the beam intensity passed beyond the razor blade. The sensitivity of this approach is very similar to position-sensing detectors (Jackson *et al.*, 1981). For the proper alignment of the razor blade with the probe beam, and for small displacements, it can be shown that the voltage change output by the photodiode is linear with the change in beam position (for Gaussian beams). However, a change in probe-beam intensity will result in an apparent signal, which is a disadvantage of this approach.

ACKNOWLEDGMENTS

The author would like to thank colleagues R. Cattolica, C. Dasch, R. Gupta, A. Tam, and R. Teets for many helpful discussions, useful ideas, and collaborations.

REFERENCES

Bialkowski, S. (1986). *Spectroscopy* **1**, 26.
Brewer, R., Bruce, C., and Mater, J. (1982). *Appl. Opt.* **21**, 4092.
Cattolica, R. and Sell, J. (1986). "Linear imaging of gas velocity using the photothermal deflection effect." In *Advances in Laser Science, Optical Science, and Engineering Series* **6** (W. C. Stwally and M. Lapp, eds.), p. 668.
Chang, T., Wilcox, N., and Tatterson, G. (1984). *Opt. Eng.* **23**, 283.
Cheng, S., Zimmermann, M., and Miles, R. (1983). *Appl. Phys. Lett.* **43**, 143.
Dasch, C. and Sell, J. (1986). *Opt. Lett.* **11**, 603.
Dovichi, N. and Harris, J. (1981). *Anal. Chem.* **53**, 689.
Drain, L. (1980). *The Laser Doppler Technique*. Wiley, New York.
Durst, F., Melling, A., and Whitelaw, J. (1981). *Principles and Practice of Laser Doppler Anemometry*, 2nd ed. Academic Press, New York.
Gustafson, E., McDaniel, J., and Byer, R. (1981). *IEEE J. Quantum Electron.* **QE-17**, 2258.
Hiller, B. and Hanson, R. (1985). *Opt. Lett.* **10**, 206.
Hiller, B., Booman, R., Hassa, C., and Hanson, R. (1985). *Rev. Sci. Instrum.* **55**, 1964.
Jackson, W., Amer, N., Boccara, A., and Fournier, D. (1981). *Appl. Opt.* **20**, 1333.
Meynart, R. (1983). *Appl. Opt.* **22**, 535.
Moosmüller, H., Herring, G., and She, C. (1984). *Opt. Lett.* **9**, 536.
Nicholaisen, S. and Bialkowski, S. (1985). *Anal. Chem.* **57**, 758.
Nickolaisen, S. and Bialkowski, S. (1986). *Anal. Chem.* **58**, 215.
Nie, Y., Hane, K., and Gupta, R. (1986). *Appl. Opt.* **25**, 3247.
Rose, A. and Gupta, R. (1985). *Opt. Lett.* **10**, 532.
Rose, A. and Gupta, R. (1986). *Opt. Commun.* **56**, 303.
Rose, A., Vyas, R., and Gupta, R. (1986). *Appl. Opt.* **25**, 4626.
Sell, J. (1984). *Appl. Opt.* **23**, 1586.
Sell, J. (1985a). *Appl. Opt.* **24**, 3725.
Sell, J. (1985b). *Appl. Opt.* **24**, 152.
Sell, J. (1987). *Appl. Opt.* **26**, 336.
Sell, J. and Cattolica, R. (1986). *Appl. Opt.* **25**, 1420.
Sontag, H. and Tam, A. (1985). *Opt. Lett.* **10**, 436.
Sontag, H., Tam, A. C., and Hess, P. (1987). *J. Chem. Phys.* **86**, 3950.
Tam, A., Zapka, W., Coufal, H., and Sullivan, B. (1983). *J. Phys. Colloq.* (*Supplement*) *C6*, **44**, 573.
Weimer, W. and Dovichi, N. (1985a). *Appl. Opt.* **24**, 2981.
Weimer, W. and Dovichi, N. (1985b) *Appl. Spectrosc.* **39**, 1009.
Zapka, W. and Tam, A. (1982a). *Appl. Phys. Lett.* **40**, 310.
Zapka, W. and Tam, A. (1982b). *Appl. Phys. Lett.* **40**, 1015.
Zharov, V. and Amer, N. (1985). "Pulsed photothermal deflection spectroscopy in a flowing media." In *Fourth International Topical Meeting on Photoacoustic, Thermal, and Related Sciences*, p. TuB9. Ville de Esterel, Quebec.

CHAPTER 8

COMBUSTION DIAGNOSTICS BY PHOTOTHERMAL DEFLECTION SPECTROSCOPY

R. GUPTA
Department of Physics
University of Arkansas
Fayetteville, AR

I. Introduction . . . 249
II. Variation of the PTDS Signal with Temperature . . . 253
III. Variation of Relevant Parameters with Temperature . . . 255
A. The Refractive Index n, and $\partial n/\partial T$ . . . 256
B. Absorption Coefficient α . . . 256
C. Density ρ, Specific Heat C_p, and the Diffusion Constant D . . . 256
IV. Observation of PTDS Signals in Flames . . . 257
V. Concentration Measurements . . . 260
VI. Temperature Measurements . . . 263
VII. Flow-Velocity Measurements . . . 264
VIII. Flame Spectroscopy . . . 265
References . . . 267

I. Introduction

The design of future combustion systems will require a thorough understanding of the combustion process. Therefore, extensive research for the purpose of understanding the combustion phenomenon is being performed in various laboratories. In order to test the theoretical combustion models, experimental diagnostic techniques must be developed. The parameters that one is generally interested in measuring are (i) the majority- and minority-species concentrations, (ii) the temperature, and (iii) the flow velocity. All of these parameters affect the chemical reactions in combustion in a rather complicated, and spatially and temporally dependent manner. Sometimes one is also interested in doing spectroscopy in a flame, particularly on radicals that can conveniently be produced only in a combustion environment.

An ideal combustion diagnostic technique should satisfy the following criteria: It should (i) be capable of making *in situ* measurements, (ii) be nonperturbing, (iii) have a high sensitivity, (iv) have a high degree of spatial resolution, (v) have a high degree of temporal resolution, and (vi) be capable of giving absolute values of the concentrations. Optical techniques

ISBN 0-12-636345-5

in general satisfy the first two criteria. Several optical techniques have been developed that satisfy many of the above criteria to varying degrees of success, for example, coherent anti-Stokes–Raman spectroscopy (CARS) for majority-species concentration measurements and laser-induced fluroescence (LIF) for minority-species concentration measurements. Photothermal deflection spectroscopy does, in principle at least, satisfy all six criteria enumerated above. Moreover, all three relevant parameters (minority-species concentration, temperature, and flow velocity) can be measured *simultaneously*. This is a very important advantage, since the reaction rates depend on all three parameters, and any stringent test of the theoretical models of combustion would require a simultaneous measurement of all three parameters. In this chapter, we review the progress that has been made in applying photothermal deflection spectroscopy (PTDS) to combustion diagnostics. Although any of the three photothermal detection schemes described in Chapter 3 (phase shift, deflection, or lensing) can be used, PTDS appears to be particularly suitable for these measurements. Therefore, we will restrict ourselves to the discussion of PTDS.

Figure 1 illustrates the basic ideas involved in using PTDS for combustion diagnostics. A dye-laser beam (pump beam) passes through the region of interest (flame in this case). The dye laser is tuned to one of the absorption lines of the molecules that are to be detected, and the molecules absorb the optical energy from the laser beam. Due to fast quenching rates in a flame, most of this energy quickly appears in the rotational–translational modes of the flame molecules. For most molecules in atmospheric pressure flames, only a negligible fraction of the energy is emitted as fluorescence. Thus the dye laser–irradiated region gets slightly heated, leading to changes in the refractive index of the medium in that region. If the density of absorbing molecules is uniform over the width of the dye-laser beam, the refractive index will have the same spatial profile as the dye-laser beam (generally assumed to be a Gaussian). Now if a probe-laser beam, generally a He–Ne beam, overlaps the pump beam, as shown in Fig. 1a, the probe beam is deflected due to the variations in the refractive index of the medium created by the pump beam. This deflection can be easily measured by a position-sensitive optical detector. The deflection of the probe beam is proportional to, among other things, the number density of absorbing molecules. If a pulsed pump laser is used, a signal similar to that sketched in Fig. 1b is obtained. The probe beam gets deflected shortly after the instant of laser firing, as discussed above, and gradually returns to its original position on the time scale of the diffusion time of the heat from the irradiated region. Thus the width of the signal depends on the thermal diffusion time constant, and consequently on the local temperature of the flame. Therefore, both the concentration of the molecules of interest and local temperature of the flame may be determined simultaneously from the

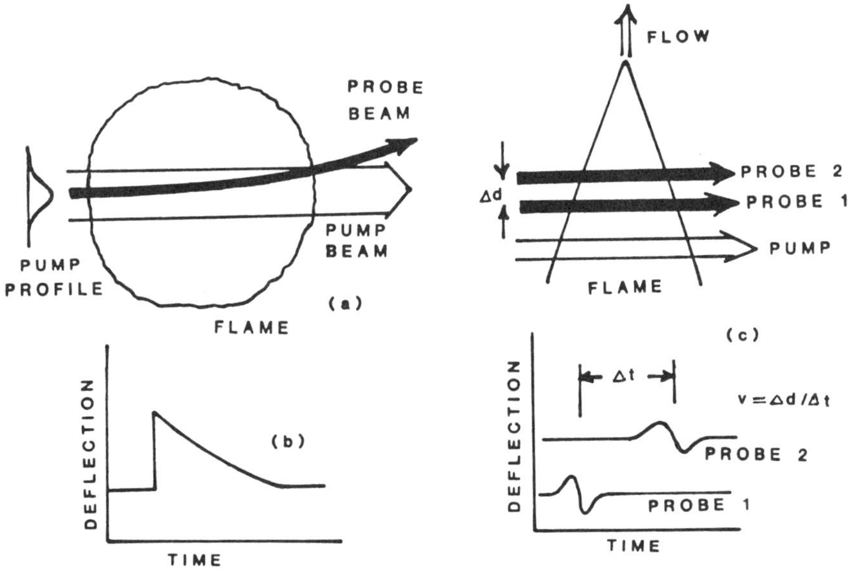

FIG. 1. (a) Deflection of a probe beam due to refractive-index gradients produced by the absorption of a pump beam. (b) Sketch of the photothermal deflection spectroscopy (PTDS) signal in a stationary medium. (c) PTDS in a flowing medium. Transit time for a photothermal pulse is measured.

size and the width of the signal, respectively. In the presence of a flow, the heat pulse produced by the absorption of the dye laser travels downstream with the medium. Of course, the heat pulse is accompanied by modifications of the refractive index. The transit time of the heat pulse between two positions of the probe beam a distance Δd apart is measured, as shown in Fig. 1c. The distances between the three beams have been greatly exaggerated for clarity. In practice, these distances are only a fraction of a mm; thus the spatial resolution is preserved. The flow velocity is derived simply from $v = \Delta d/\Delta t$, where Δt is the measured transit time. In this case, the shape of the signal is given by the gradient of the spatial profile of the pump beam, broadened by the thermal diffusion effects. Therefore, the concentration can still be measured from the size of the signal, and the temperature can be measured from the broadening of the signal. In the following paragraphs, we shall discuss why PTDS is a particularly suitable technique for combustion diagnostics.

Species-concentration measurements. The other techniques that are sensitive enough to be suitable for minority-species concentration measurements (e.g., LIF) appear not to be able to give absolute measurements due to problems associated with collisions. In these techniques, collisions are an unavoidable complication whose effect is difficult to quantitatively account for in a combustion environment. On the other hand, PTDS *relies* on quenching collisions (and it is relatively unaffected by other types of collisions). It is not necessary to know the quenching rates, as long as they are much faster than radiative rates (which is almost always the case in atmospheric pressure flames). This point can perhaps be clarified most easily by using a numerical example. If quenching is fast enough that, say, 99% of the molecules in the excited state relax nonradiatively into thermal modes, then the PTDS signal would be almost identical to what it would be if the quenching were 99.9% or 99.99%. However, for this same situation, the fluorescence signal would vary by orders of magnitude. Although special techniques are being developed to circumvent the problem of quenching in fluorescence (e.g., saturated LIF), quenching collisions continue to present a serious complication for absolute measurements. In this sense, PTDS is capable of giving absolute concentrations without quantitative knowledge of the quenching.

Temperature measurements. The only other nonintrusive methods of temperature measurement in a flame that have been proposed or demonstrated (CARS, LIF) rely on a measurement of Boltzmann distribution among rotational sublevels of the molecules. Therefore, one must assume that the thermal equilibrium exists, which is not always found to be the case. Moreover, in order to obtain the temporal resolution, the entire Boltzmann distribution must be measured in a single shot, which may not be possible. Even if it is possible, it generally requires an expensive optical multichannel analyzer. PTDS, on the other hand, is capable of giving a *direct* measurement of the kinetic temperature of a flame with a high temporal resolution.

Flow-velocity measurements. Other techniques of flow-velocity measurements in flames (demonstrated ones as well as those proposed) rely on Doppler shift measurements. Several of these techniques require seeding of the medium. Seeding may not always be possible in a practical combustion device, and there are questions whether the seeding alters the combustion parameters. Particle seeding is particularly troublesome because particles may sometimes lag behind the gas flow, particularly in regions of high acceleration in a turbulent flow. Techniques based on Doppler shift measurements are difficult to use for low velocities. PTDS, on the other hand, is free from these objections and limitations. Using PTDS, velocities as low as a few cm/s can be measured, with the ultimate lower limit being imposed by the thermal diffusion effects.

The PTDS technique satisfies all six of the criteria for an "ideal" combustion diagnostic technique. The technique is suitable for *in situ* measurements. It is nonperturbing because only miniscule temperature changes are induced by the pump beam. It has a high sensitivity and therefore is particularly suitable for minority species. A high degree of spatial resolution can be achieved by crossing the pump and probe beams near right angles. The region from which the signal is observed can thus be localized. A typical spatial resolution of 10^{-4} cm^3 is easily obtained. Finally, temporal resolution is achieved by using a pulsed pump laser. A PTDS signal is obtained typically in less than 100 microseconds. Therefore, the signal can be acquired at a rate of about 10 KHz, provided that a laser with this repetition rate is available. Absolute measurements of the concentration are possible for reasons mentioned above. Finally, PTDS permits a *simultaneous* measurement of species concentration, temperature, and flow velocity.

II. Variation of the PTDS Signal with Temperature

The photothermal deflection signal due to pulsed excitation is given by (see Chapter 3),

$$\phi_T(x, t) = -\frac{1}{n_0}\frac{\partial n}{\partial T}\frac{8\alpha E_0}{\sqrt{2\pi}\,\rho C_p}\frac{(x - v_x t)}{(a^2 + 8Dt)^{3/2}} \times e^{-[2(x - v_x t)^2/(a^2 + 8Dt)]}, \qquad (1)$$

where ϕ_T is the deflection angle of the probe beam. The pump beam propagates along the z-axis, and the probe beam propagates along the y-axis. The pump beam intersects the origin of the coordinate system, and the probe beam is displaced in the x-direction by a variable distance x from the origin; E_0 is the energy in the pump-laser pulse, and a is the $(1/e^2)$-radius of the pump beam; v_x is the flow velocity of the medium, assumed to be the entirely in the x-direction; α is the absorption coefficient of the medium, and the medium is assumed to be weakly absorbing; ρ, C_p, and D are, respectively, the density, specific heat at constant pressure, and the diffusivity of the medium; n_0 is the refractive index of the medium at the ambient temperature T. For combustion diagnostics, pulsed PTDS is most useful because of the requirement of temporal resolution. Therefore only pulsed PTDS will be discussed here. Moreover, only transverse PTDS will be considered here because it corresponds to the highest spatial resolution. In cases where it is critical to gain S/N at the expense of spatial resolution, an arbitrary angle may be used, as discussed by Rose, Vyas,

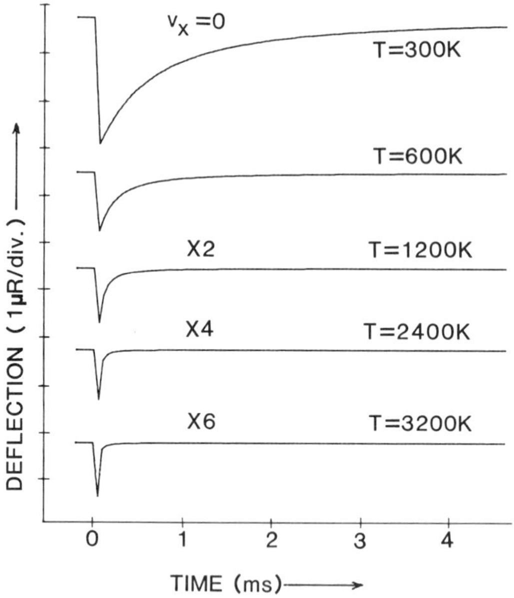

FIG. 2. Dependence of the photothermal deflection signal, in a stationary medium, on the temperature of the medium. In this calculation, pump-laser radius $a = 0.33$ mm, pump–probe distance $x = a/2$, pump energy $E_0 = 1$ mJ, pulse width $t_0 = 1$ μs, absorption coefficient $\alpha = 0.39$ m^{-1} (corresponding to 1000-ppm NO_2), and the diffusion coefficient D is assumed to vary as $T^{1.7}$. Curves at elevated temperatures have been expanded by the indicated factors for clarity.

and Gupta (1986). Finally, in order to simplify a more complicated expression to Eq. (1), we have assumed that the pump-laser pulse is very short ($t_0 \leq 10$ μs). For longer pulses, appropriate expressions can be found in Chapter 3.

Rose, Vyas, and Gupta (1986) have calculated the variation of the PTDS signals with the ambient temperature of the medium, as predicted by Eq. (1), and these are shown in Figs. 2 and 3. The medium is assumed to be N_2 seeded with 1000-ppm NO_2 ($\alpha = 0.39$ m^{-1}). Figure 2 shows the PTDS signals in a stationary medium for various temperatures up to 3200 K. In this calculation, the change in the absorption coefficient of NO_2 with temperature has not been taken into account. We note that the signal gets much narrower and weaker as the temperature increases. The diffusion constant has been assumed to increase as $T^{1.7}$ in these plots, as derived from the data of Rutherford, Roos, and Kaminsky (1969), in the 200 to 1100-K range. (See below for a fuller discussion of the variation of various parameters with temperature.) The signal gets narrower at elevated temper-

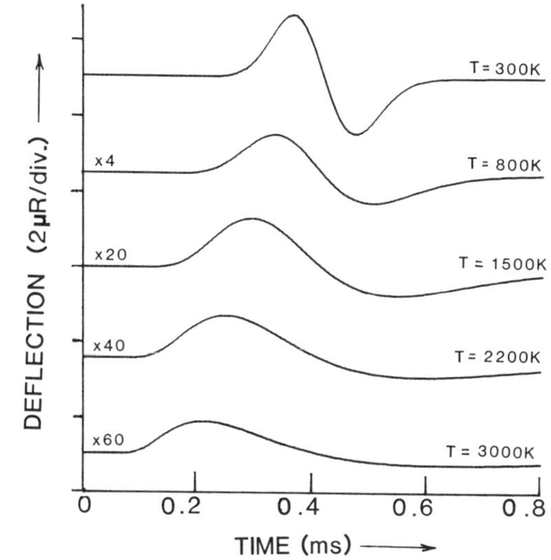

FIG. 3. Dependence of the photothermal deflection signal, in a flowing medium, on the temperature of the medium. Curves at elevated temperatures have been expanded by the factors indicated. The flow velocity of the medium was taken to be 4 m/s, and the probe–pump distance was 1.5 mm. All the other parameters are the same as in Fig. 2.

atures because the heat is able to diffuse through the probe beam much faster. Figure 3 shows similar curves for $v_x = 4$ m/s. We note that in this case the signal becomes weaker and broader as the temperature increases. The increase in D with temperature manifests itself as broadening of the curves for $v_x \neq 0$ (for nonoverlapping beams), while it manifests itself as narrowing for $v_x = 0$. As the diffusion constant becomes large at elevated temperatures, the PTDS signal becomes more and more asymmetric (the second peak being much broader compared with the first) due to the variation in the thermal diffusion rate during the duration of the signal.

III. Variation of Relevant Parameters with Temperature

In order to be able to reliably predict the size and the shape of the PTDS signals at flame temperatures using Eq. (1), one must know the variation of the relevant parameters with temperature. An examination of Eq. (1) shows that the parameters that are of importance for this purpose are: n, $\partial n/\partial T$, α, ρ, C_p, and D. We shall consider each of these parameters below.

A. The Refractive Index n, and $\partial n / \partial T$

The variation of n and $\partial n/\partial T$ with the ambient temperature of the flame can be derived simply by using the ideal-gas law. However, for gases, n is so very nearly equal to unity that it can conveniently be replaced by 1 in Eq. (1). The variation of $\partial n/\partial T$ is given by

$$\frac{\partial n}{\partial T} = \frac{(n_0 - 1)T_0}{T^2}, \tag{2}$$

where n_0 is the refractive index at temperature T_0.

B. Absorption Coefficient α

The absorption coefficient α depends on the density of molecules in the particular rotational-vibrational level involved in the excitation process and on the Einstein B-coefficient of the transition. While the B-coefficient is unaffected by the temperature of the medium, the number density of molecules in a particular rotational-vibrational level changes with the temperature. If the vibrational and rotational constants of the molecule are known, the change in the number density can be predicted by using Boltzmann distribution (assuming thermal equilibrium). Sometimes it is possible to choose a particular rotational-vibrational level such that the number density in that level has only a slight dependence on temperature.

The absorption coefficient also has a spectral profile that may change with temperature. The effect of the change in spectral profile may be significant for accurate measurements of the concentration if the temperature changes are large and the pump-laser spectral profile is very narrow.

C. Density ρ, Specific Heat C_p, and the Diffusion Constant D

These parameters are related to each other by

$$D = \frac{k}{\rho C_p}, \tag{3}$$

where k is the thermal conductivity of the medium. Tabulated values of several of these parameters can be found (Hilsenrath *et al.*, 1955; Powell,

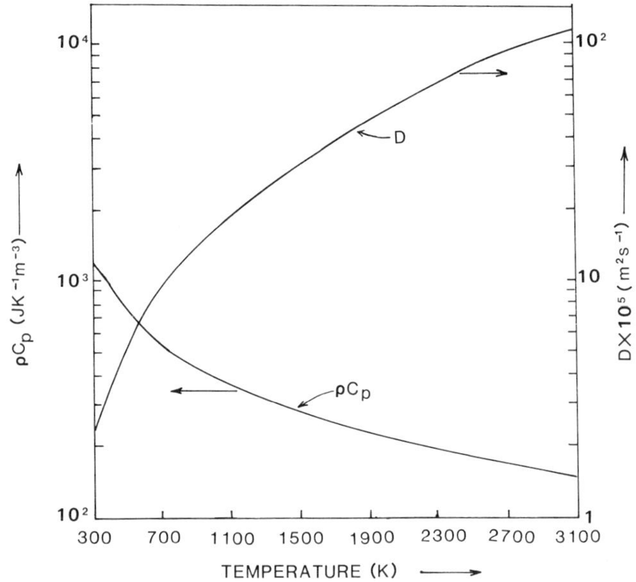

FIG. 4. Variation of diffusivity and ρC_p with temperature plotted on a semilog scale.

Ho, and Liby, 1966; Touloukian *et al.*, 1970) for a wide range of temperature for many gases and gas mixtures. These values are generally derived from a combination of experiment and theory. Figure 4 shows the variation of D and ρC_p with temperature for N_2 on a semilog scale. For this purpose, the values of ρ and C_p were taken from Hilsenrath *et al.* (1955), and values of D were derived by using Eq. (3) and the values of k listed by Powell, Ho, and Liby (1966). The variation of ρC_p and D over a wide temperature range cannot be fitted to a simple power law. (Apparently, the variation of D as $T^{1.7}$, assumed by Rose, Vyas, and Gupta (1986), is not valid over a wide temperature range.)

IV. Observation of PTDS Signals in Flames

In this section, some of the experimental observations of the PTDS in flames will be described. The basic experimental setup used in these experiments is shown in Fig. 5. The OH radicals produced in the combustion of hydrocarbons or hydrogen were excited by the frequency-doubled radiation from a Chromatix CMX-4 flash-lamp-pumped dye laser. The laser

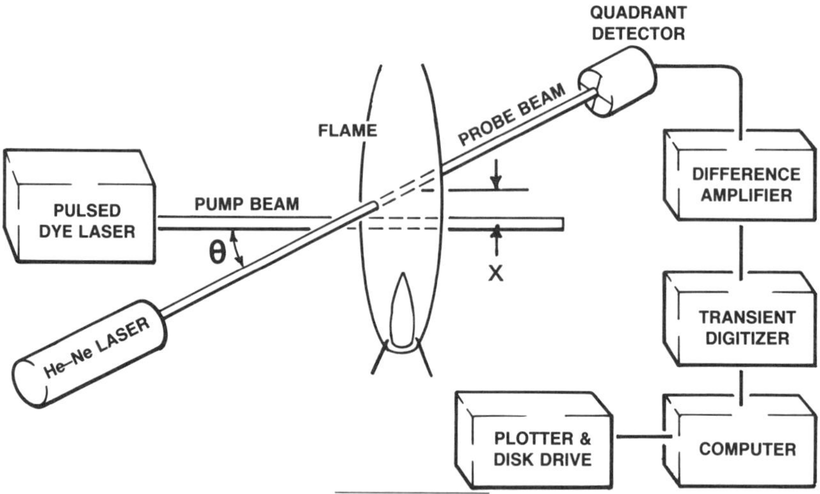

FIG. 5. Schematic illustration of a typical PTDS experiment in a flame. The pump and the probe beams made an angle θ and were separated by a variable distance x in the vertical direction.

was tuned either to the $Q_2(5)$ line of OH at 309.14 nm or to the $Q_1(8)$ line at 309.24 nm. This laser gave a $\sim$ 1-μs long pulse of radiation. A 0.8-mW He-Ne laser at 632.8 nm provided the probe beam. The probe beam could be translated up or down to vary the distance x between the two beams. Deflection of the probe beam was monitored by a quadrant detector. The difference signal from two diametrically opposite quadrants was amplified and measured. The probe beam was arranged to produce a null signal at the detector in its quiescent position. Shortly after the firing of the pump laser, the probe beam suffered a transient deflection that was recorded as a difference signal from the quadrant detector. The difference signal was digitized by a transient digitizer that was interfaced to a microcomputer for signal averaging and data storage.

Figure 6 shows typical PTDS signals in a flame (Rose, Salamo, and Gupta, 1984). For this experiment, a stainless-steel flat-flame burner with methane, oxygen, and nitrogen was used. The burner, $\sim$ 6 cm in diameter, produced a faint blue pancake-shaped $\sim$ 1.5-mm-thick flame. The pump laser was tuned to the $Q_1(8)$ line (309.2 nm) of OH produced in the combustion of methane. To maximize the S/N ratio, collinear ($\theta = 0$) pump and probe beams (low spatial resolution) were used in this experiment. The deflection of the probe beam has been plotted as a function of time after the pump-beam firing for various pump–probe beam distances. The top curve corresponds to overlapping pump and probe beams (maxi-

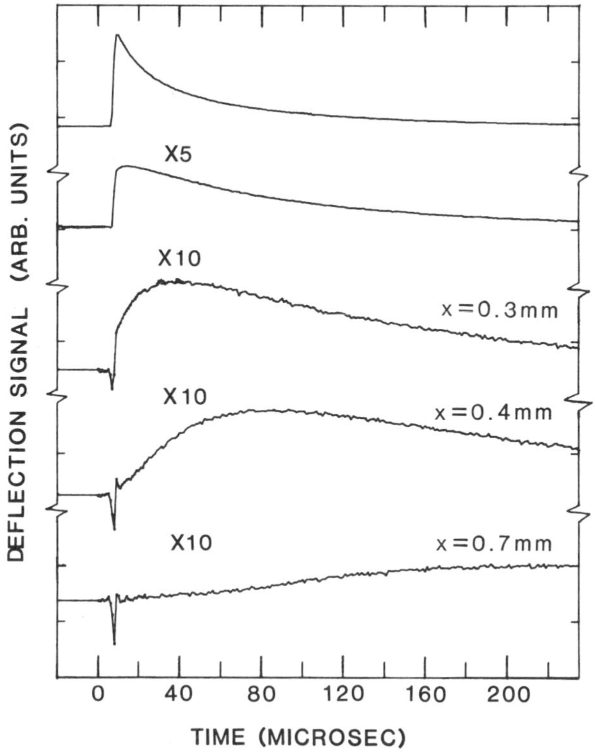

FIG. 6. Deflection of the probe beam as a function of time after the pump-laser firing for various pump–probe beam distances x. The distance x was estimated from the measurements of acoustic travel times; x for top two curves could not be estimated. Laser fired 5.3 μs after the trigger pulse. The second curve from the top has been magnified by a factor of 5, and the three bottom curves have been magnified by a factor of 10. The broad curves represent the photothermal deflection signal, while the sharp spike in the bottom three curves is the photoacoustic deflection signal.

mum signal amplitude), and successive curves toward the bottom of the diagram correspond to increasing pump–probe beam distances, as indicated on the diagram. In the top curve, the probe beam suffers a deflection at the instant of pump-laser firing due to photothermal effect and returns to its original position on the time scale of the thermal diffusion time. In this burner, the flow rate of the gases was very small, therefore the observed signal shape is essentially that which would be observed in a stationary medium. Moreover, when the two beams begin to separate, the signal is primarily produced by the diffusion of heat from the dye-laser-irradiated region to the probe beam, and consequently, it arrives at a later time. By

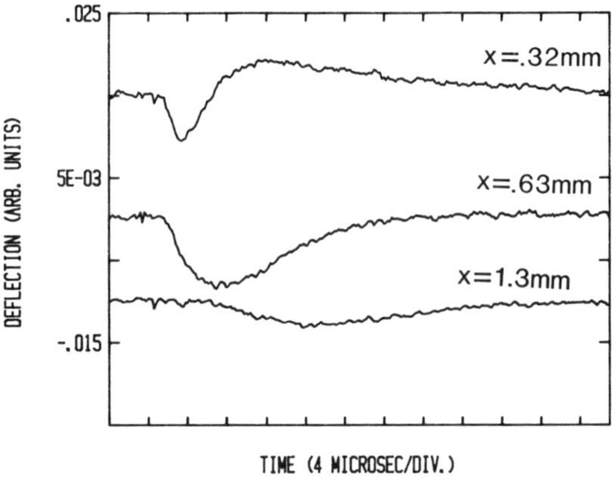

FIG. 7. Photothermal deflection signals in a flowing medium for three separations x of the pump and the probe beams. Signals were generated in OH in a H_2/O_2 glass-blowing torch.

the time the two beams are separated by about 0.7 mm, the signal is very broad and small, and its peak occurs at ~ 200 μs after the pump-laser firing. As the two beams are separated, one observes a sharp spike develop shortly after the dye-laser firing. This spike is due to a pressure pulse produced by the heating and the thermal expansion of the pump-laser-irradiated region (photoacoustic effect) (Rose, Salamo, and Gupta, 1984).

Figure 7 shows PTDS signals in a flowing medium (Rose, 1986). In this case, the burner used was a H_2/O_2 glass-blowing torch. The measured flow velocity (see Section VII) at the center of the flame was ~ 55 m/s. The pump and the probe beams crossed at an angle of 54° near the center of the flame to give good spatial resolution (~ 10^{-4} cm^3). The pump laser was tuned to the $Q_2(5)$ line of OH at 309.1 nm. Figure 7 shows three curves for three pump–probe separations, as indicated on the diagram. The probe beam was downstream from the pump beam as shown in Fig. 5. The signal shapes are typical of those observed in a flowing medium and show asymmetry characteristic of signals at high temperature (see Fig. 3). By the time the separation between the beams is ~ 1.3 mm, the signal has become very weak and broad due to a high value of the diffusion constant at elevated temperatures.

V. Concentration Measurements

Minority-species concentration measurements were the first reported application of PTDS to combustion diagnostics (Rose *et al.*, 1982). Kizirnis *et al.* (1984) have measured the concentration profile of OH in a flame, and

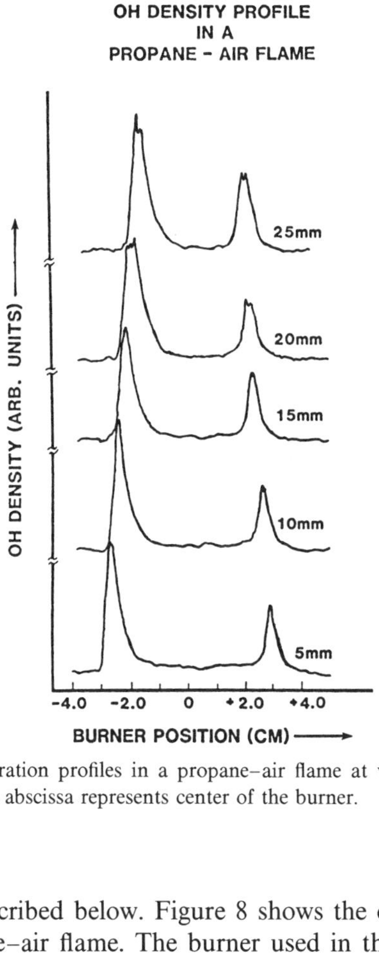

FIG. 8. OH concentration profiles in a propane–air flame at various heights above the burner head. Zero of the abscissa represents center of the burner.

their results are described below. Figure 8 shows the concentration profile of OH in a propane–air flame. The burner used in this experiment was a 3-inch-diameter stainless-steel burner, with its top surface being a perforated stainless-steel screen. A fuel-rich mixture of propane and air was used. The flow velocity of the gases was very small, and the observed PTDS signals had the shape of those in a stationary medium (Fig. 6). OH was excited on the $Q_1(7)$ line at 308.96 nm by the frequency-doubled radiation from a Nd : YAG pumped dye laser. The output of the difference amplifier was connected to a boxcar integrator that was triggered by the Nd : YAG pulse. The boxcar gate was set at 20 μs to envelope most of the PTDS signal, and the output of the boxcar was recorded on a chart recorder as the burner was translated under the pump and the probe beams. Therefore, each curve in Fig. 8 represents the relative OH concentration as a function of the lateral position of the flame. Different curves represent concentration profiles at different heights above the burner surface, as indicated on the

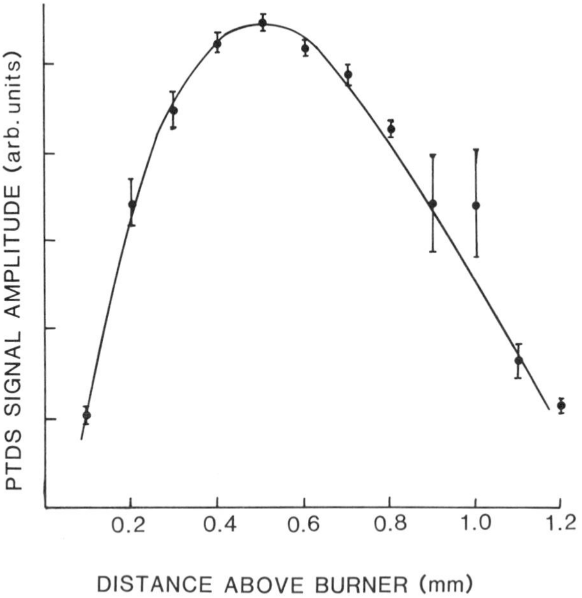

FIG. 9. NH_2 concentration profile in an ammonia flame. PTDS signal amplitude has been plotted as a function of height above the burner.

diagram. The pump and the probe beams crossed at an angle of 12°, giving a moderate spatial resolution of 3×10^{-3} cm^3. Note that the OH density was highest near the edges of the flame. As one goes higher up in the flame, the OH peaks get closer together because of the conical shape of the flame. The asymmetric OH density was consistent from run to run and was believed to be characteristic of that burner. Strictly speaking, these curves must be corrected for variations in velocity and temperature in order to represent true relative concentration. However, the flow velocity was very small, and the temperature in a plane parallel to the burner surface was nearly uniform (as measured by the photoacoustic technique (Kizirnis *et al.*, 1984)). Therefore, the curves in Fig. 8 do represent very nearly the OH concentration profiles.

As another example of the use of PTDS to measure relative concentrations, Fig. 9 shows the variation of the concentration of NH_2 in a flat-flame burner (Williams and Gupta, 1988) as a function of the height above the burner surface. This burner was discussed earlier in Section IV. An oxygen-rich mixture of NH_3, O_2, and N_2 was burned. Three unresolved rotational transitions of the (0, 9, 0–0, 0, 0) vibronic band of NH_2 at 597.72 nm were excited by the Chromatix CMX-4 dye laser. Amplitude of the

PTDS signal as a function of the distance above the burner surface is shown. The concentration peaks about 0.5 mm above the surface. The NH_2 concentration is constant in the horizontal direction. As mentioned previously, the flow velocity of gases in this burner was extremely small, and it did not affect the signal size appreciably. The curve in Fig. 9 has not been corrected for variations in the temperature of the flame with distance above the burner surface, and therefore, the concentration profile it represents is only an approximate one.

Absolute concentration measurements are in principle possible using PTDS; however, such measurements in flames have not yet been reported.

VI. Temperature Measurements

It was pointed out in the Introduction that the temperature of a flame may be measured by using PTDS. The temperature measurement can be performed via a measurement of the thermal diffusion constant, which depends on the temperature as shown in Fig. 4. In a stationary medium,

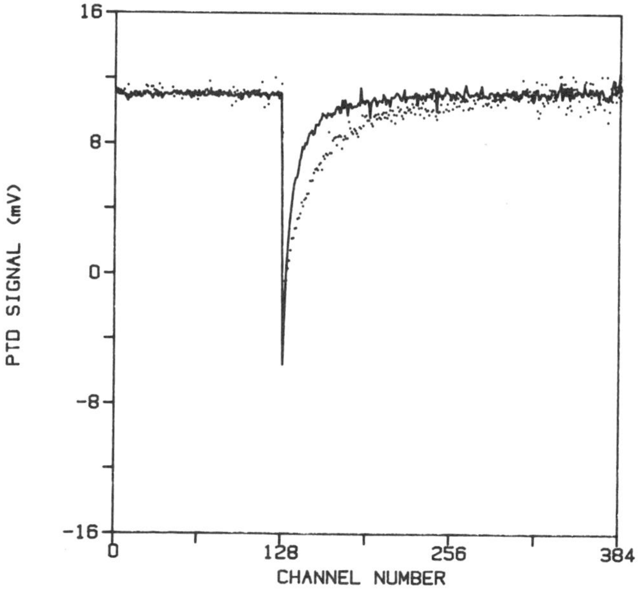

FIG. 10. PTDS signals generated in NH_2 at two different heights above the burner surface. The solid curve corresponds to 1 mm above the burner, whereas the dotted curve corresponds to 0.2 mm. One channel corresponds to 2 μs.

the diffusion constant can be measured by a measurement of the width of the PTDS signal, whereas in a flowing medium, it can be measured from the broadening of the signal with time. Figure 10 shows the PTDS signal in NH_2 (Williams and Gupta, 1988) at two different heights above the flat-flame burner described in Section IV. The solid curve is for a height of 1 mm above the burner surface, whereas the dotted curve corresponds to 0.2 mm above the burner surface. We note a marked change in the widths of the curves indicating different temperatures at the two positions. An absolute measurement of the temperature from the widths of the curves, however, requires a pump laser with clean Gaussian spatial profile, and exact knowledge of the pump-beam radius, flow velocity, etc. Such measurements are possible; however, they have not yet been reported.

VII. Flow-Velocity Measurements

Use of PTDS for flow-velocity measurements in a flame was demonstrated by Rose and Gupta (1985), and their experiment is described below. In this experiment, a H_2/O_2 glass-blowing torch, described earlier in Section IV, was used. The OH produced in the combustion was excited on the $Q_2(5)$ line at 309.1 nm. The pump and the probe beams crossed at 54° for good spatial resolution ($\sim 10^{-4}$ cm^3). Signals similar to those shown in Fig. 7 were obtained. Local flow velocity was derived from the transit time of the thermal pulse between $x = 0.32$ mm and $x = 0.63$ mm (Fig. 7). A small correction for the thermal diffusion was applied. The lateral profile of the flow velocity of the flame gases about 40 mm above the torch tip was measured and is shown in Fig. 11. The velocity profile of this flame fits a parabola very well, except near the center of the flame, as shown in the figure. This indicates that approximately a laminar flow existed in this torch. There was, however, considerable turbulence at the measurement points near the center of the flame (which was directly above the tip of the inner cone), which is also indicated by larger error bars on these points in Fig. 11. The least-squares-fitted value of v_x was found to be 55.4 m/s in good agreement with a value 55 m/s, obtained by an estimate based on the flow rates of H_2 and O_2, the area of cross section of the flame at the point of observation, and the temperature of the flame. Rose and Gupta (1986) have also measured the velocity profile of a sooting oxy-acetylene flame by generating PTDS signals on the soot particle.

The above-mentioned experiments clearly demonstrate the usefulness of PTDS for flow-velocity measurements in flame. Using a more powerful pump laser and two probe beams, a high temporal resolution (in addition to

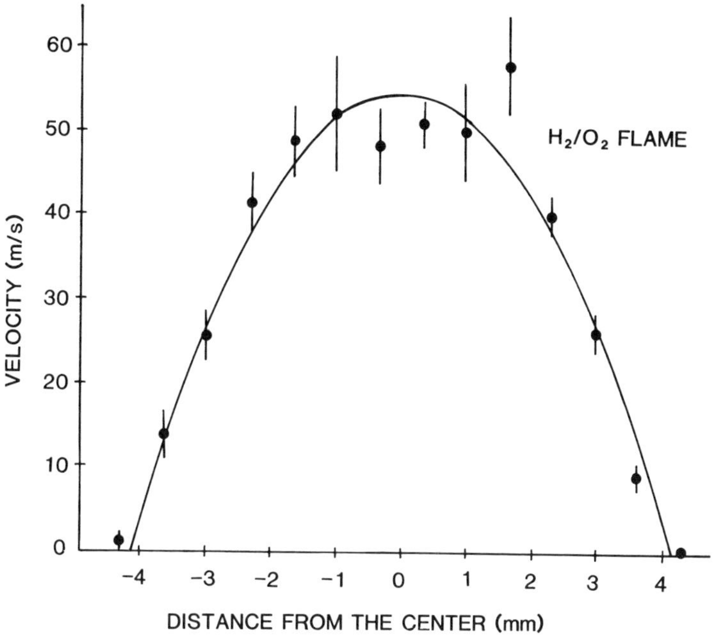

FIG. 11. Velocity profile of a laminar H_2/O_2 flame, 40 mm above the tip of the torch. The inside diameter of the tip was 2.5 mm, and it was operated with H_2 and O_2 flow rates of 98.6 and 49.2 cm^3/s, respectively, representing an equivalence ratio of 1. The experimental data fitted a parabolic curve quite well. The error bars represent one standard deviation of the mean of the three runs.

the spatial resolution) can be achieved in a straightforward manner (Rose and Gupta, 1985). The entire velocity profile can also be obtained in a single shot by using a multichannel device as proposed by Rose and Gupta (1985), or as demonstrated by Sell and Cattolica (1986) in a cold gas jet.

VIII. Flame Spectroscopy

The PTDS technique is essentially the absorption spectroscopy technique with the following two added advantages over the conventional absorption spectroscopy: (i) While the conventional absorption spectroscopy is a line-of-sight technique, PTDS can give a very high spatial resolution. (ii) PTDS is a null-background technique. For small absorptions, while the signal is superimposed on a large background in conventional absorption

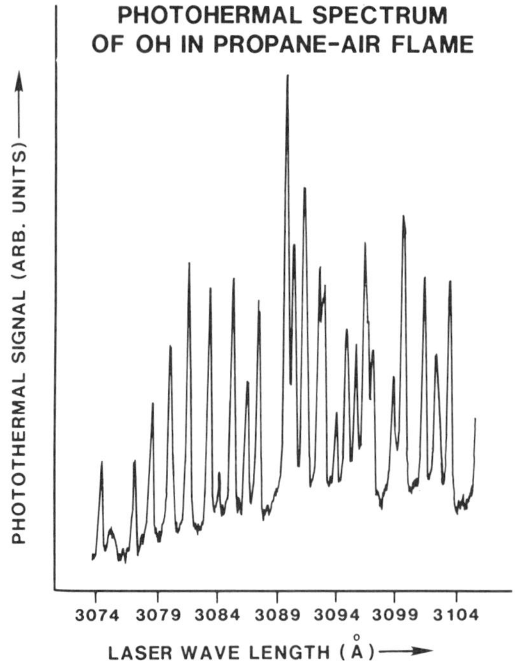

FIG. 12. Absorption spectrum of OH molecules. PTDS signal strength was plotted as the pump-laser frequency was scanned.

spectroscopy (and the background contributes to the noise), the PTDS signal does not ride on any background. It is this factor that makes PTDS a particularly suitable technique for spectroscopy, particularly for those radicals and transient species that can only be conveniently produced in flames and plasmas, and for those species that do not fluoresce strongly in the collision-dominated conditions of a flame. To demonstrate that PTDS can be used as a sensitive spectroscopic technique in flames, the PTDS spectrum of OH produced in a propane–air flame (described in Section V) is shown in Fig. 12 (Kizirnis *et al.*, 1984). The PTDS signal was recorded on a strip chart recorder by using a boxcar averager while the dye-laser frequency was scanned. All features in Fig. 12 can be identified with known features in the OH spectrum. Although OH is a poor example in this regard (since it does fluoresce strongly in a flame), Fig. 12 does show the value of PTDS as a sensitive spectroscopic technique in a combustion environment.

ACKNOWLEDGEMENTS

It is a pleasure to acknowledge the collaboration of Allen Rose and Karen Williams in the experimental results presented in this chapter. Partial support for this work was provided by the U.S. Air Force Wright Aeronautical Laboratories.

REFERENCES

Hilsenrath, J. *et al.*, (1955). *Tables of Thermal Properties of Gases*, N.B.S. Circular No. **564**. U.S. Govt. Printing Office, Washington, D.C.

Kizirnis, S. W., Brecha, R. J., Ganguli, B. N., Goss, L. P., and Gupta, R. (1984). *Appl. Opt.* **23**, 3873.

Powell, R. W., Ho, C. Y., and Liby, P. E. (1966). *Thermal Conductivity of Selected Materials*, National Standard Reference Data Series, NBS-**8**. U.S. Govt. Printing Office, Washington, D.C.

Rose, A. (1986). *The Development of Pulsed Photoacoustic and Photothermal Deflection Spectroscopy as Diagnostic Tools for Combustion*. Ph.D. dissertation, University of Arkansas.

Rose, A., and Gupta, R. (1985). *Opt. Lett.* **10**, 532.

Rose, A., and Gupta, R. (1986). *Opt. Commun.* **56**, 303.

Rose, A., Pyrum, J. D., Muzny, C., Salamo, G. J., and Gupta, R. (1982). *Appl. Opt.* **21**, 2663.

Rose, A., Salamo, G. J., and Gupta, R. (1984). *Appl. Opt.* **23**, 781.

Rose, A., Vyas, R., and Gupta, R. (1986). *Appl. Opt.* **25**, 4626.

Rutherford, W. M., Roos, W. J., and Kaminsky, K. J. (1969). *J. Chem. Phys.* **50**, 5359.

Sell, J. A., and Cattolica, R. J. (1986). *Appl. Opt.* **25**, 1420.

Touloukian, Y. S., *et al.* (1970). *Thermophysical Properties of Matter*. IFI/Plenum Press, New York.

Williams, K. and Gupta, R. (1988). *Appl. Opt.* (To be submitted.)

CHAPTER 9

PHOTOTHERMAL CHARACTERIZATION OF ELECTROCHEMICAL SYSTEMS

ANDREAS MANDELIS
Photoacoustic and Photothermal Sciences Laboratory
Department of Mechanical Engineering
University of Toronto
Toronto, Ontario, Canada

I. Introduction 269
II. Out-of-Cell Characterization of Electrodes 271
III. *In situ* Characterization of Electrodes 273
A. Passive Electrode Inspection 273
B. Working-Electrode Surface Monitoring 275
IV. Electrochemical Interface Reactions 277
V. Corrosion and Thin-Film Growth Processes 285
VI. Diffusion of Electrochemical Species 290
VII. Energy-Transfer Physics at Photoelectrochemical Interfaces 293
VIII. Conclusions and Future Directions 306
References 307

I. Introduction

The increasing need for new accurate, sensitive, and nondestructive methods for probing the scientifically and technologically important electrode–electrolyte interface, has forced electrochemists in the last decade to turn their attention to emerging photothermal techniques as a very promising diagnostic tool for a wide variety of physicochemical processes. Over the years since the first photothermal applications to electrochemical systems (Brilmyer *et al.*, 1977; Nordal and Kanstad, 1978; Fujishima, Brilmyer, and Bard, 1977; Gray and Bard, 1978), several photothermal detection techniques have enjoyed various degrees of popularity in electrochemistry. Major established methods include photothermal spectroscopy (PTS) (Brilmyer *et al.*, 1977); photoacoustic spectroscopy (PAS) (Rosencwaig, 1980), both in the microphone–gas-coupled cell and the piezoelectric transducer configurations; photothermal deflection spec-

ISBN 0-12-636345-5

troscopy (PDS) or mirage effect (Boccara, Fournier, and Badoz, 1980); and variants of these techniques. New promising photothermal detection principles are starting to be applied to electrochemical systems, such as photopyroelectric spectroscopy (P^2ES) (Mandelis, 1984). Thus, the arsenal of photothermal tools available for probing the electrochemical interface is further enriched, and just as importantly, the necessary evolutionary momentum in the quest for more sensitive, more nonintrusive and more remote-sensing interfacial techniques is maintained. At the time of the writing of this chapter, all indicators show that PDS appears to be the singlemost popular detection method among the established techniques, thanks to its remote noncontact sensing character.

In this chapter we present a comprehensive review of major contributions—milestones in the development of photothermal characterization of electrochemical systems. It is assumed throughout that the reader is familiar with the physical principles involved in the major photothermal detection techniques; therefore, only brief descriptions will be given when, and if, required for contextual continuity and clarity. The chapter proceeds from simple (and most straightforward) applications to single components of electrochemical systems, namely dry-working-electrode materials and fabricated electrodes examined photothermally in isolation outside the electrochemical environment (Section II), to *in situ* probing of the working electrode in the presence of an electrolyte and with or without simultaneous electrochemical activity (Section III). Section IV considers photothermal investigation of electrode chemical reactions *in situ*. Combinations of analytical techniques are typical of many of these studies as the complexity of the electrochemical system increases. As a result of electrochemical activity, growth of corrosive surface thin films often occurs on the working electrode, following matrix decomposition or electrochemical species deposition from the liquid phase. Section V reviews the successes of photothermal methods in diagnosing and characterizing such corrosive thin layers, many of which lead to electrode passivation. Photothermal detection of electrochemical concentration gradients in the electrolyte away from the interface is possible by means of PDS. This capability has resulted in a growing amount of theoretical and experimental studies of electrochemical product out-diffusion from working electrodes and is reviewed in Section VI. Section VII is concerned with the physics of fundamental problems in optoelectronic energy transfer and conversion across the electrochemical interface in the light of recent results obtained through the use of photothermal probes, usually in conjunction with (one or more) other analytical techniques. In this area the need for quantitative methods of interfacial energetics is great, and the section reviews the progress achieved to date.

Finally, the chapter closes with a perspective on future directions in photothermal electrochemical research.

II. Out-of-Cell Characterization of Electrodes

Most studies of remote (i.e., not *in situ*) photothermal characterization of electrodes before or after their exposure to the electrochemical environment appeared in the early years of photothermal detection and are almost exclusively photoacoustic. The emergence of PDS as an easy to implement, viable noncontact remote-sensing technique in this decade had no doubt a big influence in the fact that few "dry"-electrode studies are being performed nowadays. The use of electrode samples in the "dry" state resulted, of course, from the fact that they could be inserted in the highly confined working space of a photoacoustic (PA) cell coupled to a microphone transducer. In what follows, we give only a brief account of major PA experimental results with electrode samples. Nordal and Kanstad (1978) and Kanstad and Nordal (1978) used a very sensitive differential photoacoustic reflection-absorption spectroscopic technique (PARAS) to monitor microphone signals from thin Al_2O_3 films on Al metal, generated by selective absorption of laser light.

The fact that photothermal evaluation of electrodes taking place out of the electrochemical cell may give altered results due to the change in the environment has been observed photoacoustically with the SnO_2 electrode on which rhodamine B (RhB) was covalently bound (Fujihira *et al.*, 1978). These studies show that out-of-cell PAS detection of electrode surface species should be performed with the utmost care to prevent environmental artifacts from complicating the interpretation.

Out-of-cell Fourier transform infrared photoacoustic spectroscopy (FTIR PAS) has been performed on a number of electrode surfaces. The PA ability to monitor molecular surface spectra of highly absorbing, massive, irregularly shaped samples seems to hold promise in monitoring surface decomposition due to aging of the surface-bound species on carbon electrodes (Childers *et al.* 1983). Gardella *et al.*, (1983) used FTIR PAS to study specific types of catalytic surface decomposition with components of electrochemical systems, such as the modification of the Ni/SiO_2 catalyst upon adsorption of CO, and the corrosion of iron.

The ability of transverse PDS to measure absolute optical absorption coefficients (Mandelis, 1983) has been exploited to characterize electrode materials and surfaces in isolation remotely and nondestructively. Palmer and Smith (1986) inserted a simple mirage-effect accessory in the IBM–Bruker 9195 FTIR spectrometer and utilized conventional (front-side illumination), as well as reverse PDS (with the back-side of the sample illuminated). A ferricyanide-containing polymer, $Fe(CN)_5$-polyvinylidene (PVP), was deposited on the Pt disk electrode and Prussian Blue was electrogenerated on the surface of the Pt thin-film-on-glass electrode in 0.1 M KCl electrolyte according to the redox reactions

$$\underset{\text{(Berlin Green)}}{[KFe^{3+}Fe^{II}(CN)_6]_{1/2}[Fe^{3+}Fe^{III}(CN)_6]_{1/2}} \xrightarrow{+(1/2)e^- + (1/2)K^+} \underset{\text{(Prussian Blue)}}{KFe^{3+}Fe^{II}(CN)_6} \xrightarrow{+e^-,\, +K^+} \underset{\text{(Everitt's Salt)}}{K_2Fe^{2+}Fe^{II}(CN)_6}. \quad (1)$$

The dried electrode was placed in the FTIR-PDS cell. Subsequent electrogeneration of Berlin Green was further effected and its spectra were obtained. From the instrumentation point of view, liquid CCl_4 had to be coupled to the FTIR-PDS spectrometer to obtain decent signal-to-noise ratios (SNR), due to its large refractive-index gradient $dn/dT \approx 3.3 \times 10^{-3}$ K^{-1} (Dovichi, Nolan, and Weimer, 1984). Palmer and Smith (1986) were able to avoid complications arising from liquid opacity at the exciting wavelengths of the FTIR spectrometer by use of the reverse PDS method. Spectra of a $Fe(CN)_5$-PVP-modified Pt electrode and of hexacyanoferrates on the Pt electrode were thus easily obtained. Those corresponding to Prussian Blue and Berlin Green were found to be in agreement with the PA spectra obtained by Childers *et al.* (1983) in the 1800–2400-cm^{-1} range. No spectral alterations due to the "dry"-electrode character of the sample were noticed (see Fujihira *et al.*, 1978), due to the fact that any deterioration of the electrode surfaces was slow on the time scale of the sample transfer and spectral measurement; however, these authors did point out the value of, and need for, *in situ* infrared (IR) spectral measurements of electrode species. Out-of-cell PDS studies are, nevertheless, valuable for spectroscopic inspection of electrode structures, as to the degree of material and electronic response qualities as functions of fabrication history. Tamor and Hetrick (1985) correlated photothermal deflection (PD) spectra of TiO_2 powder photoelectrodes with crystal structure and degree of electronic conductivity. PDS responses from electrochromic WO_3 powder layers on quartz and Pt substrates were also obtained to monitor electronic charge transfer between powder and substrate, which is considered to be the mechanism for WO_3 electrochromism. The advantage of PDS over other spectroscopies, here, lay in the ability to obtain spectra from noncontinuous solids (powders), a feature commonly shared with other photothermal techniques.

III. *In situ* Characterization of Electrodes

A. Passive Electrode Inspection

Few *in situ* studies have been reported with electrodes in the electrochemical environment of a working device without actual electrochemistry occurring simultaneously. Of particular interest is the photocorrosion study of a CdTe single crystal in the presence of an acidified $SnCl_2$ electrolyte performed by Mendoza-Alvarez *et al.* (1983). In that work, a CdTe single crystal was irradiated with superbandgap light from a 1-mW He–Ne laser, and the spectrum was probed via PDS detection of optical absorption from a monochromatized tungsten source. The photocorrosion reaction gave rise to an out-diffusion of Cd^{2+} ions from the crystal surface and the formation of a tellurium surface layer. This well-known mechanism was found to be consistent also with spectra obtained from a CdTe thin film grown on SnO_2 thin layer previously deposited on glass (Figs. 1A, B). Thus, it was deduced that the CdTe thin films are present nonstoichiometrically with excess tellurium on the surface. It was further shown that thin CdTe films deposited directly on glass are stoichiometric with PDS spectra similar to that for the nonphotocorroded single crystal, Fig. 1B. This work thus demonstrated the use of PDS in characterizing the chemical composition and quality of thin-film CdTe solar cells through the monitoring of photocorrosion processes in single crystalline *n*-CdTe.

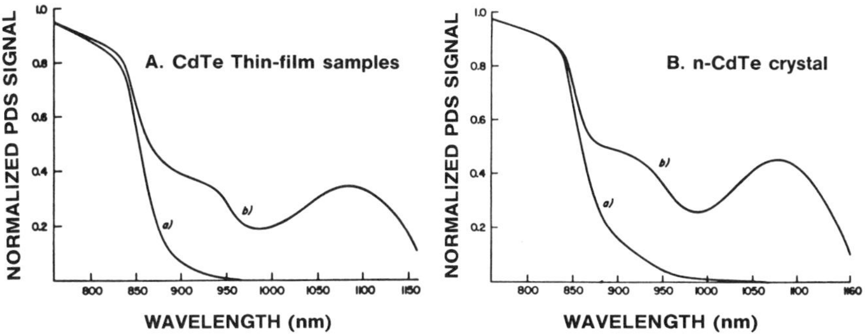

FIG. 1. A) Normalized PDS spectra from a *n*-CdTe single crystal with $n \approx 10^{13}\ cm^{-3}$: a) fresh crystal; b) photocorroded crystal, after immersion in an $SnCl_2$ acidified electrolyte and illumination for ca. 15 min. B) Normalized PDS spectra from: a) a CdTe thin film deposited on glass; b) a CdTe thin film grown on an SnO_2 layer previously deposited on glass. Deionized water was used as the electrolyte. (From Mendoza-Alvarez *et al.*, 1983.)

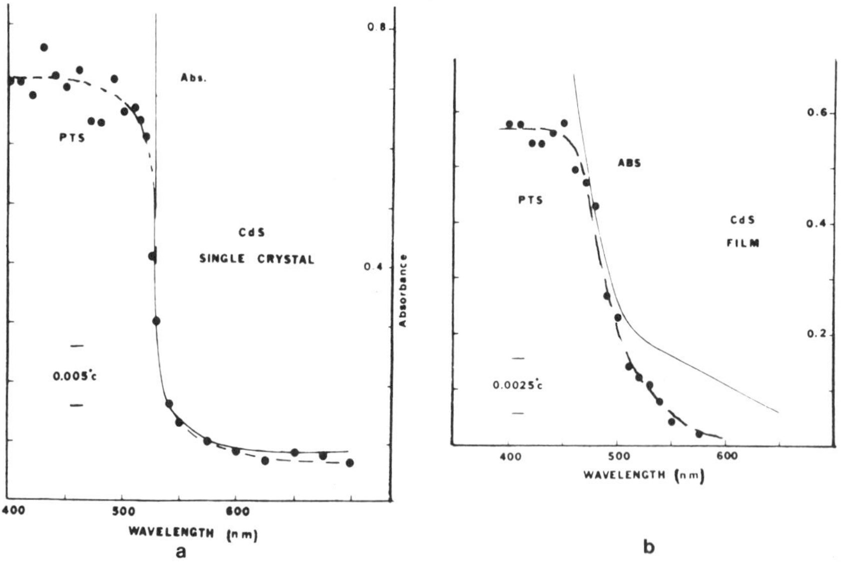

FIG. 2. Comparison between photothermal (PT) and optical absorption spectra of CdS: a) single crystal, and b) vacuum-evaporated thin film. (From Brilmyer *et al.*, 1977.)

PAS has also been used for *in situ* characterization of electrodes via back-surface illumination. In the passive mode, Takaue, Matsunaga, and Hosokawa, (1983) obtained line scans of flawed electrodes 0.1 mm thick, using a He–Ne laser and variable chopping frequency, thus demonstrating the ability of scanning photoacoustic microscopy (SPAM) to perform depth profiling of electrodes *in situ*. The PTS technique has also been used extensively for *in situ* electrode characterization. This technique was introduced by Brilmyer *et al.* (1977) and measures optical heat generation of material surfaces via a sensitive thermistor in direct contact with the surface. Typically, such devices would be black thermistors with nominal resistance of 2 kΩ, a sensitivity of 78 Ω/°K, and a time constant of 0.4 s in still water. To minimize direct optical absorption at the thermistor, usually a reflective coating (e.g., silver) is deposited on its surface. The thermistor is fixed with some adhesive on the sample surface, outside the light path. PTS presents an advantage over PAS as it is free of acoustic noise problems. The ability of PTS to perform spectroscopic studies of materials (Brilmyer *et al.*, 1977; Tom *et al.*, 1979) has found applications in semiconductor crystalline and thin-film photoelectrode studies in the passive *in situ* mode. Figure 2 shows typical PTS spectra for a CdS single crystal (a) and vacuum-evaporated CdS thin film (b). Although these particular spectra were obtained in air, similar spectra of somewhat smaller amplitude were also reported in water (Fujishima *et al.*, 1980a). PTS spectra of semiconducting powders on solid metal electrodes have been further obtained (Brilmyer *et al.*, 1977). The main disadvantage of PTS, besides its contact nature, appears to be the slow response associated with the long time constant of the thermistor. This feature forced the original studies to concentrate on dc photothermal responses, with the well-known associated signal drift problems (Brilmyer *et al.*, 1977), while later studies with ac optical excitation were only performed with very long modulation periods, e.g., "on" 20 sec, "off" 40 sec (Fujishima *et al.*, 1980a), or up to 8 Hz (Fujishima *et al.*, 1980b). These are some of the reasons for which PTS has not enjoyed, so far, the popularity of other photothermal techniques.

B. Working-Electrode Surface Monitoring

Metallic working electrodes have been extensively studied *in situ* using rear-side "open-membrane" detection PAS (Masuda, Fujishima, and Honda, 1980). Similar arrangements using a piezoelectric transducer instead of a microphone have also been reported (Oda, Sawada, and Kamada, 1978). Fujishima *et al.* (1980b) obtained PTS spectra of a Au electrode before and after electrolysis at -0.20 V vs. SCE in 1 M $HClO_4$ containing 5×10^{-3} M Cu^{2+} ions; and of a Au electrode in 5×10^{-3} M Cu^{2+}, 0.1 N Na_2SO_4 electrode. The spectrum of the electrolyzed Au electrode was shown to agree with that of a Cu plate, thus indicating that the Au electrode was covered with a Cu layer. These same authors further performed combined cyclic PA and voltammetric measurements of the same system (Fujishima, Masuda, and Honda, 1979; Masuda, Fujishima, and Honda, 1980). Figure 3 shows results consistent with those obtained spectroscopically: Light absorption by the Cu layer at 514.5 nm was larger than that by pure Au at that wavelength. The reduction current around -0.1 V vs. SCE in the cathodic sweep was due to the electrochemical deposition of Cu. The PA signal also began to increase at approximately the same potential due to increased absorption by the Cu layer. On the other hand, the anodic current peak at 0.20 V vs. SCE was attributed to the dissolution of the Cu layer. The PA signal simultaneously began to decrease, returning to the value for the bare Au surface. Using the same cyclic combination of the two techniques, these authors further studied the formation of an oxide layer on the Au electrode in 1 M $HClO_4$ solution via anodic polarization, with the PA signal change effected through increased absorption by the oxide layer. On monitoring the total amount of charge passed through the electrode during the anodic scan, the thickness of the (assumed) Au_2O_3 layer was

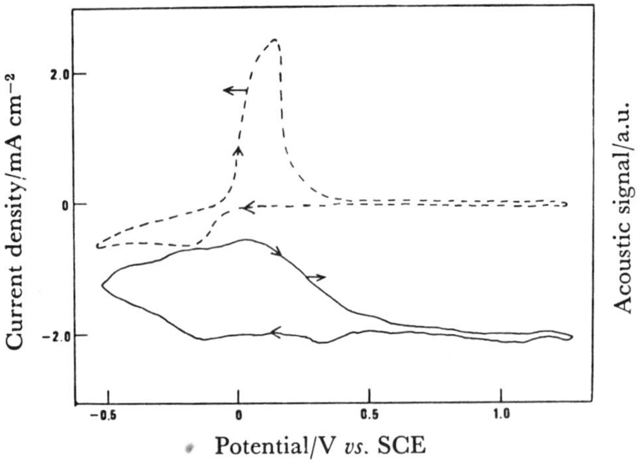

FIG. 3. PA signal vs. potential (—) and current vs. potential (- - - -) curves of Au electrode in the presence of Cu^{2+} ions. Concentrations: 5×10^{-3} M Cu^{2+}, 1 M $HClO_4$; potential sweep rate: 40 s/V; $\lambda = 514.5$ nm. (From Masuda, Fujishima, and Honda, 1980.)

calculated to be 5 Å, which served as a reference for the sensitivity of the PA probe to oxide growth.

Almost simultaneously, the same group of workers performed similar electrochemical measurements of essentially the same system using PAS as the photothermal probe (Fujishima, Masuda, and Honda, 1979). Similar sensitivity limits were reported (5 Å Au_2O_3 in 1 M $HClO_4$ electrolyte).

Direct probing of the working electrode–electrolyte interface can be accomplished via *in situ* PDS, an advantage over the remote, rear-side PAS detection. The PDS signal is, however, sensitive not only to temperature gradients, but also to concentration gradients in the electrolyte (Mandelis and Royce, 1984). Such concentration gradients often appear adjacent to the electrode surface as the result of electrochemical reaction product formation, and the PDS probe can be used to study the kinetics of such reactions. In practice, however, both types of gradients are superimposed leading to two distinct but spatially overlapping contributions to the refractive-index gradient. Dorthe-Merle, Morand, and Maurin (1985) investigated the electrochemistry of a Pt electrode in 1 M H_2SO_4 electrolyte without and with the presence of 10^{-3} M KI. They used a Pt-wire counter-electrode (CE) and pulsed potential in order to limit the formation of electrode surface species. The concentration-gradient component of the PDS deflection in the 1 M H_2SO_4 electrolyte was monitored at 0.08 Hz, the pulsation frequency of the potential, while the temperature-gradient component was monitored at 20 Hz, the light-modulation frequency. No electrochemical reaction occurs without I^- ions in solution, and the observed strong PDS signal at 0.08 Hz at a few microns from the surface was attributed to the establishment of a diffuse layer. A 10-fold enhancement of this signal when I^- ions were present attested to the extension of the diffuse layer following electrochemistry, i.e., I_2 molecules created at the electrode surface. The signal at 20 Hz was found to be 10^4 times smaller than the one due to the I_2 concentration gradient, and PDS of the surface after potential pulsing corroborated the growth of a spectral band at 430 nm characteristic of iodine.

Roger, Fournier, and Boccara (1983) used an ingenious variation of PDS, based on light polarization modulation to study Cu deposition on Pt electrodes with a much higher sensitivity than the conventional PDS. This technique is, in essence, similar to the principle of PARAS (Section II), in that it measures the differential PDS signal generated by unequal absorption of the exciting light polarized parallel or perpendicular to the plane of incidence, in the presence of a thin-grown surface layer on a Pt electrode. The polarization modulation of the incident Ar^+ laser beam was achieved by using an electro-optic modulator, with the Pt electrode in contact with 1 N H_2SO_4 solution containing 10^{-3} M Cu^{2+} ions. Transverse PDS detection of sensitivity corresponding to 10^{-1} monolayer of Cu on Pt was claimed. It is also interesting to note that these authors reported the presence of a huge transient beam deflection related to the ionic concentration gradient that appears during the Cu deposition on (or dissolution from) the Pt electrode. This effect can, in principle, be separated out by the method employed by Dorthe-Merle, Morand, and Maurin (1985).

IV. Electrochemical Interface Reactions

Chemical reactions catalyzed in the presence of various types of electrodes have been investigated photothermally since the early days of photothermal spectroscopy. Gray and Bard (1978) performed a few key experiments that established the utility of microphone–gas-coupled PAS as a spectroscopic technique suitable for electrode reaction studies. These authors monitored the heterogeneous photocatalytic decomposition of acetic acid on suspended platinized anatase (Pt/TiO_2) powder under irradiation (photo-Kolbe reaction) to form methane and CO_2 in an enclosed PA chamber (Kraeutler and Bard, 1978)

$$CH_3COOH \xrightarrow[h\nu]{Pt/TiO_2} CH_4 + CO_2. \tag{2}$$

Gray and Bard (1978) further studied the oxygen uptake by rubrene (tetraphenylnaphthacene) supported on MgO powder to form the endoperoxide according to the chemical reaction:

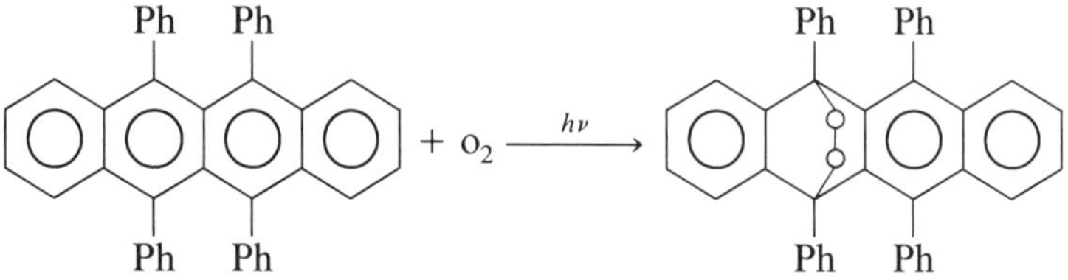

in an enclosed microphonic PA cell, under irradiation in the 250–650-nm region. They concluded that PAS was unique, albeit diffuse reflectance was found to be just as sensitive spectroscopically, in detecting chemical reactions at electrode surfaces that result in production or uptake of gases, with the sensitivity limits obtained being at the $1.2 \times 10^{-9}\,\mathrm{cm}^3$ peak-to-peak gas volume change per irradiation period, or 5.3×10^{-14} mol of gas consumed or released per cycle. The PA ability to monitor photochemical reactions at the powder electrode interface was later exploited by Mandelis (1982) to determine the nonradiative quantum efficiency of the $S_1^0 \rightarrow S_0^0$ transition in the photo-excited uranyl formate monohydrate, $UO_2(HCOO)_2 \cdot H_2O$.

The PA technique with piezoelectric detection has also been used to monitor interfacial electrochemistry. Malpas and Bard (1980) implemented PA detection on a piezoelectric stress-detection electrochemical system, which utilized a working electrode coupled to a piezoelectric ceramic disk. These authors had previously used the piezoelectric technique without optical excitation to detect the derivative of the stress on a Pt electrode surface as a function of the applied potential (Malpas, Fredlein, and Bard, 1979a, b), thus obtaining information from such electrode surface processes as adsorption and thin-film formation. The PA implementation of this system constitutes one of the first *in situ* investigations of working-electrode reactions, besides the rear-surface qualitative microphonic PA studies of Masuda, Fujishima, and Honda (1980) and Fujishima *et al.* (1980b). Malpas and Bard applied their technique to the study of the reduction of heptyl viologen (HV^{2+}) bromide on a Pt electrode surface of an electrochemical cell with a Pt wire CE and an SCE reference electrode (RE). Reduction of this species leads to the formation of

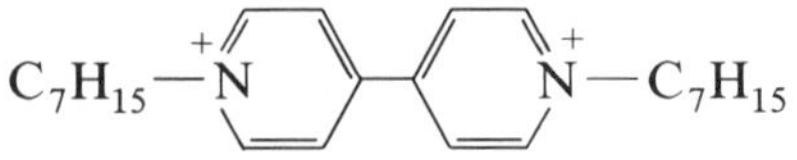

HV^{2+}

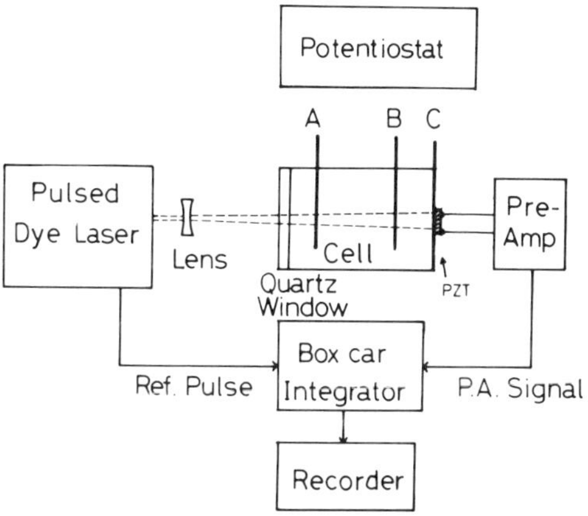

FIG. 4. Block diagram of experimental apparatus for pulsed-laser piezoelectric PA detection of heptyl viologen on Pt electrodes: A) Counter-electrode; B) Reference Ag electrode; C) Working Pt-foil electrode. (From Sawada and Bard, 1982–83.)

i.e., a film of the deeply colored salt, HVBr, on the electrode surface according to the reaction:

$$HV^{2+} + Br^{-} + e^{-} \rightleftarrows HVBr. \qquad (3)$$

The temporal PA response of the Pt electrode–heptyl viologen system to a step-function type of applied potential was found to be controlled by surface tension change effects at early times at the electrode–electrolyte interface. This information regarding the coloration response time of this system is crucial for its application to electrochromic display devices. The sensitivity of this technique is limited by the low power of the light source, a xenon-lamp and monochromator arrangement, because the PA signal is directly proportional to the incident light intensity. A much increased sensitivity was demonstrated later on (Sawada and Bard, 1982–83) by the use of a pulsed tunable dye laser (rhodamine G) with the same electrochemical system (Fig. 4). In that work the *in situ* detection of monolayer coverage on the Pt electrode, due to reaction Eq. (3), was reported corresponding to a surface charge of ca. 2.5×10^{-6} Coul/cm^2 or 2×10^{13} molecules/cm^2 at the detection limit of unity SNR. Malpas and Bard (1980) further performed piezoelectric PAS experiments on the electrochemical reaction involving the coloration of WO_3 on reduction and its bleaching upon subsequent oxidation, using the Xe-lamp source:

$$WO_3 + xH^{+} + zH_2O + xe^{-} \rightleftarrows H_x(H_2O)_zWO_3. \qquad (4)$$

Eq. (4) is an anodic reaction of a ca. 300-Å-thick WO_3 film grown on the W piezoelectric electrode by repeated cycling between +1.3 and −0.25 V vs. SCE in 1 M H_2SO_4 for approx. 2h. The product is responsible for the blue coloration of the WO_3 film. PAS characterization upon anodic bleaching and reduction of the $H_x(H_2O)_z WO_3$ layer to the original pale yellow color showed a signal increase at ≈380 nm, above the optical gap of amorphous WO_3 as expected from the restoration of the WO_3 layer. These effects were also corroborated by the cell-current responses, which were monitored simultaneously.

Another heterogeneous photocatalytic reaction studied by piezoelectric PAS using a lead–zirconate–titanate (PZT) transducer is the MV^{2+} reduction of the CdS—EDTA–MV^{2+} system with the CdS in the form of suspension (Morishita, Fujishima, and Honda, 1983). The value of this work in electrochemistry is in the analytic strength of PZT PAS in clarifying the electronic transfer mechanism at the CdS–EDTA–MV^{2+} interface and its potential applications to electrochemical systems utilizing CdS colloidal electrodes (Dunn, Aikawa, and Bard, 1981). Under Ar^+ laser irradiation at 514.5 nm, increasing time response PA signals were obtained from the CdS–EDTA–MV^{2+} system with saturation constant ≤1 min, due to light absorption by $MV^{\dot{+}}$, produced by the photocatalytic reaction, as shown in Fig. 5.

In situ PZT PAS combined with differential capacity measurements of the adsorption of tetrabutylammonium ions at a gold electrode–electrolyte interface were performed by Kusu *et al.* (1983–84). Piezoelectric detection was shown to be particularly well suited for adsorption studies at light-absorbing electrodes, such as carbon electrodes. These authors were further able to monitor absorption accompanying Faradaic reactions, which are difficult to study by use of differential capacity measurements only.

In situ PZT PAS combined with photoconductive spectroscopy (PCS) and electrochemical detection of chemical reactions at cadmium electrodes (0.5 mm thick), was used by Wun, Milgaten, and Hwang (1981) and Wun-Fogle, Milgaten, and Hwang (1982). Both CW and pulsed-laser excitation modes were used to study the passivation of the cadmium electrode, according to the overall reaction (Wun, Milgaten, and Hwang, 1981):

$$Cd + 2OH^- \rightleftarrows Cd(OH)_2 + 2e^-. \tag{5}$$

This electrode is used in NiCd batteries, and its electrochemical reactivity can be limited by passivation upon discharge. There has been much speculation that a metastable CdO film at open-circuit in KOH might be responsible for the Cd electrode passivation, Eq. (5). Prior to that study, no definite evidence of CdO presence and its role in passivation existed. The

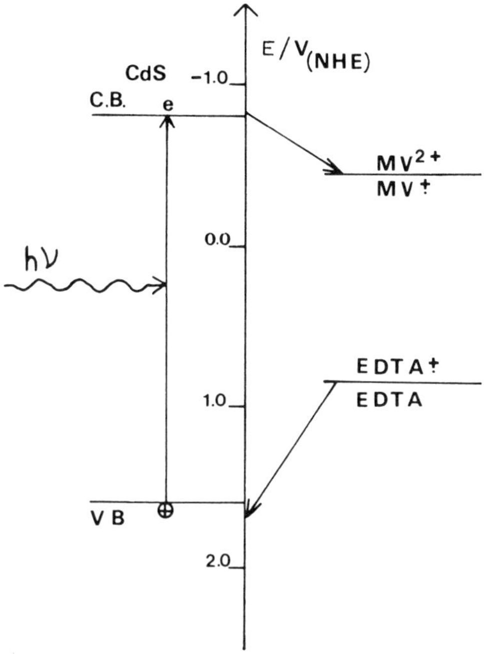

FIG. 5. Energy diagram of CdS–EDTA–MV^{2+} system constructed by use of PZT PAS. (From Morishita, Fujishima, and Honda, 1983.)

combined PAS and PCS were used to distinguish the CdO spectrum from those of the charged and discharged electrodes. Both the CW and pulsed PAS spectra showed a strong CdO signal after discharge, with a 50-Å film detection limit. The PCS detection limit was estimated at 5 Å. The identity of this electrochemical product was established from the well-known band gap of CdO at ca. 500 nm. The PAS results further showed that a physical or chemical process associated with some other species was also involved in the passivation. No such species was identified; however, the back-surface sensing nature of PZT PAS should have been considered, in that the resulting PA signals represent averages over material depths much thicker than typical PCS probe depths. Therefore, there remains a possibility that the thermal diffusion-length-controlled PA signal (Rosencwaig, 1980) might also monitor bulk electrode structural changes upon the onset of passivation, without a corresponding PCS change and without necessarily the formation of new electron-active chemical species, to which the PCS signal might be expected to be sensitive. Microphone PAS has been applied to *in situ* working-semiconductor-electrode electrochemical reaction studies in

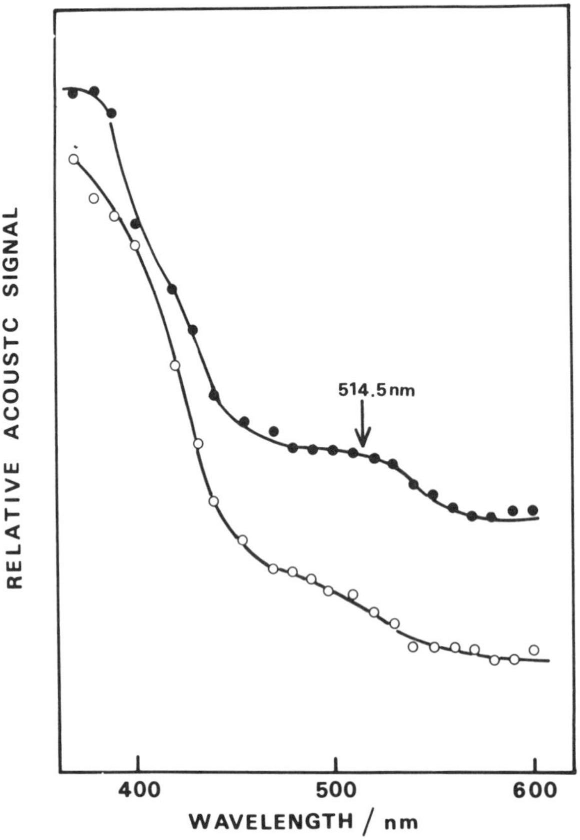

FIG. 6. Microphonic PA spectra of ZnO electrode in 0.2 M KNO_3 + 0.01 M Pb^{2+} solution before (○) and after (●) photoelectrolysis at 1.5 V vs. SCE. Rear-surface detection. (From Masuda *et al.*, 1981.)

the rear-side detection mode briefly discussed in Section III.B. Masuda *et al.* (1981) used two light sources, one as the constant-intensity exciting beam (1-kW Xe arc lamp) and the other as the intensity-modulated probe beam (Ar^+ laser 514.5-nm line), to obtain PA spectra from a *n*-type polycrystalline ZnO electrode subject to the photoanodic reaction of Pb^{2+} in a standard three-electrode configuration. Figure 6 shows PA spectra of ZnO before and after photoelectrolysis. The photoanodic reaction of Pb^{2+} ions at the ZnO electrode is known to lead to the formation of PbO_2, which absorbs visible light, as shown by the enhanced spectrum of Fig. 6. Combined photocurrent and PA cycling between −0.5 V and +1.0 V vs. SCE revealed a two-step photoanodic reaction of the ZnO electrode, consistent with the following mechanism:

$$ZnO \xrightarrow{h\nu} ZnO + e^- + H^+ \quad \text{(excitation step)} \tag{6a}$$

$$Pb^{2+} + 2h^+ \longrightarrow Pb^{4+} \quad \text{(first-wave reaction)} \tag{6b}$$

and

$$2ZnO + 2h^+ + Pb^{2+} \longrightarrow PbO_2 + 2Zn^{2+} \quad \text{(second-wave reaction)} \tag{6c}$$

where h^+ indicates a photogenerated hole.

Masuda, Fujishima, and Honda (1982a) repeated the ZnO-electrode measurements in a different electrolyte (1 M KCl) and added a rhodamine 6G tunable dye laser (550–660 nm) to the Ar^+ probe-laser instrumentation. After electrolysis at −1.6 V vs. SCE, the ZnO electrode decomposed, forming a deposited Zn layer as follows:

$$ZnO + 2H^+ + 2e^- \xrightarrow{h\nu} Zn + H_2O. \tag{7}$$

After photocurrent and PA cycling between −1.5 V and 0 V vs. SCE, the PA signal did not return to its initial value upon anodic scanning, thus implying that the deposited zinc layer could not completely dissolve and remained partially on the polycrystalline ZnO surface. These workers further performed PA studies of the same nature on CdS single-crystal electrodes (Masuda, Fujishima, and Honda, 1982a, b) before and after electrolysis at −2.0 V vs. SCE in 1 M KCl solution. Their results were consistent with the photodecomposition reaction:

$$CdS + 2e^- \xrightarrow{h\nu} Cd + S^{2-}. \tag{8}$$

Perhaps due to its slow response coupled with its contact nature, back-surface PTS has been much less popular than PAS in monitoring electrodic reactions. To the best of our knowledge, this technique has only been used to study the enhanced (anomalous) reaction rate for oxygen reduction at a n-TiO_2 rotating disk photoelectrode in a 1 M KOH electrolyte (Decker, Juliao, and Abramovich, 1979). The observed anomalously high heating of the crystal was monitored with a Copper–Constantan thermocouple and was attributed to surface states at the TiO_2–electrolyte interface.

In recent years, photopyroelectric spectroscopy (P^2ES) is becoming progressively popular as a remote detection method, combining fast response and high sensitivity (Coufal and Mandelis, 1987). P^2ES exploits the pyroelectric character of thin-polymer-film detectors, which exhibit spontaneous, rapid polarization as a result of a minute temperature change across the film. As a result, a static charge redistribution occurs across the detector that manifests itself as a voltage. When properly amplified, this pyroelectric

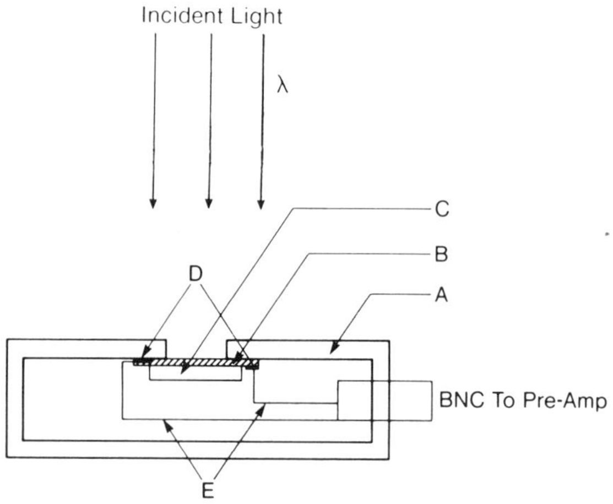

(a)

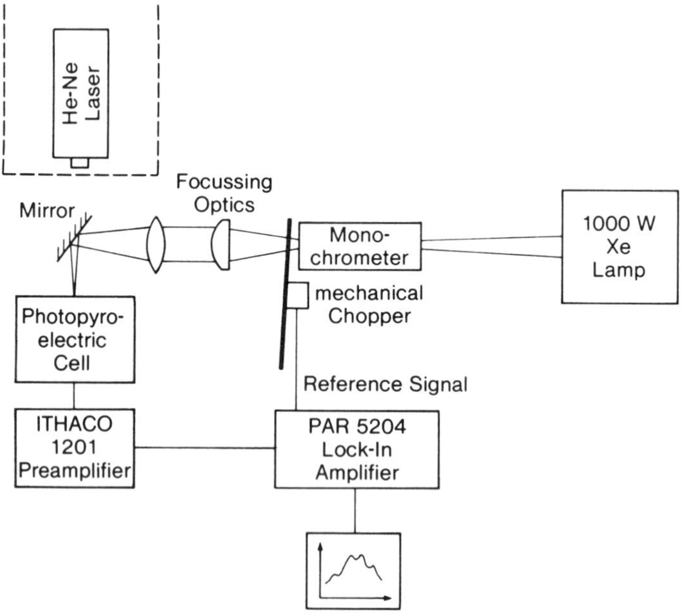

(b)

FIG. 7. a) P^2E cell with a 0.5″ × 0.25″ PVDF sample holder. A) Teflon (or metal) cell; B) PVDF film; C) Teflon support; D) Solderable copper foil electrodes with conductive adhesive; F) Ag wires. b) P^2E experimental apparatus. (From Mandelis, 1984.)

voltage carries optical and thermal information from samples in intimate contact with the thin-film detector (Mandelis and Zver, 1985). By using a thin (28 μm) pyroelectric polyvinylidene difluoride (PVDF or PVF_2) film, with a simple experimental setup (Fig. 7), Mandelis (1984) first applied P^2ES to the study of the kinetics of the electrode reaction

$$CuO + 2H^+ \rightarrow Cu^{2+} + H_2O. \tag{9}$$

A 0.34-mm-thick Cu plate was oxidized thermally and mounted between the holder opening and the PVDF film, i.e., between A and B in Fig. 7. Simple physical contact of the pyroelectric with the sample was sufficient to produce strong P^2E signals using a He–Ne laser. The kinetics of Eq. (9) was monitored *in situ*, upon its initiation, by introducing into the open chamber a 0.1 M HCl solution. The time-dependent amplitude of the lock-in-detected signal at 10 Hz was recorded and was found to be consistent with the first-order kinetic behavior

$$[CuO](t) = [CuO](0)\exp(-kt) \tag{10}$$

with

$$k = (1.55 \pm 0.6) \times 10^{-3}\, s^{-1}.$$

The future looks promising with respect to electrochemical reaction monitoring applications of P^2ES. The potential of thin pyroelectric films as acoustic-noise and electrolyte-bulk concentration gradient-insensitive electrodes, covered with thin layers of electrochemically active components (metals, semiconductors), appears quite exciting for truly *in situ* spectroscopic CW (Mandelis and Zver, 1985) and time-resolved studies (Coufal, 1986).

V. Corrosion and Thin-Film Growth Processes

Photothermal probes are particularly suited for thin-film growth studies at the electrochemical interface, due to their high sensitivity and, in particular, the fact that photothermal signals are linear functions of the reflectivity change of the surface on which thin-film growth occurs, which, in turn, is proportional to the film thickness (McIntyre and Aspnes, 1971). Therefore, photothermal studies of electrochemically formed thin surface films have flourished, with a number of them being quantitative film-thickness measurements. The main advantage of the photothermal probes is the relative ease with which such studies can be undertaken, along with the monolayer

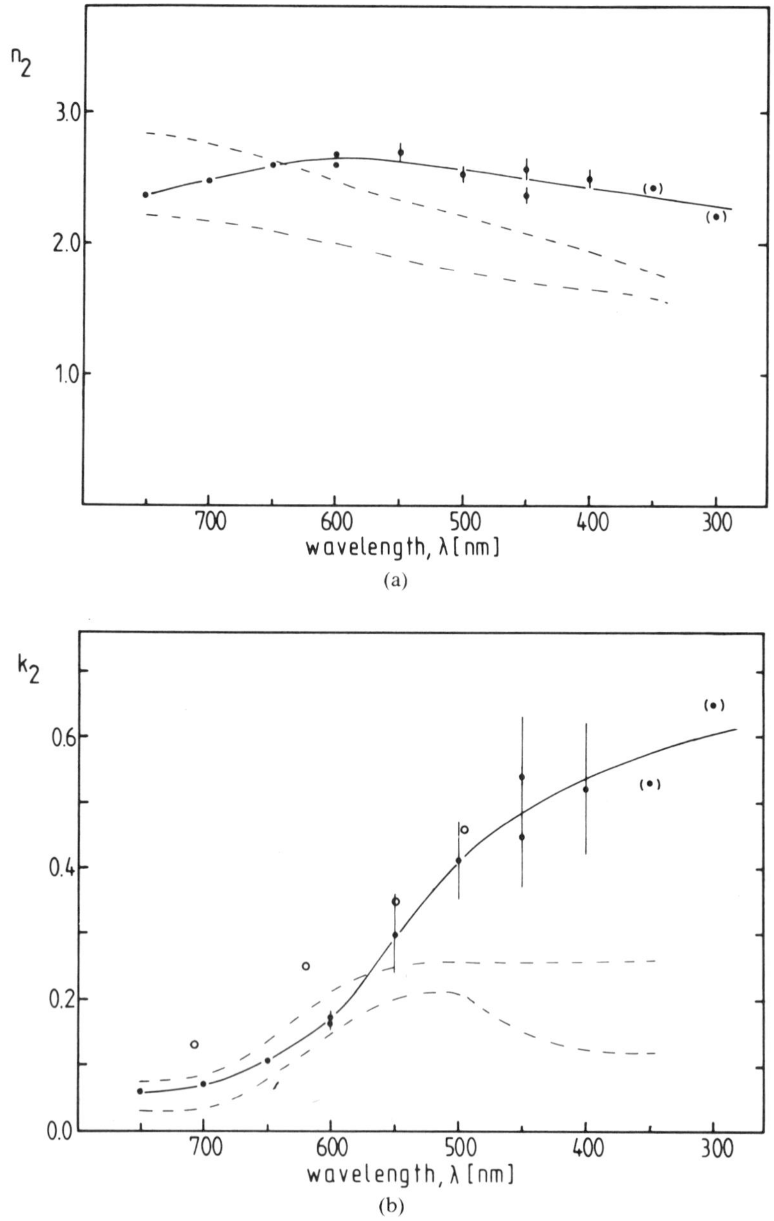

FIG. 8. a) Index of refraction $n_2(\lambda)$ and b) absorption constant $k_2(\lambda)$ of PbO_2 films (thickness $\leq 1\ \mu m$) electrodeposited on Pt as determined *in situ* from PA interference signals (●). For comparison: k_2 for β-PbO_2 films on SnO_2 from transmission spectroscopy (○) [Mindt, W. (1969). *J. Electrochem. Soc.* **116**, 1076] and range of optical constants of 80–280 nm PbO_2 films on Au from *in situ* reflection spectroscopy (broken lines) [Naegele, K. D., and Plieth, W. J. (1980). *Electrochem. Acta* **25**, 241]. (From Dohrmann and Sander, 1986.)

sensitivity, features not matched by either transmission or diffuse reflectance spectroscopies. Quantitative foundations for the photoacoustic detection of thin surface films have been developed by Mandelis, Siu, and Ho (1984), based on a two-layer PA model: The thin-film thickness L can be calculated from the spectral positions λ_1 and λ_2 of two consecutive thermal wave interference maxima (or minima) using the relation:

$$L = \frac{2\pi + \phi_{12}(\lambda_1) - \phi_{12}(\lambda_2)}{4\pi\left[\dfrac{n_1(\lambda_2)}{\lambda_2} - \dfrac{n_1(\lambda_1)}{\lambda_1}\right]}; \qquad \lambda_1 > \lambda_2, \tag{11}$$

where n_1 is the real part of the index of refraction of the thin film, and $\phi_{12}(\lambda)$ is the phase angle (in polar coordinates) of the complex optical reflection coefficient $R_{12}(\lambda)$ at the thin-film–substrate interface, at the wavelength λ. This theory was further shown to be in quantitative agreement with experimental PA spectra obtained from Si wafers oxidized with surface thin SiO_2 films. Dohrmann and Sander (1986) used a similar PA interferometric approach for quantitative studies of PbO_2 films grown anodically on a Pt electrode by constant-current deposition from $Pb(ClO_4)_2$ in aqueous $HClO_4$. Film thickness was independently determined coulometrically and was used as a known parameter for the PA interference spectra, in a fashion entirely analogous to the procedure used by Mandelis, Siu, and Ho (1984). Upon expressing the amplitude of the complex optical reflection coefficient at the $(i–j)$ interface in the form

$$R_{ij} = \frac{(n_j - n_i)^2 + k_i^2}{(n_j + n_i)^2 + k_j^2}, \tag{12}$$

where $k_i(\lambda)$ is the optical extinction constant (i.e., the imaginary part of the complex refractive index), Dohrmann and Sander (1986) used PA data from the clean Pt electrode before $(i = 1, j = 3)$ and after $(i = 1, j = 2)$ the deposition of the opaque PbO_2 layer. A best fit of the theoretical expressions to the PA wave-interference data yielded $n_2(\lambda)$ and $k_2(\lambda)$ spectra for the PbO_2 films as shown in Fig. 8. Thus, PA interferometry was shown to be a quantitative substitute for ellipsometry for *in situ* electrochemical applications where the latter cannot be used.

Dohrmann, Sander, and Strehblow (1983) further utilized the thin-film thickness dependence of surface-reflectivity changes ΔR (McIntyre and Aspnes, 1971) to monitor the growth of oxide films on noble metal electrodes photoacoustically. These authors showed that the difference PA signal ΔP of the layer-covered electrode and that of the layer-free electrode is given by:

$$\frac{\Delta P}{P_0 - P} = -\frac{\Delta R}{R}, \tag{13}$$

where P_0 is the PA signal from the electrode covered with an appropriate calibration standard, and R is the reflectivity of the layer-free surface at the electrode–electrolyte interface. Equation (13) further assumes that all non-radiative quantum yields are equal and that $R_0 = 0$ for the standard (an ideal blackbody). By monitoring PA spectra in oxide-free metal electrodes made of Cu, Ag, Au, and Pt, these workers were able to identify spectroscopically from ΔP signals the presence of oxides grown electrochemically on these metal surfaces. It is important to note that Sander, Strehblow, and Dohrmann (1981) were able to identify both the type of oxide (or hydroxide) species on a Cu electrode *and* the chemical compositions (e.g., Cu_2O, CuO, $Cu(OH)_2$) via combined information from simultaneous PA and potentiodynamic measurements, as well as ESCA data. The contribution of the PA information to the overall picture was largely similar in nature to that obtained by Masuda, Fujishima, and Honda (1980), Fig. 3; that is, the correlation of abrupt PA signal step changes with cyclic potentiodynamic-current density peaks as diagnostics for formation or dissolution of various thin surface layers. The electrochemical potential position of the step, when combined with ESCA findings, was used in identifying the chemical structures and stoichiometry of the layer in question. Yamagishi, Moritani, and Nakai (1980, 1981) also used a differential PA technique similar, in principle, to that of Dohrmann, Sander, and Strehblow (1983) to measure anodic film growth on *p*-GaAs wafers and *p*-Bi_2Te_3 layered semiconductors. The technique utilized a piezoelectric transducer as the PA detector on the rear-surface of the working electrode, and the change in transmittance through the sample, $\Delta T = -\Delta R$, was measured photoacoustically. The PA signal from Bi_2Te_3 exhibited a periodically varying structure as a function of anodization time, which reflected the layered crystal structure of the semiconductor. No further identification of the chemical composition of the anodic thin film was attempted. This study ultimately showed that the PA technique can be used as a substitute for the differential reflectance technique, especially where reflectance/transmittance changes too small to be observed by the latter method are involved. PA signal saturation due to a PbO_2 layer growth during the photoanodic reaction Eq. (6) at +1.65 V vs. SCE, coupled with the photoelectrochemical decomposition of ZnO, was observed by Masuda *et al.*, (1981). These authors used the saturation-characteristic time with the amount of charge passed through the electrode to estimate the PbO_2 thickness (200 nm).

This brief review of PA applications to surface film growth and corrosion would not be complete without mentioning the work of Ogura, Fujishima, and Honda (1984), who attempted to resolve the controversial issue of the single- or double-layer nature of the oxide film responsible for the passivation of Fe electrodes. These workers cleverly used combined cyclic PZT PAS and potentiostatic measurements under variable potential ranges between -1.2 V and $+0.8$ V vs. SCE, with the Fe electrode in 0.1 M potassium phosphate and 0.05 M sodium borate solution of pH 8.0. From cathodic and anodic scan data, they showed that the results supported the existence of two layers on iron, γFe_2O_3 and Fe_3O_4 (reduction potentials -0.4 V and -0.9 V vs. SCE, respectively). The PA signal thus helped resolve the controversy by providing a resolved Fe_3O_4-related feature at -0.9 V, whereas the cathodic voltammogram could not resolve such a feature because of the overlapping large cathodic current of hydrogen evolution.

Other photothermal techniques besides PAS have also been applied successfully to problems of corrosion and/or thin-film growth *in situ*. PTS was used to study nonelectrochemical, vacuum-evaporated thin films of zinc phthalocyanine (ZnPC) and proved to be quite sensitive to film thicknesses of 200–5000 Å (Brilmyer and Bard, 1980). These authors further used PTS to characterize the thin-film precipitate $HV^{+}\cdot Br^{-}$ produced from the Pt electrode reaction Eq. (3), as a result of continuous pulsing of the electrode between -0.2 V and -0.55 V vs. SCE at 1 Hz. Photothermal temperature saturation effects after controlled deposition at -0.5 V vs. SCE were observed, similar to PA saturation phenomena related to the PbO_2 growth reported by Masuda *et al.* (1981). Brilmyer and Bard (1980) estimated saturation to occur at several hundred monolayers of precipitate. Similar PTS experiments were also conducted for the electrodeposition of Cu from a $CuSO_4$ solution. The sensitivity of PTS to thin-film thickness, however, was found to be no better than 75–100 monolayers (about one order of magnitude lower than PAS), which, when considered together with the slow response of the thermistor, may explain the relative lack of popularity of this technique for corrosion and film-growth analysis.

PDS, on the other hand, has enjoyed good success in the quantitative monitoring of thin-photocorrosion-film growth on ZnSe. The electrochemical reaction:

$$ZnSe + 2h^{+} \xrightarrow{h\nu \geq E_g} Zn^{2+} + Se \tag{14}$$

results in the dissolution of the photoelectrode and growth of amorphous Se layer at the electrochemical interface (Gautron *et al.*, 1979). Transverse PDS was used with superbandgap excitation (460 nm) in a pH 4 electrolyte (Royce *et al.*, 1982) and the thin-Se-film absorptance (αl) was monitored through

$$S = A[1 - \exp(-\alpha l)], \tag{15}$$

where A is a constant, l is the film thickness, and S is the PDS signal in the thermally thin limit. An independent measurement of the optical absorption coefficient α for amorphous Se allowed the determination of the (effective) thickness l. The growth rate of the photocorrosion films was thus found to exhibit a parabolic dependence on time, suggesting a diffusion-controlled reaction. After 18-h exposure to radiation and photocorrosion, the Se film thickness was estimated to be $(0.32 \pm 0.1) \times 10^{-4}$ cm for the pH 4 and $(0.85 \pm 0.2) \times 10^{-4}$ for a pH 7 electrolyte. These types of measurements hold good promise for more detailed studies of photocorrosion mechanisms and sample quantum efficiencies (see Section VII).

VI. Diffusion of Electrochemical Species

Photothermal laser beam deflection (or mirage effect) has the unique characteristic ability among photothermal techniques to detect concentration gradients in an electrolyte as a result of electrodic reactions. Frequently, concentration gradients are present together with temperature gradients, in which case the decoupling of the two effects on the probe-beam deflection requires detection at different modulation frequencies, taking advantage of the widely different characteristic time constants between thermal and mass diffusion (Dorthe-Merle, Morand, and Maurin, 1985). Another way of separation of the mass diffusion component is mirage-effect detection of electrochemical interface species gradients without optical excitation and heating of the electrode surface. Under these conditions, time domain detection appears to have several advantages over the conventional frequency domain technique, the main one being the ease of interpretation of diffusion processes in terms of thermal and mass transfer response delay times at the probe-beam offset location. A detailed theoretical analysis of the time domain situation with a constant concentration of electrode surface species, the assumed result of an electrochemical reaction, has been presented by Mandelis and Royce (1984). These authors showed that the probe-laser beam contains information about the diffusing species, both in the time evolution of the deflection $\Theta(z, t)$ *and* in the shape of the beam

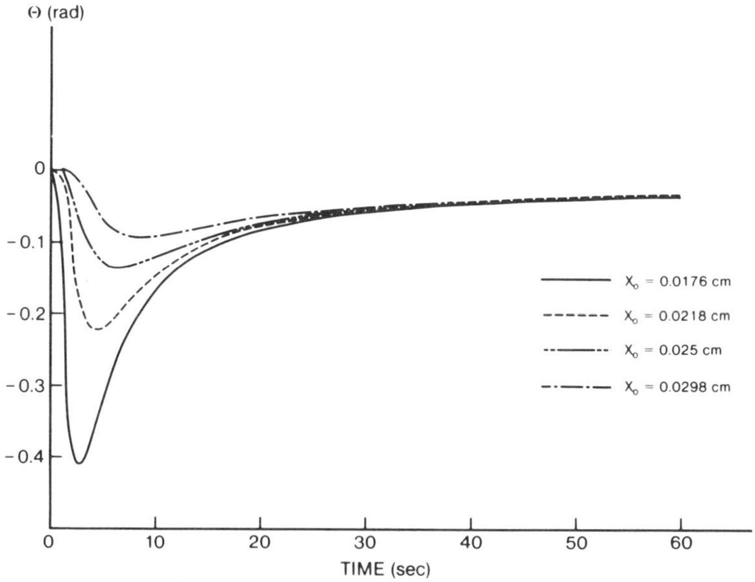

FIG. 9. Probe-laser beam deflection according to Eq. (17). $A = 5.9 \times 10^{-4}$; $D = 1.49 \times 10^{-5}$ cm^2/s; $z = 3.0$ cm. (From Mandelis and Royce, 1984.)

spot in the direction x, perpendicular to the plane of the electrode, $W_x(z, t)$. For a concentration profile

$$C(x, t) = C_0 \operatorname{erfc}\left(\frac{x}{2\sqrt{Dt}}\right), \tag{16}$$

the beam deflection along the z-dimension parallel to the plane of the electrode is given by (Fig. 9):

$$\Theta(z, t) = -\sqrt{m}\, z\left[1 + \frac{\sqrt{m}\, x_0 z^2}{12Dt}\right]\exp\left(-\frac{x_0^2}{4Dt}\right), \tag{17}$$

and the spot size along the x-direction:

$$W_x^2(z, t) = W_{0x}^2\left[\cosh^2(\sqrt{m}\, z) + \left(\frac{1}{mz_{0x}^2}\right)\sinh^2(\sqrt{m}\, z)\right], \tag{18}$$

where D is the diffusion coefficient of the electrochemically generated species, $m = m(t) = A^2/\pi Dt$, with $A \approx (C_0/n_0)\partial n/\partial C|_{C=C_0}$; n_0 is the refractive index of the pure electrolyte without the product species; x_0 is the beam offset distance when $C = 0$; and $W_{0x}(z_{0x})$ is the laser-beam spotsize at the beam waist (magnitude of the complex beam radius). The theoretical predictions of the model by Mandelis and Royce (1984) regarding the offset distance x_0-time delay for maximum deflection relationship, Eq. (17) and Fig. 9, have been found to be in agreement with experiment in the case of the diffusion of ferri-ferrocyanide redox species, 50 mM $Fe(CN)_6^{3-}$/50 mM $Fe(CN_6^4)^-$ in 0.5 M K_2SO_4, produced electrochemically via a step voltage applied to a Pt electrode (Royce, Voss, and Bocarsly, 1983). These workers observed a linear dependence of Δt (time since the application of the voltage step and the maximum beam deflection) on x_0^2:

$$\Delta t = x_0^2/4D \tag{19}$$

as predicted from Eq. (17). Therefore, the diffusion coefficient for $Fe(CN)_6^{3-/4-}$ in 0.5 M K_2SO_4 was found to be $(9.4 \pm 0.2) \times 10^{-6}$ cm^2/s, in agreement with published values. A small opposite signal was also observed prior to the deflection associated with the concentration gradient. This signal was attributed to the rapid thermal wave resulting from the electrochemical Peltier effect, in view of the fact that $(\partial n/\partial C)$ was found to be $(7.2 \pm 0.1) \times 10^{-2}$/mole for $Fe(CN)_6^{4-}$, and $(5.2 \pm 0.1) \times 10^{-2}$/mole for $Fe(CN)_6^{3-}$, whereas the $(\partial n/\partial T)\Delta T$ associated with the Peltier heat was three to four orders of magnitude smaller than the $(\partial n/\partial C)\Delta C$ for each species.

Electrochemical diffusion as the result of electrode potential modulation (0.5 Hz) in the absence of optical excitation has been reported by Pawliszyn *et al.* (1986a). These authors used a sensitive differential deflection experimental arrangement (Pawliszyn, Weber, and Dignam, 1985) to monitor concentration gradients generated by nonequilibrium chemical processes. A cell containing a Pt WE, Pt CE and silver wire RE, quartz windows, and a He–Ne laser probe beam were used with two redox systems, i.e., oxygen and *p*-benzoquinone. Both are known to undergo a reversible electrochemical reduction to produce anion radicals with products superoxide and benzoquinone anion radicals, respectively, after a one-electron reduction in dimethyl sulfoxide (ME_2SO). The potential was modulated at the rate of 500 mV/s, and $(\partial C/\partial x)$ was produced by the electrochemical reaction

$$\mathrm{O} + ne \rightleftarrows R. \tag{20}$$

That study concluded that the high-sensitivity deflection detection apparatus was capable of tracing concentration gradients down to

$$\left(\frac{\partial C}{\partial x}\right)_{\min} = \Theta_{\min}(n/d)/(\partial n/\partial C) \approx 10^{-5}\ \mathrm{M/m}, \tag{21}$$

where Θ_{min} is the minimum deflection, and d is the distance parallel to the surface of the electrode over which the refractive index gradient exists. The limit established in Eq. (21) was obtained from the reduction of *p*-benzoquinone. It is important to note that the sensitivity of the laser-beam deflection technique was found to be at least one order of magnitude higher than cyclic voltammetry performed intermittently at the same electrode to check the dissolved oxygen concentration due to loss from its saturated Me_2SO solution. Pawliszyn *et al.* (1986b) further investigated the spectroscopic properties of electrochemical species using chopped light from a 1000-W Xe lamp dispersed through a monochromator, and were thus able to selectively monitor individual species in *p*-benzoquinone (BQ) in Me_2SO. The rest of the experimental system was the same as before (Pawliszyn *et al.*, 1986a), with the electrode potential modulated with a sweep rate of 5 mV/s. In these experiments, concentration gradients were generated above the Pt electrode during cyclic voltammetric experiments. Either the excitation wavelength was varied in order to collect the absorption spectrum of the product, or signal changes due to concentration variations at a fixed wavelength were monitored. Spectra of two consecutive BQ reduction products were obtained. The products were believed to be $BQ^{-}\cdot$ and $BQH\cdot$ radicals from the spectral positions of the peaks at ca. 435 and 460 nm ($BQ^{-}\cdot$), and 400 nm ($BQH\cdot$), that were formed according to the reaction sequence:

$$BQ + e^{-} \leftrightarrows BQ^{-}\cdot \tag{22a}$$

$$BQ^{-}\cdot + H^{+} \leftrightarrows BQH\cdot \tag{22b}$$

$$BQH\cdot + e^{-} \leftrightarrows BQH^{-} \tag{22c}$$

and

$$BQH^{-} + H^{+} \rightarrow BQH_2. \tag{22d}$$

The signal phase shift attributed to concentration gradients was also used to distinguish between a signal originating at the electrode surface and that due to a dissolved product, thus providing spatially resolved information: $\partial C/\partial x$ at the intersection volume between pump and probe beams is expected to contribute to the modulated signal characterized by zero phase shift.

VII. Energy-Transfer Physics at Photoelectrochemical Interfaces

Photothermal studies of the physics of energy transfer and conversion at electrode–electrolyte interfaces have been performed exclusively with semiconducting photoelectrodes, primarily due to the well-known electrochemical behavior and energy band structures of many semiconductors (Morrison, 1980). From the point of view of a photoelectrochemical (PEC) cell device, two important parameters have been studied in some detail using photothermal probes: The intrinsic quantum efficiency of the photoelectrode reaction, η_Q, and the energy conversion efficiency of the PEC cell, η_E. The former quantity is defined as the number of electrons generated per unit time for each photon absorbed at the photoelectrode:

$$\eta_Q = \frac{i(\text{electrons/s})}{I(\text{photons/s})}. \tag{23}$$

The latter quantity can be defined as the ratio of the chemical or electrical power output to the optical power input to the photoelectrode

$$\eta_E = \frac{W_{out}}{W_{in}}. \tag{24}$$

Fujishima *et al.* (1980a) developed a thermodynamic model for the enthalpy change ΔH_c during an electrochemical reaction in terms of the heat evolved ($= T\Delta S$, where ΔS is the entropy change), and related their calculations to the η_Q and the generated overpotential $(V - V_{FB})$, as follows:

$$E\frac{\Delta T}{\Delta T_{oc}} = [(Q_{sc} + T\Delta S)/It] + \eta_Q(V - V_{FB}), \tag{25}$$

where E is the energy (eV/photon) of an optical pulse of duration t, exciting the electrode, which absorbs an average intensity I (photon/s); V_{FB} is the flatband potential of the semiconductor (Morrison, 1980); $(Q_{sc} + T\Delta S)/It$ is the total heat change in the PEC system (usually evolved) at the photoanode; ΔT_{oc} is the electrode temperature change at open-circuit, and Q_{sc} is the heat evolved in the semiconductor via nonradiative recombination and other radiationless processes. Equation (25) shows that η_Q can be obtained experimentally from the slope of the (theoretically) straight line of the $E(\Delta T/\Delta T_{oc})$ vs. $(V - V_{FB})$ plot under constant illumination conditions, while the thermal loss term, $(Q_{sc} + T\Delta S)/It$ may be estimated from the intercept with the $V = V_{FB}$ axis, Fig. 10. Fujishima *et al.* (1980a) applied their model to the PEC systems $CdS/Fe(CN)_6^{3-}/Fe(CN)_6^{4-}$ and TiO_2/H_2SO_4, using PTS detection. In the CdS experiment, the photo-oxidation of $K_4Fe(CN)_6$ at the CdS photoanode was monitored. The solution used was 0.1 M $K_4Fe(CN)_6$, 0.001 M $K_3Fe(CN)_6$ and 0.2 M Na_2SO_4 as the electrolyte. Pt CE and SCE RE were

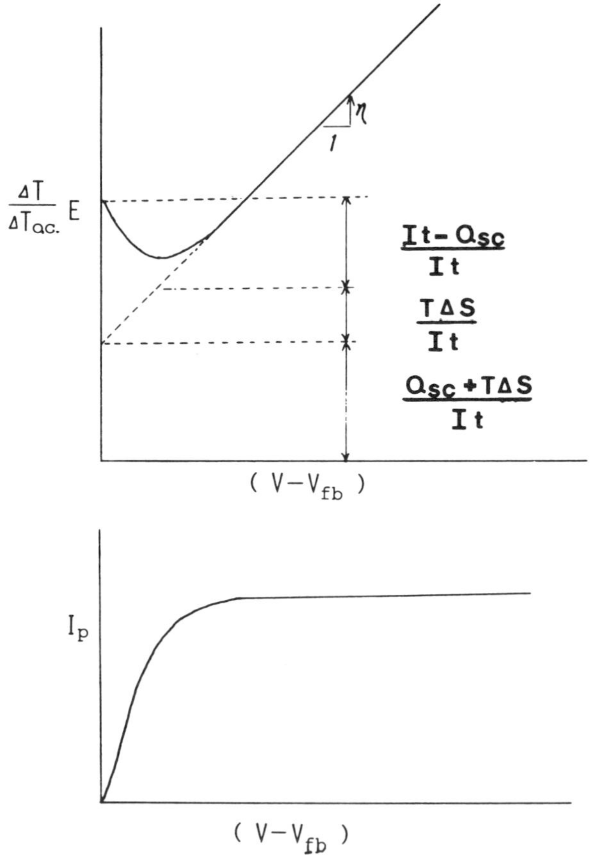

FIG. 10. (*Top*) Theoretical behavior of the PTS signal vs. potential, according to Eq. (25). The solid line represents typical semiconductor electrode (nonideal) behavior. (*Bottom*) Theoretical photocurrent vs. potential behavior. (From Fujishima *et al.*, 1980a.)

utilized. Under the conditions of Eq. (25) with $E = 2.5$ eV, η_Q was found to be unity, with $(Q_{sc} + T\Delta S)/It \approx 1.0$ eV. The monochromatic energy conversion efficiency, η_E, can be found from the intercept as:

$$\eta_E = \left(\frac{EIt - Q_{sc}}{EIt}\right) \times 100(\%), \tag{26}$$

or 40% for the CdS electrode. Similarly, three different TiO_2 photoanodes in 1 M H_2SO_4 yielded the η_Q values of 0.7, 0.3, and 0.1. The η_E was 37% for the $\eta_Q = 0.7$ electrode. The largest η_Q was found to be accompanied by the largest η_E, as expected. Although the linear theory developed by those authors did not account for the PTS signal minimum observed experimentally, it proved successful in estimating the system η_Q and η_E quantities. The nonzero values for η_E encountered with both CdS and TiO_2 single-crystalline electrodes were attributed by Fijishima *et al.* (1980a) to two possible mechanisms for energy dissipation during the oxidation of the redox species: a) via an electron donation–hole filling in the valence band to produce a vibrationally excited species, whose energy is quickly dissipated in the solution as heat; or b) an isoenergetic electron transfer from the reduced species in solution to a surface state, followed by electron–hole recombination in the valence band. In this case, heating would be expected to occur in the semiconductor itself. The PTS measurements alone could not distinguish between the two mechanisms, and the experimental precision of $\pm 1\%$ was found to limit the use of PTS to the more efficient PEC systems. Corrections in temperature, however, due to the endothermic entropy change (electrochemical Peltier effect) on the CdS photoanode were implemented through Eq. (25) theoretically and earlier, experimentally (Fujishima, Brilmyer, and Bard, 1977), for a similar PEC cell, by comparing the magnitudes of heating and cooling on reference carbon and CdS electrodes. This correction amounted to a rigid upward vertical shift of the straight portion of the uncorrected line in the $E(\Delta T/\Delta T_{oc})$ vs. $(V - V_{FB})$ plot, thus yielding the (corrected) Fig. 10.

The same PTS methodology was used further (Fujishima, Maeda, and Honda, 1980c) to calculate η_Q for PEC reactions on various semiconductor

TABLE I[a]

COMPARISON OF INTRINSIC QUANTUM EFFICIENCIES, η_Q AT 340-nm IRRADIATION

Electrode	η_t	η_a	Electrolyte
n-CdS s	1.00	1.00	Na_2SO_3
n-CdS p	0.67	0.67–0.70	Na_2SO_3
n-CdSe s	0.42	0.40–0.42	Na_2S
*n*GaP s	0.85	0.62	Na_2S
n-GaAs s	0.80	0.47–0.49	Na_2S
n-TiO_2 s	0.70	0.70	H_2SO_4
n-ZnO p	0.85	0.76–0.78	Na_2SO_4
n-MoS_2 s	1.00	0.54	Na_2SO_4
p-GaP s	0.80	0.60–0.66	H_2SO_4
p-GaAs s	0.80–1.00	0.56–0.59	H_2SO_4

[a] Fujishima, Maeda, and Honda (1980c).
η_t: Quantum efficiency by temperature measurement.
η_a: Quantum efficiency by actinometry.
p Polycrystal.
s Single crystal.

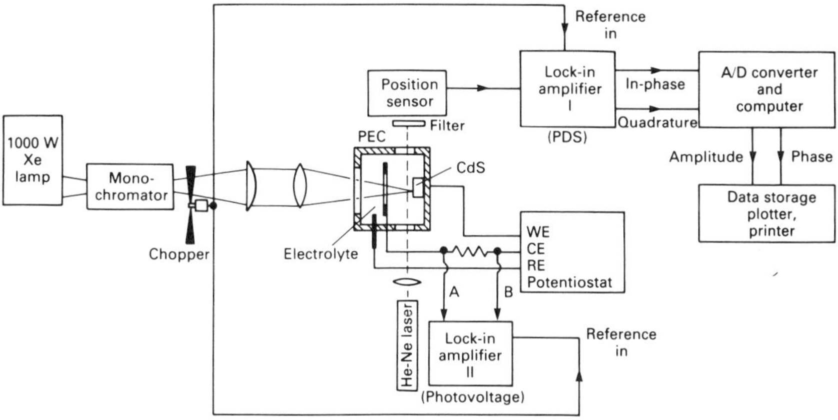

FIG. 11. Experimental apparatus for combined PDS and PCS measurements. (From Wagner, Wong, and Mandelis, 1986.)

electrodes. The intrinsic quantum efficiencies calculated by PTS, were compared with those obtained by conventional chemical actinometric methods and are shown in Table I.

Recently, PDS has emerged as a very suitable technique for quantum efficiency studies at the semiconductor electrode–electrolyte interface. Combined PDS and PCS studies of single crystalline *n*-CdS–polysulphide electrolyte (1 M OH^-/1 M S^{2-}/1 M S) interfaces (Wagner, Wong, and Mandelis, 1986) have yielded *in situ* self-consistent qualitative descriptions of energy-conversion mechanisms. Figure 11 shows the experimental apparatus for the combined PCS and PDS PEC studies. The spectroscopic data were supplemented by Mott–Schottky measurements of the space–charge layer capacitance, which yielded values for the CdS-effective doping density (5.02×10^{14} cm^{-3}) and V_{FB} (-1.52 V to -1.20 V vs. SCE). The particular electrolyte used in that work was chosen because of its well-known ability (Ellis, Kaiser, and Wrighton, 1976) to quench the photoanodic dissolution of the CdS reaction:

$$CdS \xrightarrow{h\nu} Cd^{2+}(aq.) + S(s) + 2e^-. \tag{27}$$

Very recently, quantitative analysis of the PDS signal dependence on various nonradiative de-excitation (i.e., heat production) mechanisms was developed by Wagner and Mandelis (1987) based on the results of Wagner, Wong, and Mandelis (1986) and the PTS model of Maeda, Fujishima, and Honda (1982). The significant thermal processes at the working-electrode–electrolyte interface are represented in the schematic diagram of Fig. 12. Quantitative expressions for the partial contributions of each of these

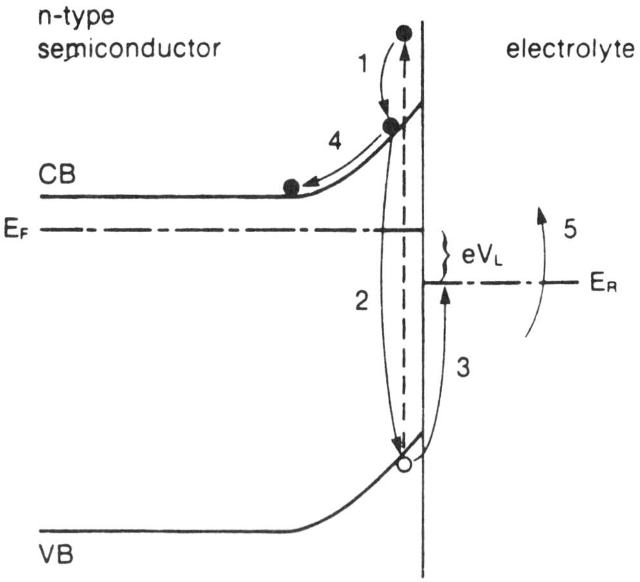

FIG. 12. Heat-generating mechanisms which contribute to the PDS signal at the WE–electrolyte interface; E_F is the semiconductor Fermi energy, E_R is the electrolyte redox level, CB(VB) is the conduction (valence) band. Refer to the text for an explanation of the five heat sources (PDS_j; $j = 1, 2, \ldots, 5$; Maeda, Fujishima, and Honda, 1982). (From Wagner and Mandelis, 1987.)

processes to the PDS signal were developed: i) Following optical absorption, nonradiative electron intraband de-excitation from higher states in the conduction band to states near the band edge yields a signal component

$$PDS_1 = KN(\hbar\omega - E_g), \tag{28}$$

where E_g is the optical band gap of the semiconductor electrode, and K is a geometry-dependent proportionality constant. All optically generated conduction-band (CB) electrons are assumed to undergo this fast transition. ii) Nonradiative interband de-excitation between the conduction and valence bands contributes a signal component

$$PDS_2 = KN(1 - \eta_Q)E_g. \tag{29}$$

iii) Electron injection from electrolyte species into the valence band (for n-type working electrode) gives

$$PDS_3 = KN\eta_Q(E_R - E_{VB}), \tag{30}$$

where E_R is the redox level of the electrolyte (vs. reference), and E_{VB} is the valence-band (V) energy level (vs. reference). iv) Carrier separation in the depletion layer, during which the carriers lose energy under the influence of the built-in field, plus the Peltier heat at the back Ohmic contact of the n-type electrode, contribute

$$\mathrm{PDS}_4 = KN\eta_Q(E_{CB} - E_R - E_L) \tag{31}$$

or

$$\mathrm{PDS}_4 = KN\eta_Q e(V - V_{FB}), \tag{32}$$

where E_{CB} is the CB energy level (vs. reference), $E_L \equiv V_L/e$, where V_L is the voltage drop across an external load attached to the cell, and e is the electronic charge, V is the applied bias (vs. reference), and V_{FB} is the flat-band potential (vs. reference). Equation (31), a relation applicable for the PEC cell under load, includes the Peltier heat for the back Ohmic contact; Eq. (32) does not include the Peltier heat, and is generally valid only for low-resistivity electrodes where the semiconductor Fermi level is close to the conduction-band edge. Equation (32) becomes valid for high-resistivity (more intrinsic) crystals by adding a term $KN\eta_Q(E_{CB}^{\mathrm{bulk}} - E_F)$, where E_F is the semiconductor Fermi energy. v) The enthalpy change of the redox reaction (expressed as a change in entropy), the so-called electrochemical Peltier heat (EPH), gives

$$\mathrm{PDS}_5 = KN\eta_Q T\Delta S, \tag{33}$$

where T is the absolute temperature, and ΔS is the change in system entropy.

When the five partial PDS signals are summed up, the total PDS signal for a PEC cell under bias can be written as:

$$\begin{aligned}\mathrm{PDS}(V) = KN\big[(\hbar\omega - E_g) + (1 - \eta_Q)E_g + \eta_Q(E_R - E_{VB}) \\ +\eta_Q(V - V_{FB}) + \eta_Q T\Delta S\big].\end{aligned} \tag{34}$$

If the cell is at open circuit, the quantum efficiency, η_Q, is zero. Thus, the PDS signal is:

$$\mathrm{PDS}_{OC} = KN\hbar\omega. \tag{35}$$

The open-circuit PDS signal provides a convenient means for normalizing the PDS signal under bias, because under open-circuit conditions all of the absorbed optical energy is converted to heat at the electrode–electrolyte interface. The concept of the internal quantum efficiency stems from the fact that the PDS signal under bias is normalized by the open-circuit PDS signal. Since neither signal is sensitive to the absolute number of photons incident on the electrode, but only to absorbed photons, PDS measurements can only yield an internal quantum efficiency. Taking into account that K and N do not depend on whether the system is at open circuit or under an applied bias, Eqs. (34) and (35) yield an expression for the normalized PDS signal under bias:

$$\mathrm{PDS}(V)/\mathrm{PDS_{OC}} = \big[(\hbar\omega - E_g) + (1-\eta_Q)E_g + \eta_Q(E_R - E_{VB}) + \eta_Q e(V - V_{FB}) + \eta_Q T\Delta S\big]/\hbar\omega. \tag{36}$$

Considering that the cell photocurrent and, hence, quantum efficiency are independent of bias for large reverse biases, we note that Eq. (36) becomes independent of η_Q for large reverse biases. Therefore, from the slope of that expression:

$$\partial[\mathrm{PDS}(V)/\mathrm{PDS_{OC}}]/\partial(V - V_{FB}) = e\eta_Q^{max}/\hbar\omega, \tag{37}$$

where η_Q^{max} is defined as the maximum internal efficiency of the PEC cell under given experimental conditions. The resistive (Joule) heating of the electrode and electrolyte were ignored in Eq. (36) for a low-resistivity crystal and nondilute electrolyte (Maeda, Fujishima, and Honda, 1982; Wagner and Mandelis, 1987). Equation (36), when combined with experimentally determined values for $(E_R - E_{VB}) + T\Delta S$ and $\eta_Q(V)$, yields results such as that of Fig. 13. Very good agreement has been found between the model of Wagner and Mandelis (1987) and experiment in all cases, as well as with data such as those of Fujishima, Maeda, and Honda (1980c) obtained via PTS, overcoming the shortcomings of the straight line approximation of Eq. (25) in the theory by Fujishima *et al.* (1980a). Maeda, Fujishima, and Honda (1982) used the nonradiative de-excitation mechanisms leading to Eq. (34) above to study in detail the energy efficiency of a ZnO photoelectrode using PTS detection. Their results are shown in Table II for four different electrolytes: 0.5 M Na_2S, 0.5 M KBr in 0.2 M Na_2SO_4, 0.5 M $Na_2S_2O_3$, and 0.5 M H_2Q, respectively. In that Table, Q is the value of the intercept of the $E(\Delta T/\Delta T_{OC})It$ vs. $(V - V_{FB})$ curve at $V = V_{FB}$, ϕ_T is the energy conversion efficiency from the PTS data, and ϕ_V is the ideal maximum conversion efficiency; E_R is as in Eq. (30). It is interesting to note that Roger *et al.* (1985) used a combination of PDS and PCS to obtain η_Q spectra for a $CuInSe_2$ thin film on Mo/Al_2O_3 and for a $CuInSe_2$ crystal in the acetonitrile-based redox electrolyte CAT at −0.4 V and −0.5 V vs. SCE, respectively: PDS gave absorptance (αl) spectra via Eq. (15), whereas

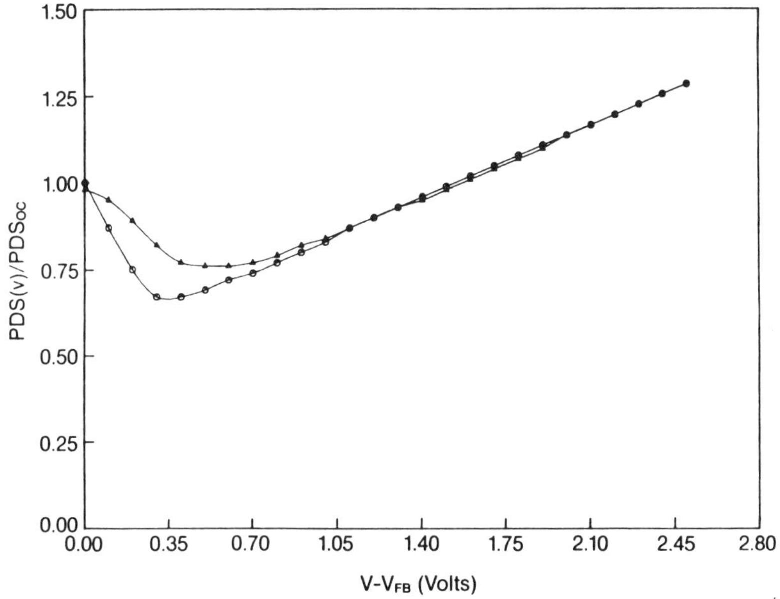

FIG. 13. Theoretical (△) and experimental (○) PDS(V)/PDS$_{OC}$ vs. ($V - V_{FB}$) curves for the n-CdS, S_2^{2-}/S^{2-} system; 505 nm excitation; experimental curve after Wagner, Wong, and Mandelis (1986). (From Wagner and Mandelis, 1987.)

the ratio of the photocurrent to the number of incident photons yielded η_Q, so that the (αl) vs. $\eta_Q(\lambda)$ curves could be compared. This method, however, would give accurate values for $\eta_Q(\lambda)$ only if the optical source is very well calibrated. Otherwise, the relative PDS measurements of Wagner and Mandelis (1987) could prove simpler and more reliable.

Photothermal investigations aiming at improved energy conversion for semiconducting electrodes of PEC cells via sensitization of the semiconductor have recently emerged and are gaining popularity. The main reasons for the need for such applications include the lack of, or difficulty associated with, other conventional spectroscopic techniques in the investigation of the largely opaque electrochemical interface with sensitized electrodes, as well as the powder nature of many semiconductors deposited as electrodes, which excludes other spectroscopic measurement possibilities.

The purpose of electrode sensitization is usually the change of the absorption characteristics of the semiconductor in order to maintain its stability against dissolution in the electrolyte, which is often the case with wide band-gap materials such as TiO_2, while improving its quantum efficiency through the increase in the density of photon-absorbing centers in

TABLE II[a]

ENERGY EFFICIENCIES FOR THE PHOTOOXIDATION OF EACH REDUCING AGENT ON ZnO POLYCRYSTAL ELECTRODE

Redox agent	(E_R vs. SCE)/V	Q/eV	ϕ_T/%	ϕ_V/%
S^{2-}	−.072	3.50	2.8	7.9
SO_3^{2-}	−0.22	3.25	10.8	10.7
$S_2O_3^{2-}$	−0.16	3.15	12.5	12.2
H_2Q	+0.08	2.85	20.8	16.1

[a] Maeda, Fujishima, and Honda (1982)

$$\phi_T = \frac{3.6 - Q}{3.6} \times 100(\%),\ \phi_V = \frac{|E_{FB} - E_R|}{3.6} \times 100(\%)$$

the matrix. Organic compounds have been used as sensitizers, such as dyes and porphyrins. These compounds, when in contact with the semiconductor surface, are capable of inducing a great variety of electronic transitions as a result of strong absorption in the visible range. Merle, Cherqaoui, and Gianotti (1983) used microphonic PAS for the investigation of out-of-cell electrochemically active electrodes consisting of sensitized semiconductor oxides (e.g., Al_2O_3 and Ta_2O_5) and insulators by adsorption of cobalt tetraphenyl porphyrin ($Co^{II}TPP$). In order to study electron exchanges that such electrodes can undergo in the presence of an electrolyte, the oxidation state of the adsorbed porphyrin was determined from the PA spectra, which were superpositions of the porphyrin and substrate absorptions. The results showed that porphyrin on SiO_2 and ZrO_2 undergoes photo-oxidation as $Co^{II}TPP$ gets transformed into $(CO^{III}TPP)^+$ upon excitation with $\lambda > 400$ nm. Al_2O_3 and Ta_2O_5 did not show evidence of photo-oxidation, whereas SnO_2, Nb_2O_5, TiO_2, ZnO, and $SrTiO_3$ exhibited both types of spectra. From their PA spectra, these workers were able to relate the band structure of the semiconductors (energy of the conduction-band edge) to the oxidation potential of the porphyrin, as shown in Fig. 14. The proposed mechanism in agreement with Fig. 14, allowing for electron transfer between the porphyrin and the semiconductor conduction band, postulated that the oxidation potential of the cobalt porphyrin in the adsorbed state ought to be higher than that in solution, thus excluding the commonly advanced isoenergetic electron-transfer interpretation. Merle, Cherqaoui, and Gianotti (1983) further used PDS for *in situ* observations of the evolution of the absorption spectrum of a sensitized electrochemical interface consisting of tetraphenyl porphyrin ($TTPH_2$) in contact with silica gel powder and an acid electrolyte. Drastic change in the absorption spectrum of the interface resulting in spectral broadening (sensitization) around 660 nm was observed upon adsorption of $TTPH_2$ on the SiO_2 gel. This change in the presence of the

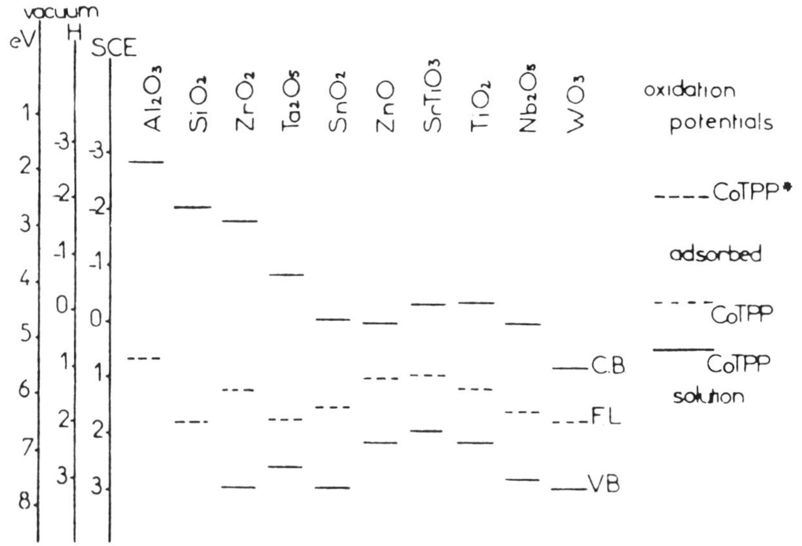

FIG. 14. Conduction-band-edge energies of several sensitized semiconductor oxide electrodes shown relative to the first oxidation potential of $Co^{II}TPP$ measured in 0.1 M *n*-Bu_4NBF_4/benzonitrile solution and corrected for the electrolyte effect at 5.27 eV vs. vacuum (0.2 V vs. SCE). (From Merle, Cherqaoui, and Gianotti, 1983.)

acidic electrolyte was interpreted as the result of protonation of the porphyrin by the acidic groups of the silica gel leading to the species H_4TPP^{2+}. This type of *in situ* monitoring of electron exchange at the electrochemical interface shows great promise in the study of the solid-state physics of the sensitized electrode and is of sound technological value at the PEC device level.

Other compounds used as sensitizers include organic dyes, the photoelectrochemical properties of which have been extensively studied photoacoustically by Iwasaki's group. The postirradiation microphonic PA spectra of ZnO powder sensitized by rhodamine B and 1,1′-diethyl-2,2′-cyanine chloride were obtained as a function of irradiation time with photons of $\lambda > 500$ nm (Iwasaki *et al.*, 1980). A decrease in the latter spectral intensity with time was interpreted as an indication of drastic decomposition of the dye chromophores following irradiation. A similar decrease of the rhodamine B–sensitized ZnO spectrum, accompanied by a blue shift of the spectral peak from 565 to 510 nm (total irradiation time: 1030 s), was attributed to the efficient N-deethylation of the dye. From the combined results, the observed photochemical activity supported the injection of a dye electron in an excited state into the conduction band of the semiconducting

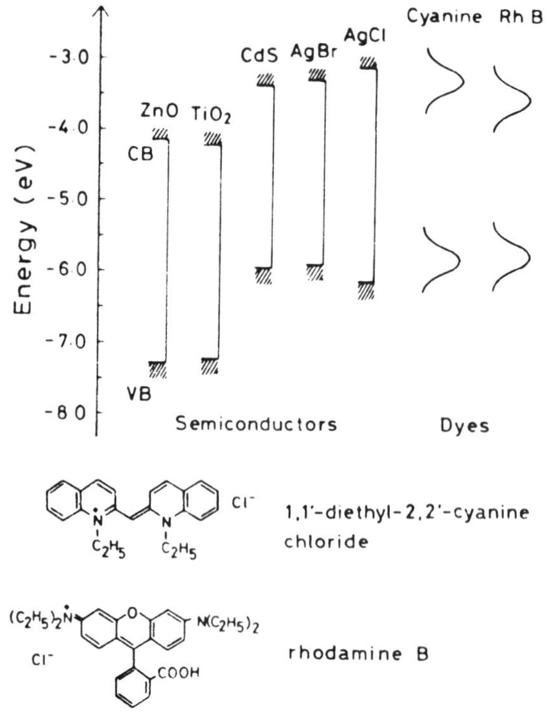

FIG. 15. Energy structure of the sensitizing dyes and correlation between energy level of the dyes and that of the conduction and valence band edges of semiconductor electrodes. (From Iwasaki *et al.*, 1981.)

electrode. The injected electron would subsequently be removed by interaction with and trapping by oxygen to form O_2^- to explain the oxidative N-dealkylation. This assignation was further supported by direct ESR evidence of the presence of the O_2^- species at $g = 2.002$ (Iwasaki *et al.*, 1981). Further experiments were performed with dye-sensitized silver halide crystals and other semiconductors with photochemical changes analogous to those of ZnO. No change was observed when nonsemiconducting solids were used (e.g., Al_2O_3) (Iwasaki *et al.*, 1981). From the energy-level structures of several dyes and their relative positions with respect to semiconductor electrode band structures (Fig. 15), these authors attributed the photosensitivity to a mechanism similar to the one advanced for ZnO, with the electron-injecting dye in the singlet-excited state. The PA technique was also able to provide spectroscopic evidence of the presence of transient intermediates of eosine adsorbed on ZnO powder during irradiation with photons of wavelength $\lambda > 500$ nm. The peak of the microphonic

PA spectrum of the intermediate species coincided with the cation radical of eosine. This identification, finally, helped complete the dye–semiconductor interface electron-transfer picture: Upon injection of an electron by the excited dye molecule into the semiconductor, the cation radical concentration would increase. As a result, a decrease in the dye concentration at the surface of the electrode and at the semiconductor–dye solution interface would be expected, following the sensitization (injection) process. This mechanism proved to be consistent with experimental observations of an SnO_2 PEC electrode–sodium fluorescein dye solution interface *in situ*, using combined PAS and fluorescence spectroscopy (Iwasaki *et al.*, 1979). A 2200-Å SnO_2 thin film layered on quartz was used as the WE in the PEC cell with a distilled water/10^{-4} M dye/0.1 N KNO_3 electrolyte, a Pt wire CE and Ag/AgCl RE. Sodium fluorescein was chosen for its high fluorescence efficiency. SnO_2 ($E_g = 3.7$ eV) was essentially transparent at the wavelength of the exciting radiation (Ar^+ laser, $\lambda = 488$-nm line), and *in situ* piezoelectric PA signals were detected simultaneously with photocurrent and fluorescence spectra. A decrease in the photocurrent above +0.5 V vs. Ag/AgCl was observed, corresponding to an increase in the dark current. This is expected from a decrease in the effective dye concentration at the electrode–solution interface via the electrolytic oxidation of the dye, in accordance with the proposed mechanism. The (assumed) singlet state of the excited dye was directly observed by fluorescence measurements at the electrochemical interface and its lifetime measured (6.9 ns) using a N_2 flash lamp. The proposed mechanism was further corroborated by the decrease in the intensity of fluorescence at the electrode–solution interface with increasing bias, since the intensity was restored after addition of 10^{-2} M hydroquinone, which is known to regenerate the dye immediately. The PA data also gave unequivocal direct evidence of the dye concentration decrease mechanism, as the signal due to direct absorption by the dye decreased reversibly when the applied voltage increased, whereas the PA signal's potential dependence disappeared upon addition of the hydroquinone.

It can thus be seen that photothermal techniques have contributed greatly toward the enhancement of our fundamental understanding of complex energy-conversion processes associated with electronic transfer and de-excitation mechanisms of photo-excited sensitized semiconductor electrode–electrolyte interfaces. The physics and chemistry of these processes is intimately related to the band structure and band-to-band electronic transition dynamics of particular semiconductor electrodes. It appears that the imaginative stress-modulated PEC technique, a kind of simultaneous optic-acoustic effect method (as opposed to the conventional serial piezoelectric optically-generated acoustic PAS), introduced by Handley, McCann, and Haneman (1982) has a strong potential for very accurate measurements of interband transition energies and photoelectronic processes and is ideally suited for *in situ* studies at the photoelectrochemical interface. The technique depends on the piezoelectric generation of a modulated strain in an optically excited semiconductor electrode, which causes modulated changes in the interband transition threshold energies detectable by PEC means. This method could be implemented directly, for instance, to monitor the excited-state dynamics of electronic transfer across the electrochemical interface when polycrystalline or amorphous semiconducting electrodes are used, having only short-range order and, therefore, exhibiting nonuniform band gaps.

VIII. Conclusions and Future Directions

This chapter has presented a comprehensive, albeit not totally encompassing, overview of the state of photothermal activity in the rich field of electrochemical systems. The many growing types of applications at all levels of system complexity, from simple qualitative spectroscopic characterizations of "dry"-electrode substrates to the study of fundamental electrochemical interface physics of working PEC cell devices *in situ*, point toward an exciting future for the photothermal sciences in electrochemistry. In retrospect, one of the major goals of this chapter was to show the *necessity* of the utilization of the various photothermal techniques in electrochemistry, where other established methods are unable to respond to the ever-increasing requirements for *in situ*, nondestructive, even noncontact characterization at operating cell levels.

As the sophistication of electrode materials increases (and new, more stringent technological requirements of the electrochemical interface arise), the future trend appears to be directed toward those above-mentioned diagnostic requirements at the operating cell level. Thus, it is very likely that photothermal techniques that can combine versatility, simplicity of implementation, high sensitivity, fast response, ruggedness, and tolerance of the electrochemical environment will become increasingly more popular in electrochemistry. Such techniques include, but are not limited to, PDS and the emerging P^2ES. The new semiconductor technologies of compound direct-gap materials (e.g., GaAs, CdS), thin amorphous films (e.g., *a*-Si:H), and quantum-well heterostructures (e.g., $Ga_xAl_{1-x}As$) offer challenging new electrochemical interfaces. As far as their physicochemical characterization is concerned, it remains an exciting prospect to see *how* photothermal science will rise to meet this challenge.

REFERENCES

Boccara, A. C., Fournier, D., and Badoz, J. (1980). *Appl. Phys. Lett.* **36**, 130.

Brilmyer, G. H. and Bard, A. J. (1980). *Anal. Chem.* **52**, 685.

Brilmyer, G. H., Fujishima, A., Santhanam, K. S. V., and Bard, A. J. (1977). *Anal. Chem.* **49**, 2057.

Childers, J. W., Crumbliss, A. L., Lugg, P. S., Palmer, R. A., Morosoff, N., and Patel, D. L. (1983). *J. Phys. Colloq.* C6, **44**, 285.

Coufal, H. (1986). *IEEE Trans. UFFC, UFFC-***33**, 507.

Coufal, H., and Mandelis, A. (1987). In *Photoacoustic and Thermal Wave Phenomena in Semiconductors* (A. Mandelis, ed.), Ch. 7. North-Holland, New York.

Decker, F., Juliao, J. F., and Abramovich, M. (1979). *Appl. Phys. Lett.* **35**, 397.

Dohrmann, J. K., and Sander, U. (1986). *Ber. Bunsenges. Phys. Chem.* **90**, 605.

Dohrmann, J. K., Sander, U., and Strehblow, H.-H. (1983). *Z. Phys. Chem. Neue Folge*, **134**, 41.

Dorthe-Merle, A.-M., Morand, J. P., and Maurin, E. (1985). In *Technical Digest 4th International Topical Meeting on Photoacoustic, Thermal and Related Sciences*, p. MD5.1. Ville d'Estérel, Québec.

Dovichi, N. J., Nolan, T. G., and Weimer, W. A. (1984). *Anal. Chem.* **56**, 1700.

Dunn, W. W., Aikawa, Y., and Bard, A. J. (1981). *J. Electrochem. Soc.* **128**, 222.

Ellis, A. B., Kaiser, S. W., and Wrighton, M. S. (1976). *J. Am. Chem. Soc.* **98**, 6855.

Fujihira, M., Osa, T., Hursh, D., and Kuwana, T. (1978). *J. Electroanal. Chem.* **88**, 285.

Fujishima, A., Brilmyer, G. H., and Bard, A. J. (1977). In *Semiconductor Liquid-Junction Solar Cells* (A. Heller, ed.), p. 172, Electrochem. Soc. (*Proc.* 77-3). Princeton, N.J.

Fujishima, A., Masuda, H., and Honda, K. (1979). *Chem. Lett.* 1063.

Fujishima, A., Maeda, Y., Honda, K., Brilmyer, G. H., and Bard, A. J. (1980a). *J. Electrochem. Soc.* **127**, 840.

Fujishima, A., Masuda, H., Honda, K., and Bard, A. J. (1980b). *Anal. Chem.* **52**, 682.

Fujishima, A., Maeda, Y., and Honda, K. (1980c). *Bull. Chem. Soc. Japan* **53**, 2735.

Gardella, J. A. Jr., Jiang, D.-Z., McKenna, W. P., and Erying, E. M. (1983). *Appl. Surf. Sci.* **15**, 36.

Gautron, J., Lemasson, P., Rabaga, F., and Triboulet, R. (1979). J. Electrochem. Soc. **126**, 1868.

Gray, R. C., and Bard, A. J. (1978). *Anal. Chem.* **50**, 1262.

Handley, L. J., McCann, J. F., and Haneman, D. (1982). *J. Appl. Phys.* **53**, 4549.

Iwasaki, T., Sawada, T., Kamada, H., Fujishima, A., and Honda, K. (1979). *J. Phys. Chem.* **83**, 2142.

Iwasaki, T., Oda, S., Kamada, H., and Honda, K. (1980). *J. Phys. Chem.* **84**, 1060.

Iwasaki, T., Oda, S., Sawada, T., and Honda, K. (1981). *Photogr. Sci. Eng.* **25**, 6.

Kanstad, S. O., and Nordal, P.-E. (1978). *Opt. Commun.* **26**, 367.

Kraeutler, B., and Bard, A. J. (1978). *J. Am. Chem. Soc.* **100**, 2239.

Kusu, F., Hoshino, H., Takamura, K. and Sawada, T. (1983–84). *J. Photoacoust.* **1**, 463.

McIntyre, J. D. E., and Aspnes, D. E. (1971). *Surf. Sci.* **28**, 417.

Maeda, Y., Fujishima, A., and Honda, K. (1982). *Bull. Chem. Soc. Japan* **55**, 3373.

Malpas, R. E., and Bard, A. J. (1980). *Anal. Chem.* **52**, 109.

Malpas, R. E., Fredlein, R. A., and Bard, A. J. (1979a). *J. Electroanal. Chem.* **98**, 171.

Malpas, R. E., Fredlein, R. A., and Bard, A. J. (1979b). *J. Electroanal. Chem.* **98**, 339.

Mandelis, A. (1982). *Chem. Phys. Lett.* **91**, 501.

Mandelis, A. (1983). *J. Appl. Phys.* **54**, 3404.

Mandelis, A. (1984). *Chem. Phys. Lett.* **108**, 388.
Mandelis, A., and Royce, B. S. H. (1984). *Appl. Opt.* **23**, 2892.
Mandelis, A., and Zver, M. M. (1985). *J. Appl. Phys.* **57**, 4421.
Mandelis, A., Siu, E., and Ho, S. (1984). *Appl. Phys. A* **33**, 153.
Masuda, H., Fujishima, A., and Honda, K. (1980). *Bull. Chem. Soc. Japan* **53**, 1542.
Masuda, H., Morishita, S., Fujishima, A., and Honda, K. (1981). *J. Electroanal. Chem.* **121**, 363.
Masuda, H., Fujishima, A., and Honda, K. (1982a). *Bull. Chem. Soc. Japan* **55**, 672.
Masuda, H., Fujishima, A., and Honda, K. (1982b). *Chem. Lett.* 1153.
Mendoza-Alvarez, J. G., Royce, B. S. H., Sanchez-Sinencio, F., Zelaya-Angel, O., Menezes, C., and Triboulet, R. (1983). *Thin Solid Films* **102**, 259.
Merle, A. M., Cherqaoui, A., and Gianotti, C. (1983). *J. Phys. Colloq.* C6, **44**, 291.
Morishita, S., Fujishima, A., and Honda, K. (1983). *J. Electroanal. Chem.* **143**, 433.
Morrison, S. R. (1980). *Electrochemistry at Semiconductor and Oxidized Metal Electrodes*. Plenum Press, New York.
Nordal, P.-E., and Kanstad, S. O. (1978). *Opt. Commun.* **24**, 95.
Oda, S., Sawada, T., and Kamada, H. (1978). *Anal. Chem.* **50**, 865.
Ogura, K., Fujishima, A., and Honda, K. (1984). *J. Electrochem. Soc.* **131**, 344.
Palmer, R. A., and Smith, M. J. (1986). *Can. J. Phys.* **64**, 1086.
Pawliszyn, J., Weber, M. F., and Dignam, M. J. (1985). *Rev. Sci. Instrum.* **56**, 1740.
Pawliszyn, J., Weber, M. F., Dignam, M. J., Mandelis, A., Venter, R. D., and Park, S.-M. (1986a). *Anal. Chem.* **58**, 236.
Pawliszyn, J., Weber, M. F., Dignam, M. J., Mandelis, A., Venter, R. D., and Park, S.-M. (1986b). *Anal. Chem.* **58**, 239.
Roger, J. P., Fournier, D., and Boccara, A. C. (1983). *J. Phys. Colloq.* C6, **44**, 313.
Roger, J. P., Fournier, D., Boccara, A. C., Noufi, R., and Cahen, D. (1985). *Thin Solid Films* **128**, 11.
Rosencwaig, A. (1980). *Photoacoustics and Photoacoustic Spectroscopy*. Wiley, New York.
Royce, B. S. H., Sanchez-Sincencio, F., Goldstein, R., Muratore, R., Williams, R., and Yim, W. M. (1982). *J. Electrochem. Soc.* **129**, 2393.
Royce, B. S. H., Voss, D., and Bocarsly, A. (1983). *J. Phys. Colloq.* C6, **44**, 325.
Sander, U., Strehblow, H.-H., and Dohrmann, J. K. (1981). *J. Phys. Chem.* **85**, 447.
Sawada, T., and Bard, A. J. (1982–83). *J. Photoacoust.* **1**, 317.
Takaue, R., Matsunaga, M., and Hosokawa, K. (1983). *Denki Kagaku*, **51**, 153 (in Japanese).
Tamor, M. A., and Hetrick, R. E. (1985). *Appl. Phys. Lett.* **46**, 460.
Tom, R., Moore, T. A., Lin, S. H., and Benin, D. (1979). *Chem. Phys. Lett.* **66**, 390.
Wagner, R. E., and Mandelis, A. (1987). In *Photoacoustic and Thermal Wave Phenomena in Semiconductors* (A. Mandelis, ed.), Ch. 14. North-Holland, New York.
Wagner, R. E., Wong, V. K. T., and Mandelis, A. (1986). *Analyst* **111**, 299.
Wun, M., Milgaten, D. J., and Hwang, W. C. (1981). In *Technical Digest. Second International Meeting on Photoacoustic Spectroscopy*, p. ThA7-1. Univ. of California, Berkeley, CA.
Wun-Fogle, M., Milgaten, D. J., and Hwang, W. C. (1982). *Appl. Opt.* **21**, 121.
Yamagishi, G., Moritani, A., and Nakai, J. (1980). *Japan J. Appl. Phys.* **19**, L711.
Yamagishi, G., Moritani, A., and Nakai, J. (1981). *Japan J. Appl. Phys.* **20**, 1423.

CHAPTER 10

PHOTOTHERMAL SPECTROSCOPY OF AEROSOLS

A. J. CAMPILLO
H.-B. LIN
Optical Sciences Division
Naval Research Laboratory
Washington, D.C.

I. Introduction . . . 309
A. Aerosols and Spectroscopy . . . 309
B. Aerosol Optics . . . 310
C. Chapter Goals . . . 314
II. Survey of Possible Photothermal Schemes . . . 315
A. Single-Particle Schemes . . . 315
B. Heat Conduction Equation for Sphere and Surrounding Gas . . . 316
III. Early Work . . . 320
IV. Photothermal Interferometry . . . 321
A. Detection of Extremely Weak Absorption in the Gas Phase . . . 321
B. Aerosol-Cloud Studies . . . 324
C. Single-Particle Studies . . . 325
V. Photothermal Modulation of Light Scattering . . . 329
A. Photothermal Modulation of Elastic Scattering . . . 329
B. Photothermal Modulation of Inelastic Scattering . . . 333
VI. Photophoresis . . . 334
VII. Conclusion . . . 340
References . . . 341

I. Introduction

A. Aerosols and Spectroscopy

Aerosols are dispersions of particulate matter suspended in the gas phase (Fuchs, 1964; Hidy and Brock, 1970; Friedlander, 1977). They range in number density from single isolated particles to dense systems (over 10^{12} cm^{-3}) and have characteristic dimensions extending from the macroscopic (ca. mm) down to the molecular level (ca. 1 nm). Gaseous molecules are generally self-attractive due to van der Waals forces and have a tendency to cluster to form small particles that may subsequently undergo

ISBN 0-12-636345-5

further agglomeration. It is quite common for a dynamically stable size distribution (Hidy and Brock, 1970) to develop as larger particles are lost through precipitation or evaporation and as new particles are formed and grow. The chemistry and physics of the resultant particles may differ dramatically from the original carrier gas or, indeed, even the bulk condensed phase. As one of the most common conditions of matter, aerosols are of importance in environmental science, meteorology, astronomy, combustion chemistry, surface science, and optics, as well as in many practical problems of technological significance.

The absorption properties of aerosols, in particular, are important in a number of applications; these include high-power laser propagation through the atmosphere, solar heating and its effects on global temperature, and particulate chemical analysis. This last application has only recently begun to be realized through the development of photothermal spectroscopies. Because there are many aerosols where volatile and reactive species are in delicate equilibrium with the surrounding gas, it would be desirable to employ *in situ* methods of analysis, such as absorption spectroscopy, which do not require premeasurement treatment such as filter collection. Several problems have, unfortunately, made particle absorption spectroscopy difficult to implement previously. First, direct transmission measurements are complicated by particle light scattering, which often obscures the minimal light absorption, and, second, background gases, such as water vapor, have wing absorptions that often dwarf the particulate contributions. As a result, absorption spectroscopy of aerosols in smog-chamber studies, for example, is usually limited to the gaseous components. However, several photothermal spectroscopies are available and others can be visualized that will allow particle-specific absorption to be measured independent of the scattering and gaseous terms. Before discussing these, it is important to understand some of the optical peculiarities of small particles, many of which may, at first glance, appear to be counterintuitive.

B. Aerosol Optics

In 1908, Gustav Mie formulated a set of light-scattering equations to account for particulate optical effects such as the variation of brilliant colors of colloidally dispersed metallic particles with size. Mie theory consists of a rigorous solution of Maxwell's equations for scattering and absorption of electromagnetic waves by a dielectric sphere. These two loss mechanisms may be related to a total loss or extinction factor and written in terms of cross-sections as

$$C_{\text{ext}} = C_{\text{sca}} + C_{\text{abs}}, \tag{1}$$

where C_{ext}, C_{sca} and C_{abs} are the extinction, scattered, and absorption cross-sections, respectively. In the case of a cloud of identical particles having density, D particles/cm^3, the ratio of transmitted to incident light intensity is given by $\exp(-DLC_{\text{ext}})$ and L is the sample length. It is customary in the literature (Kerker, 1969; Pendorf, 1958; Bohren and Huffman, 1983) to introduce a parameter called the cross-section efficiency, Q, which is defined as the cross-section per unit area; for spherical droplets, Q is $C/\pi a^2$. Here a is the droplet radius. A term called the size parameter is also often used and is $x = ka = 2\pi a/\lambda$. The cross-section efficiencies may be evaluated by integrating the time-averaged Poynting vector found from Mie's solution. Kerker (1969) has evaluated the integral and writes the following results:

$$Q_{\text{sca}} = \frac{2}{x^2}\sum_{n=1}^{\infty}(2n+1)\left(|a_n|^2 + |b_n|^2\right), \tag{2}$$

$$Q_{\text{ext}} = \frac{2}{x^2}\sum_{n=1}^{\infty}(2n+1)\left[\text{Re}(a_n + b_n)\right], \tag{3}$$

$$Q_{\text{abs}} = Q_{\text{ext}} - Q_{\text{sca}}. \tag{4}$$

The functions a_n and b_n are given by

$$a_n = \frac{\Psi_n(x)\Psi_n'(mx) - m\Psi_n(mx)\Psi_n'(x)}{\xi_n(x)\Psi_n'(mx) - m\Psi_n(mx)\xi_n'(x)}, \tag{5a}$$

$$b_n = \frac{m\Psi_n(x)\Psi_n'(mx) - \Psi_n(mx)\Psi_n'(x)}{m\xi_n(x)\Psi_n'(mx) - \Psi_n(mx)\xi_n'(x)}, \tag{5b}$$

where $\Psi_n(x)$ is the Ricatti–Bessel function of order n, $\xi_n(x)$ is the Hankel function of the first kind of order n, and the primes represent derivatives. The quantity m is the ratio of the particle complex index of refraction to the index of the surrounding medium. The a_n and b_n correspond to TM and TE modes, respectively. The behavior of Q_{sca} and consequently Q_{ext} show interesting variations as a function of x, especially near $x = 1$.

As x and mx become small, one may make significant simplifications in Eqs. (2)–(4) by expanding a_n and b_n in series and taking leading terms. Pendorf (1958) carried out this expansion to the tenth order of x. From this work, one may take the lowest-order terms in x, yielding

$$Q_{\text{sca}} = \frac{8}{3}x^4\left|\frac{m^2 - 1}{m^2 + 2}\right|^2, \tag{6}$$

$$Q_{\text{abs}} = \text{Im}\left[-4x\left(\frac{m^2 - 1}{m^2 + 2}\right)\right]. \tag{7}$$

These efficiency factors are identical to the Rayleigh theory, which is based on the assumption of small x. From Eq. (6), we see that scattering varies as size of the particle to the sixth power and as wavelength to the fourth power. The size dependence results in larger particles scattering disproportionately more per unit mass than gases or smaller particles. This sixth-power dependence breaks down as x approaches 1 and Q asymptotically approaches 2 for $x \gg 1$. The wavelength dependence results in the apparent blue color of sky and ocean. For a given x and m, Eq. (7) is simply a constant times the particle size. Thus, the absorption for particles in the Rayleigh approximation is proportional to the mass of the particle. This is strictly true for spherical particles and generally true for those of arbitrary shape. There are exceptions for particles of dimensions equal to natural resonances or probed by light at resonant frequencies, e.g., plasmon or polariton.

As the size parameter approaches 1, many interesting features appear in the scattering and extinction efficiencies. In the Rayleigh regime, the angular distribution of scattered light is nearly evenly distributed over all angles. However, as a increases, a significant amount of light is diffracted into a small-angle frontal lobe. The larger particles also scatter a significant amount of light via specular reflection off the back surface into the backward direction. As $x \gg 1$, the two terms result in Q approaching 2. This surprised early researchers, who termed this the "extinction paradox," since large particles were exhibiting a scattering cross-section twice that of their geometrical value. In the range of $0.1 < x < 200$, examination of the angular distribution of the scattering shows that besides the frontal lobe and backscattered light, there exists a complex distribution of lobes and nulls, whose positions and strengths are sensitive functions of the size and the m of the particle. Some of these features can be explained by refracted light rays passing through the sphere glancing off the far surface and eventually reemerging through a third point. These rays interfere with specularly scattered rays and account for rainbow effects in clouds.

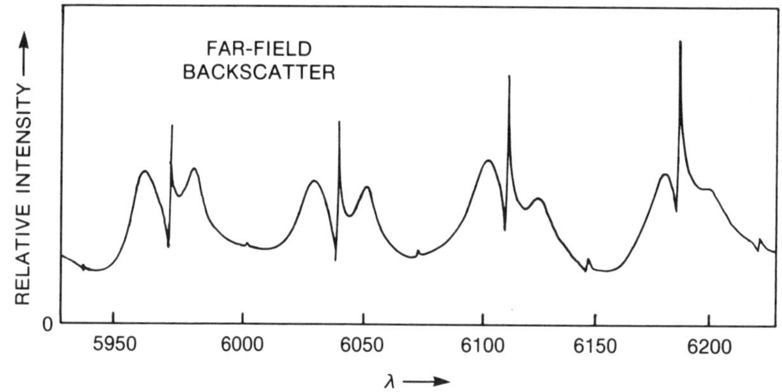

FIG. 1. Wavelength dependence of the far-field backscatter from a single 11.4-μm silicone oil droplet. The sharp features are due to surface-wave resonances. (After Ashkin and Dziedzic, 1981).

The electromagnetic-wave analog to whispering-gallery acoustic waves (Rayleigh, 1914) account for many of the interesting features normally observed in Mie scattering (Mie, 1908; Kerker, 1969; Pendorf, 1958; Bohren and Huffman, 1983). Surface waves formed by grazing-incidence light at the boundary of two materials with dissimilar indices of refraction can, by skimming along the interface, propagate over long distances with little loss. In cylindrical- or spherical-cavity geometries, resonances occur when an integral number of wavelengths equals the circumference. These waves have an evanescent tail extending into the surrounding gas, which weakly spews radiation tangentially to the surface. Interference of this component with light reflected or refracted from the particle accounts for the ripple structure observed in the extinction cross-section (see Fig. 1) and the glory phenomenon of backscattering. The naturally occurring mirrorless cavity present in aerosol droplets may be used to provide the necessary feedback for stimulated processes (Tzeng *et al.*, 1984; Snow, Qian, and Chang, 1985; Qian and Chang, 1986; Lin *et al.*, 1986). At resonance, the fields within the particle are enhanced dramatically. Figure 2, from Chylek, Pendleton, and Pinnick (1985), shows the fields internal to a particle of $x = 40$ and $m = 1.47$, where the highest peaks are enhanced by a factor of about 10^6 over that of the incident plane wave. In the case of weakly absorbing particles, the enhanced fields play an important role in the effective absorption of the particle. Consequently, single-particle absorption can have spectral features where none exist in the bulk.

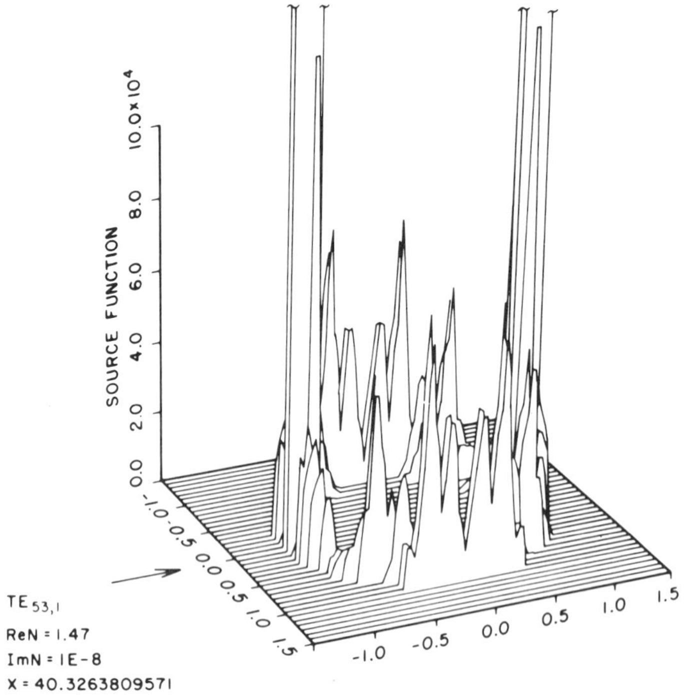

FIG. 2. Internal field distribution (source function) in equatorial plane of a sphere with size parameter $x = 40.33$ and refractive index $m = 1.47$ at the $TE_{53,1}$ mode resonance condition. The source function is enhanced by a factor of 10^6 over that of the incident plane wave in the forward and backward scattering directions. The arrow indicates the direction of incoming (unpolarized) radiation (From Chylek *et al.*, 1985).

C. CHAPTER GOALS

In general, many of the spectroscopies discussed in other chapters of this volume may be used to study aerosol clouds. Often, after a short thermal buildup time, the particulate absorption may be treated as a term that increases the effective "gaseous" absorption. Therein lies a fundamental problem, since often the gaseous absorption is greater than that of the particles and many dominate the signal. In this chapter, we shall emphasize spectroscopies that have the potential to be or are inherently "particle specific," i.e., provide a signal independent of the surrounding gas. Many of these rely in some way on the peculiar heat pattern that develops around the absorbing particle bathed in a radiation field. Consequently, many of these spectroscopies exhibit "single-particle" sensitivity as well.

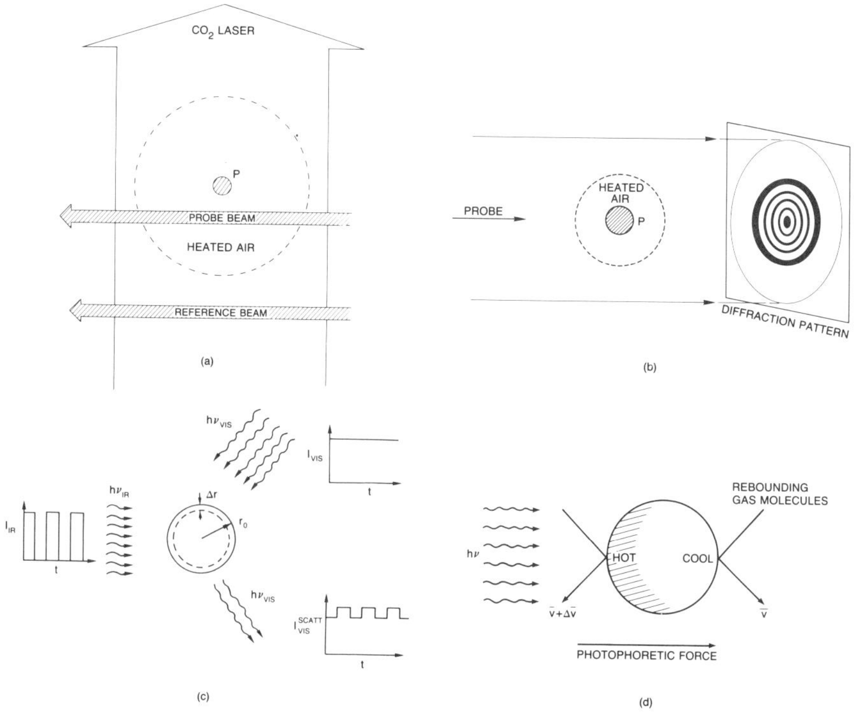

FIG. 3. Survey of some possible single-particle photothermal spectroscopies: (a) interferometry, (b) photothermal Fraunhofer diffraction, (c) photothermal modulation of light scattering and (d) photophoresis (see text for explanation).

II. Survey of Possible Photothermal Schemes

A. Single - Particle Schemes

Possible photothermal schemes rely on probing thermal changes occurring within or in the immediate vicinity of the particle. As the particle absorbs radiation, it heats and begins to conduct energy to the surrounding gas. A bubble of hot gas forms around the particle, expanding radially at a rate $\sqrt{4\kappa t}$. Here κ is the thermal diffusivity of the gas, and t is the time after the incident radiation is turned on. The hot air bubble represents a region of lower gas density and lower index of refraction. Figures 3a and b represent two of the possible schemes that exploit the localized low-index region. In (a), interferometry is employed to probe the index change of this region by comparing the phase difference between a probe beam and a reference beam. In the geometry shown, the gaseous signal is also present in the reference beam and so is effectively subtracted out. In (b), a form of photothermal Fraunhofer diffraction, a variation in the single-particle far-field diffraction (Airy) pattern in the presence of the hot air bubble is observed. Also possible are microscopic analogues of photothermal beam deflection and thermal lensing, which also rely on the localized bubble. In the microscopic cases, the signals should be particle specific by nature. In photothermal modulation of Mie scattering (PMMS), a modulated infrared (IR) beam induces periodic thermal changes in the particle and surrounding region, resulting in a time-varying Q_{sca} component (Fig. 3c). When probed by a second visible CW light source, the scattered light also contains a time-varying component reflecting the magnitude of C_{abs}. The spectroscopy in (d), called photophoresis, is based on a radiometric force resulting from interaction of an asymmetrically heated particle surface and the surrounding fluid. Spectroscopies shown in (a), (c), and (d) have been studied in some detail, and have been applied to particle spectroscopy, and are described further in the following sections. Recently, Lin and Campillo (unpublished) have observed (b) but have not determined the limits of this spectroscopy. The above survey is not exhaustive but includes promising schemes proposed by various authors.

B. Heat Conduction Equation for Sphere and Surrounding Gas

In order to quantify some of the spectroscopies outlined above, we discuss in this section some aspects of the theory of heat conduction from a particle. The heat transfer between a particle of radius a, heat capacity C_p, density ρ_p, and the surrounding air characterized by a thermal conductivity k_a, heat capacity C_a, density ρ_a is governed by the heat equation (Carslaw and Jaeger, 1959)

$$\frac{\partial T(r,t)}{\partial t} = \kappa \nabla^2 T(r,t), \tag{8}$$

where $\kappa = k_a/\rho_a C_a$ is the thermal diffusivity, and T is the temperature, r is the distance from particle center, and t is time. Use of this equation implies continuum fluid behavior of the air for dimensions of physical interest. Departure from Eq. (8) occurs as the mean free path of the gas becomes comparable to particle dimensions and/or as the temperature of the particle increases above ambient. Losses due to radiation are normally negligible if the temperature rise in the particle is kept small ($< 1°C$). However, as the particle size decreases or if surrounding gas pressure decreases, heat transfer by radiation begins to play a role. For the case in which a is of the same order of magnitude as the mean free path, a correction must be made to the solution obtained from Eq. (8). In this transition between free molecular and continuum regimes, energy transport can be partly described by a purely kinetic equation and partly by a diffusion equation. The difficulties in solving the Boltzmann equation, particularly when a linearized collision term (Brock, 1982) is not sufficient to describe the transition regime, have lead researchers to apply semiempirical interpolations. Wagner (1982) has reviewed interpolations of the solution of the heat equation for the transition regime. The procedures result in a modification, k_a^*, of the thermal conductivity k_a and hence of κ, the diffusivity:

$$k_a^* = \beta k_a, \tag{9a}$$

$$\beta = \left[\frac{0.377\,K_{n+1}}{K_{n+1}} + \frac{4K_n}{3\alpha_T}\right], \tag{9b}$$

where K_n is the Knudsen number and is equal to the gas mean free path divided by a, and α_T is the empirical thermal accommodation coefficient (usually set to 1). This correction results in a reduced k_a. In the case of air (mean free path is 0.06 μm), the correction for a 0.06-μm particle is 0.685 and for a 0.01-μm particle is 0.12.

The case of a particle illuminated by a periodically chopped laser beam may be evaluated by solving the heat equation under the appropriate boundary conditions. The particle is first assumed heated by the laser for $P/2$ sec, then the radiation is turned off for $P/2$ sec. This cycle is repeated N times. The general solution to Eq. (8) is spherically symmetric and has the boundary conditions

$$\lim_{r \to \infty} T(r, t) = T_0, \qquad T(a, t) = T_s(t), \tag{10}$$

and the initial condition $T(r, 0) = f(y)$, where $y = r - a$ has been given by Carslaw and Jaeger (1959) to be

$$\begin{aligned} r[T(r, t) - T_0] &= \frac{2a}{\pi} \int\int_0^\infty dy'\, d\alpha\, f(y') \sin \alpha y' \sin \alpha y \exp(-\kappa \alpha^2 t) \\ &\quad + \frac{2a}{\sqrt{\pi}} \int_{\frac{r-a}{\sqrt{4kt}}}^\infty d\mu\, T_s\left[t - \frac{(r-a)^2}{4k\mu^2}\right] \exp(-\mu^2). \end{aligned} \tag{11}$$

We assume that the particle surface temperature is uniform, since particle thermal relaxation times, τ ($\tau = a^2 \rho_p c_p / 3k_a$), are on the order of nanoseconds (Chan, 1975), whereas the time to reach thermal equilibrium by conduction to the surrounding gas for micrometer-sized particles is of the order of milliseconds.

Initially, the particle is assumed in thermal equilibrium with the carrier gas. Thus, at $t = 0$, $f(y) = 0$, $T_s(t)$ is found from conservation of energy as applied by Chan (1975). One sets the power leaving the surface of the particle by thermal conduction plus the time rate of change of the heat content of the particle equal to the power absorbed by the particle. Thus,

$$I_0 Q_{\text{abs}}(a)\pi a^2 = 4\pi a^2 k_{\text{a}} \frac{\partial T(r,t)}{\partial r}\bigg|_{r=a} + \frac{4}{3}\pi a^3 \rho_{\text{p}} C_{\text{p}} \frac{\partial T(r,t)}{\partial T}\bigg|_{r=a}, \quad (12)$$

where I_0 is the heating laser intensity, and $Q_{\text{abs}}(a)$ is the Mie absorption efficiency factor discussed in the previous section. Equation (12) may be rewritten as

$$\Delta T = -a\frac{\partial T}{\partial r}\bigg|_{r=a} + \tau\frac{\partial T}{\partial t}\bigg|_{r=a}, \quad (13)$$

where $\Delta T = I_0 Q_{\text{abs}}(a)a/(4k_{\text{a}})$ is the maximum temperature of the particle above ambient, and τ is the thermal rise time of the surrounding gas. As an example, consider a 0.1-μm particle ($Q_{\text{abs}} = 0.1$) in air at STP and assuming $I_0 = 100$ W/cm^2. In this case, $\Delta T = 0.1$ K and $\tau = 0.13$ μsec. The surface temperature approaches equilibrium as it absorbs radiation as $T_s(t) = \Delta t[1 - \exp(-t/\tau)]$. For practical applications with excitation-chopping frequencies of < 100 Hz, that is, for times of the order of milliseconds, one finds that the particle temperature is essentially a constant ΔT when exposed to the excitation energy.

Fluckiger, Lin, and Marlow (1985) have evaluated Eq. (11) with $f(y) = 0$ and $T = 0$ and derived:

$$T(r,t) = \frac{a\Delta T}{r}\operatorname{erfc}\left(\frac{r-a}{\sqrt{4\kappa\tau}}\right) + \frac{a\Delta T}{2r}\exp\left(-\frac{t}{\tau}\right)\left\{\exp\left[\left[\frac{(r-a)^2}{4\kappa\tau}\right] + 1\right]\right.$$
$$\left.\cdot\left[w\left[1 + i\frac{(r-a)}{2\sqrt{\kappa\tau}}\right] + w\left[-1 + i\frac{(r-a)}{2\sqrt{\kappa\tau}}\right]\right]\right\}, \quad (14)$$

where $\operatorname{erfc}(y)$ is the complementary error function and $w(z) = \exp(-z^2)\operatorname{erfc}(-iz)$. The last term in Eq. (14) represents the thermal transients that die out on a time scale of τ, that is, on a microsecond time scale. If the beam is chopped on a millisecond time scale, the transient may be ignored. The temperature profile around a single particle may then be represented by

$$\Delta T(r, t) = \frac{a\Delta T}{r}\operatorname{erfc}\left(\frac{r-a}{\sqrt{4\kappa t}}\right). \tag{15}$$

The asymptotic r^{-1} behavior apparent in Eq. (15) is consistent with a solution of the heat equation in spherical geometry. If this temperature profile is used as an initial condition to find the temperature during the half-period the heating laser is off, then

$$\Delta T_{\text{off}}(r, t) = \frac{a\Delta T}{r}\left\{\operatorname{erf}\frac{r-a}{\sqrt{4\kappa t}} - \operatorname{erf}\left[\frac{r-a}{\sqrt{4\kappa(t+P/2)}}\right]\right\}. \tag{16}$$

Subsequently, one may deduce the functional form of the temperature profile around a particle as it is exposed to modulated IR radiation for any later half-period. The general form is given (Fluckinger, Lin, and Marlow, 1985) by

$$T(r, t) = \frac{a\Delta T}{r}\sum_{n=1}^{N}(-1)^{n+1}\operatorname{erfc}\left\{\frac{r-a}{\sqrt{4\kappa[t+(N-n)(P/2)]}}\right\}, \tag{17}$$

where N is the number of half-periods, and P is the chopper period.

In deriving Eq. (17) it was necessary to assume that heat transfer was by conduction only. Heated aerosols will also dissipate their energy by radiation, convection, and in the case of volatile droplets, by vaporization (Armstrong, 1984; Baker, 1975; Davis, 1983; Sageev and Seinfeld, 1984). Radiation and convection were discussed earlier. Vaporization removes a significant amount of energy, and this process may represent a potential source of error in photothermal spectroscopy, since a portion of the absorbed radiation does not show up as heat in the surrounding gas. In general, however, this problem arises when dealing with neat liquids such as water, which when exposed to a constant CW heating field will continue to evaporate. In such cases, errors as large as 25% may be realized. Fortunately, the most important applications involving volatile material center around water droplets containing salt (atmospheric chemistry, acid-rain problem, etc.). Unlike the neat liquid, saltwater droplets will, when exposed to a CW heating field, only partially evaporate and eventually stabilize at a smaller diameter. As the particle shrinks, the salt concentration increases and causes the partial pressure of water at the surface to fall to a level equal to that of the surrounding gas, thereby halting evaporation. When the heating laser is turned off, the particle recondenses to its former size. Such particles will therefore not show an error due to volatility when periodically heated, because energy lost during one part of the heating cycle will then be regained during the remainder of the cycle.

III. Early Work

The earliest photothermal work in aerosols arose as a result of the thermal blooming problem encountered in the propagation of high-power radiation in the atmosphere. Thermal blooming may occur in clear air but is far more severe in the presence of atmospheric aerosols. Chan (1975) showed that the "effective absorption coefficient" is time dependent. In the case of aerosol absorption, there is a delay in heating the air due to the relative slowness of thermal conduction from particle to air. By solving the heat conduction equation, he derived Eq. (14), and by assuming various aerosol types and size distributions was able to numerically determine the

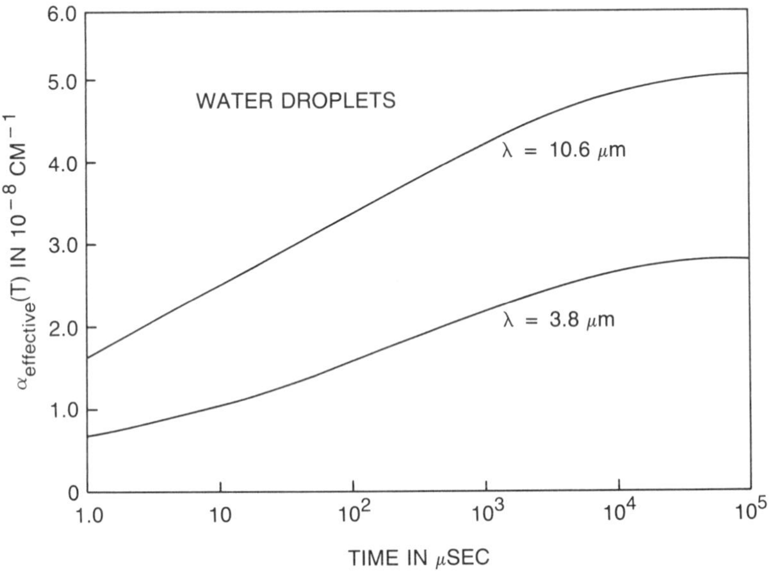

FIG. 4. Effective absorption coefficient $\alpha_{\mathrm{eff}}(t)$ as a function of t for a typical polydisperse distribution of water droplets irradiated at 3.8 and 10.6 μm. The time scale is in μsec. (From Chan, 1975).

effect of laser pulse length on the degree of blooming. Figure 4 shows the effective absorption coefficient as a function of time for water droplets irradiated at 3.8 and 10.6 μm. As can be seen, characteristic blooming times are in milliseconds. Other researchers have extended the theory in order to treat pulsed-laser excitation as well as CW and to include the effects of volatility (i.e., mass transport) both for pure water droplets (Armstrong, 1984; Baker, 1975; Davis, 1983; Sageev and Seinfeld, 1984) and aqueous salt droplets (Sageev and Seinfeld, 1984).

The first attempts to perform *in situ* photothermal spectroscopy were by Bruce and Pinnick (1977) and by Terhune and Anderson (1977) using photoacoustic detectors on quartz dust and smokes, respectively. In aerosol applications, the photoacoustic signal is independent of the scattering distribution, overcoming an important earlier obstacle. Unfortunately, any gaseous absorption that is present also contributes to the signal, and this may often overwhelm that of the particulate matter. However, when the particulate density is sufficiently high or when studying nonvolatile species in dry air, photoacoustic spectroscopy is quite useful. Yasa *et al.* (1979), using photoacoustic spectroscopy, showed that the absorbing species in many urban aerosols are mainly particulate carbon in graphite form.

IV. Photothermal Interferometry

A. Detection of Extremely Weak Absorptions in the Gas Phase

To date, the most sensitive absorption measurements in gases (10^{-10} cm^{-1}) have been accomplished through laser interferometric (Davis and Petuchowski, 1981) measurements of photothermally induced index-of-refraction changes. Although the photothermal refraction change following laser heating is small, it accounts for effects such as blooming and beam deflection. Small indices are most appropriately detected by using coherent laser interferometry. One such scheme, called phase fluctuation optical heterodyne (PFLOH) detection, has been demonstrated in pulsed and CW form (Davis and Petuchowski, 1981; Davis, 1980; Campillo *et al.*, 1980) and appears especially promising. The anticipated absorption sensitivity of this scheme appears to be $< 10^{-11}$ cm^{-1}, superior to the best values obtained (in the range 10^{-9} cm^{-1}) by using photoacoustic or other forms of photothermal spectroscopy. A comprehensive theory describing the use of the PFLOH technique in studies of molecular relaxation, thermal conduction, and extremely weak absorptions in the gas phase is given by Davis and Petuchowski (1981). The main drawback of interferometric schemes is their complexity when compared to other photothermal spectroscopies. However, in a laboratory environment, this minor inconvenience is well worth the additional sensitivity.

If an exciting laser is modulated at frequency, ω_m, it can be shown that the corresponding Fourier component of the induced index of refraction $\Delta n(t)$ is approximated by

$$\Delta n(t) = \frac{(n-1)I_0\alpha \sin(\omega_m t)}{2\omega_m C_p T \rho_a}. \tag{18}$$

Here I_0 is the CW intensity of the heating laser, n is the index of refraction, α is the absorption coefficient of the trace species, C_p is the specific heat at constant pressure, and T is the temperature. Equation 18 is derived by assuming that I_0 is nonsaturating, that the thermal conduction time is $> \omega_m^{-1}$, and that the effects of sample cell walls are minimal. For samples in nitrogen, typical values are $\omega_m = 2\pi(27)$ sec^{-1}, $T = 293$ K, $(n-1) = 2.92 \times 10^{-4}$, $\rho_a = 1.165 \times 10^{-3}$ g cm^{-3}, $C_p = 1.006$ J K^{-1} g^{-1}, and $I_0 = 250$ W cm^{-2}. One ppb ethylene in 1-atm nitrogen excited by the $P_{10}(14)$ line of a CO_2 laser ($\alpha = 32.14$ cm^{-1} atm^{-1}) will result in an induced index-change amplitude of $\Delta n = 2 \times 10^{-11}$.

Indices of this magnitude are easily detected by using a stable single-frequency probe beam, usually from a He–Ne laser source, in conjunction with a Mach–Zehnder interferometer (Davis, 1980) containing the sample in one of its arms (see Fig. 5, from Lin and Campillo, 1985). The probe beam is split into a reference component and a component to be passed through the sample. The sample probe is consequently phase modulated by the time-varying index of refraction and later recombined with the reference beam and heterodyned in a photodiode. The diode acts as a square law detector and yields a low-frequency signal at the difference frequency corresponding to ω_m. The voltage amplitude V_M of this component is related to the magnitude of the index of refraction by $V_M = V_s \sin[2\pi L \Delta n(t)/\lambda]$. Here V_s is an experimentally determined electrical calibration constant, L is the sample length, and λ is the wavelength of the probe laser. In practice, the minimum measurable Δn is limited by the interferometer vibration, sample turbulence, window absorption, or ultimately the photon noise of the He–Ne laser. The latter term is given by $\Delta n_{min} = (c/2\pi L\nu)(2h\nu\Delta f/\eta p)^{1/2}$, where Δf is the bandwidth of the detection process, p is the probe laser power, L is the sample length, and η is

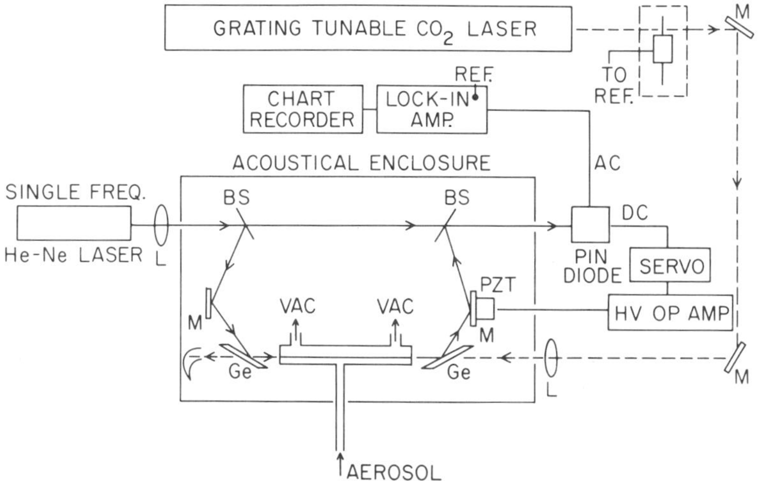

FIG. 5. Schematic of experimental PFLOH detector used to measure scatter-free absorption spectra of flowing aerosols.

the quantum efficiency of the detector. For a 0.25-mW laser, $L = 10$ cm, and $\eta = 0.6$, the minimum detectable Δn is $5 \times 10^{-14}(\Delta f)^{1/2}(\mathrm{Hz})^{-1/2}$.

In Fig. 5, the heating laser is a grating tunable CO_2 laser. The beam from this laser is amplitude modulated using a beam chopper and enters and leaves the interferometer via two Ge Brewster windows. Although transparent at 10.6 μm, the Ge windows form two of the high-reflectivity mirrors of the interferometer. Both CO_2 and He–Ne beams overlap coaxially in the sample region. As the interferometer yields an optimum (linear) response midway between destructive and constructive interference, this phase condition is maintained by using a servo control system that drives a transducer displacing one of the interferometer mirrors. The PFLOH device was vibrationally and acoustically isolated by means of an air suspension table and an aluminum box enclosure. In the device shown in Fig. 5, the limiting sensitivity was determined to be 10^{-8} cm^{-1} (ca. 300-ppt C_2H_4; see Fig. 6) and was limited by mechanical vibrations of the interferometer components. These vibrations can be reduced by at least two orders of magnitude by using a modified Jamin interferometer configuration as demonstrated by Davis and Petuchowski (1981). A Fabry–Perot interferometer (Campillo *et al.*, 1982) may also be used to take advantage of the increased sensitivity due to the finesse of the device. Besides achieving ultimate sensitivity, it is important to minimize the effect of interfering absorbing species. Thus, tunable lasers are often used to isolate the species of interest. Use of the Stark effect for fine tuning a nearby absorption line into coincidence with a

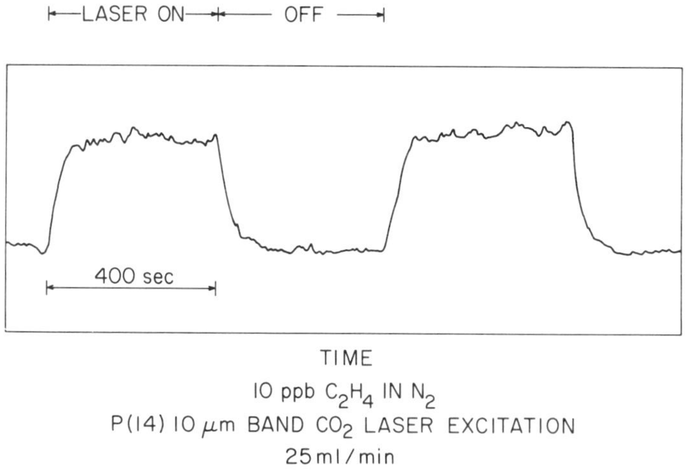

FIG. 6. PFLOH detector signal from N_2 gas stream containing 10-ppb C_2H_4. Characteristic system noise, due largely to vibration, can be seen during the laser-off periods. Signal fluctuations during the laser-on periods reflect variations in CO_2 laser power and are not indicative of the detection process. (From Lin and Campillo, 1985).

fixed-frequency laser has been employed in the PFLOH configuration with excellent results (Campillo *et al.*, 1980). This approach has advantages over the use of tunable lasers in separating a trace gas possessing a permanent electric dipole moment from other interfering species. Another approach that also appears to have promise consists of coupling a PFLOH detector to a simple gas chromatograph (Lin, Gaffney, and Campillo, 1981).

B. Aerosol - Cloud Studies

Submicrometer particles heated by a CO_2 laser are in the Rayleigh regime and, as shown in Section I.B., the quantity a^2Q_{abs} is proportional to the mass of the particle. Since the PFLOH signal gives an estimate of the total absorption, proportional to a^2Q_{abs}, the signal from an ensemble of small particles is proportional to their total mass (Lin and Campillo, 1985). An experimental study confirmed this approximation by measuring the PFLOH signal vs. ammonium sulfate mass (Fluckiger, Lin, and Marlow, 1985). In general, the formalism describing an ensemble of Rayleigh absorbers observed at $t \gg \omega_m^{-1}$ will be very similar to that describing trace gas detection, since the energy transferred to the surrounding gas very nearly equals the light energy absorbed. In this case, α in Eq. (18) can be replaced by the quantity $\alpha + A_a C$, where A_a is the specific absorption coefficient of the aerosol in $m^2 g^{-1}$, and C is the particulate mass concentration in g/m^3.

PFLOH aerosol detection capabilities (Lin and Campillo, 1985) in the submicrometer size range were explored using the apparatus shown in Fig. 5. Submicrometer aerosol particles, generally a log-normal size distribution centered on a mean particle size, were generated from aqueous salt solution by using an atomizer (Liu and Lee, 1975). The particles generated by the atomizer are generally wet and highly electrostatically charged, and therefore, drying and charge neutralization are normally necessary. The charged aerosols were neutralized by a Kr-85 neutralizer (TSI model 3012). The dried and neutralized aerosol was introduced to the capillary tube shown in Fig. 5. A partial vacuum applied to the vacuum port near the ends of the cell ensures that particles do not migrate to the Ge windows. PFLOH spectra for 9- and 2-mg m^{-3} mass concentrations of sulfate are displayed in Fig. 7. Several interesting features should be noted. First, there is a broad absorption feature that increases with decreased wave number due to a particulate absorption band centered at ca. 1100 cm^{-1}, and, second, there are several sharp features due to gaseous absorptions. The gaseous features are more obvious at low mass concentration. The line at 976 cm^{-1} is due to water-vapor absorption (at ca. 50% relative humidity), and the line at 1085 cm^{-1} is due to ammonia gas (ca. 500 ppb). Note that the data were obtained *in situ* on a continuously flowing aerosol and that simultaneous measurements of water-vapor and ammonia-gas concentration as well as sulfate mass loading are obtained. This quantitative *in situ* capability should find application in direct laboratory aerosol-chemistry studies. Lin and Campillo (1985) measured a specific absorption coefficient for sulfate at 1086 cm^{-1} of 0.5 m^2 g^{-1} and were able to detect sulfate levels as low as 100 μg m^{-3}. Measurements of levels below this value were not possible due to background water-vapor absorption. However, by a straightforward PFLOH geometry reconfiguration, it is possible to subtract the gaseous signals as shown in the next section.

C. Single-Particle Studies

PFLOH geometries which rely on the discrete nature of the particles allow the gas signal component to be effectively subtracted from the particle signal. One geometry, which monitors the absorption of a particle in a gas flow at one wavelength is shown in Fig. 8. The CW CO_2 laser and the He–Ne probe beams are arranged in a 90° crossbeam geometry in the

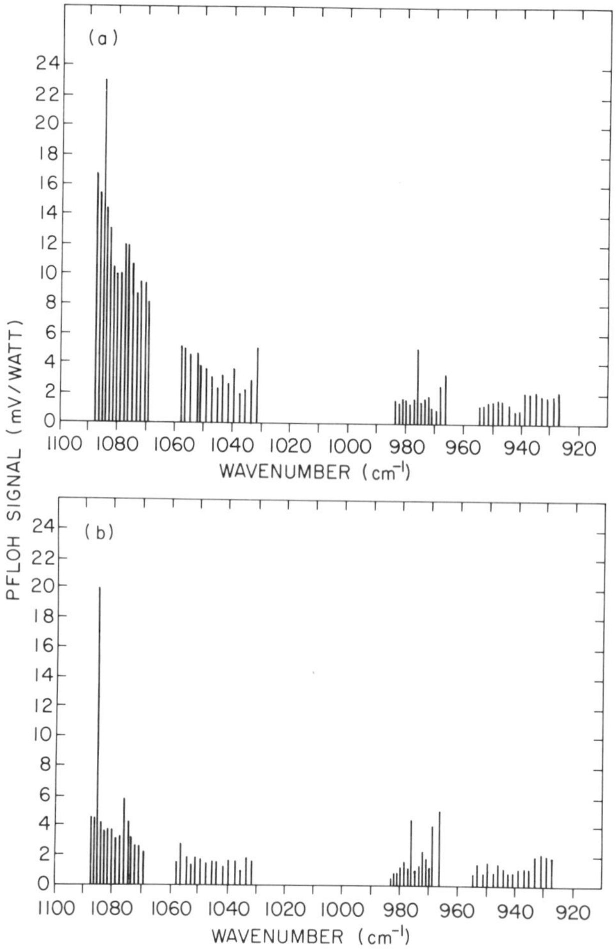

FIG. 7. PFLOH-detected 920–1090-cm^{-1} absorption spectra of flowing $(NH_4)_2SO_4$ aerosol. The sulfate mass concentration loading is (a) 9 mg m^{-3} and (b) 2 mg m^{-3}. The broad absorption feature centered at 1100 cm^{-1} is due to particulate sulfate, while sharp features at 976 cm^{-1} (water vapor ca. 50% relative humidity) and 1085 cm^{-1} (500 ppb of ammonia gas) are due to gaseous components. (From Lin and Campillo, 1985).

horizontal plane. The aerosol stream is introduced vertically through the intersection of the two beams. The background gas signal, being constant in time, is effectively subtracted by the servo/piezoelectric transducer (PZT) system, which maintains a phase quadrature relationship between probe and reference beams. As single particles pass through the crossed beams, phase fluctuations are photothermally induced. The leading edge of the

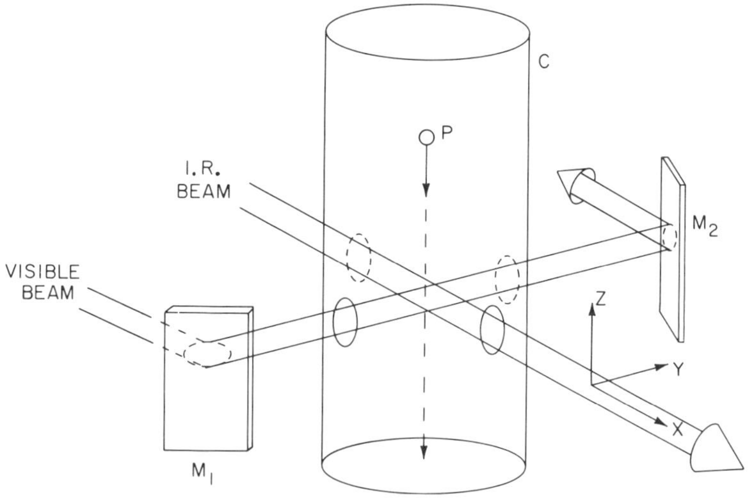

FIG. 8. Modified PFLOH geometry for single-particle flow stream detection.

phase perturbation is a ramp function that follows the relation

$$\frac{\Delta n}{\Delta t} = \frac{4}{3}(n - 1)a^3\rho_p A I_0 \left(C_p T \rho_a w^2 L\right)^{-1}. \tag{19}$$

Here w is the radius of the probe He–Ne laser beam and the other parameters as defined previously. Typical traces are shown in Fig. 9. Particles of carbon in the 1- to 10-μm size range were easily detected, and the slope of the leading edge was found to agree with Eq. (19). The slope was monitored rather than the maximum value of Δn, because the residence time of the particle in this geometry is largely determined by photophoresis and exhibited a complex dependence on size.

A more promising approach is to perform interferometry on a single particle levitated in a quadrupole trap (Lin and Campillo, 1985). It can be shown that the phase shift of an optical ray passing through the hot air zone around the particle, as in Fig. 3a, is given by

$$\Delta\phi(R) = \pi a^2 Q_{abs} I_0 \ln\left[\left(\frac{\kappa t}{R^2}\right)^{1/2} + \left(\frac{\kappa t}{R^2} - 1\right)^{1/2}\right](\lambda k_a T)^{-1}, \tag{20}$$

where R is the closest radial approach distance of the ray to the particle. If a 1-μm radius particle having $Q_{abs} \approx 1$ is exposed to a heating source of 10 mW/cm^2 for 1 sec, a phase shift of 400 μrad will be produced. This is well

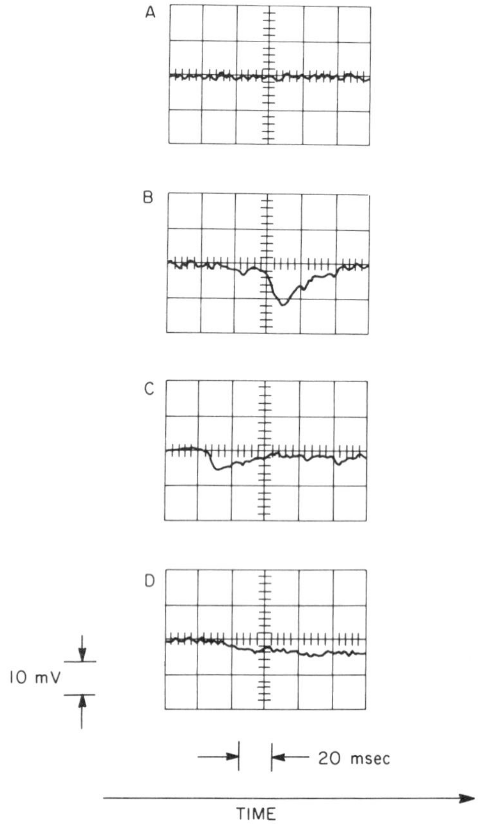

FIG. 9. PFLOH signals from the experimental geometry shown in Fig. 8. (A) No particles are present. Gaseous absorption contributes only to the background dc level in contrast to particle absorption, which appears as a phase fluctuation. (B), (C), (D) 7-, 5-, and 1.5-μm particles, respectively, detected as they pass through the interrogation region. (From Lin and Campillo, 1985).

within the limits of the heterodyne interferometer discussed previously (1 to 10 μrad). A blackbody IR source coupled to a monochromator with a 10-cm^{-1} bandwidth is quite capable of generating in excess of 10 mW/cm^2 and would be tunable over most of the mid IR. This scheme, therefore, has the potential for extremely broadband single-particle absorption spectroscopy. Since the scheme is universal, i.e., may be applied to particles regardless of shape, it has distinct advantages over PMMS and photophoresis discussed in the following sections. As of this date, however, use of this high-sensitivity geometry has not yet been reported in the literature.

V. Photothermal Modulation of Light Scattering

A. Photothermal Modulation of Elastic Scattering

In this form of Mie scattering, originally proposed and demonstrated by Campillo, Dodge, and Lin (1981) and Campillo and Lin (1983), two light beams are used, one for particle heating and another for elastic light scattering (see Fig. 3b). If the heating (absorbed) beam is amplitude modulated while the visible (nonabsorbed but scattered) probe beam is continuous, one finds in analyzing the visible Mie scattered light, that there is superimposed upon the dc signal a time-varying component at the frequency of the modulated heating beam whose amplitude is proportional to the particle-absorption and heating-beam intensity. Most importantly, there is usually no first-order background contribution to the signal from gas absorption, and so particle specific spectroscopy may be performed. There are several mechanisms that will lead to such a photothermal modulation of Mie scattered light: (1) photophoresis (see Section VI) of the particles within the Gaussian intensity profile of the visible probe beam; (2) macroscopic thermal lensing due to both gas and particle absorption and subsequent distortion of the far-field Mie scattering profile; (3) change in Q_{sca} caused by addition of a localized refracting heated gas shell surrounding the particle and index changes in the particle itself; and (4) photothermally induced physical changes (e.g., size, shape, phase changes) in the particles. For volatile particles, such as saltwater droplets, the dominant mechanism affecting scattering light is size-induced changes (Campillo, Dodge, and Lin, 1981) such as evaporation and condensation as illustrated in Fig. 3b. For other aerosol species and experimental conditions, mechanism (1)–(3) can be expected to play a contributing role. Because to date PMMS has principally been applied to volatile particles, we shall confine attention to this mechanism in this section.

We first describe the essential physics of this form of photothermally modulated Mie scattering for volatile particles containing water-soluble salts. This is a very common atmospheric case, since the bulk of the world's aerosol consists of water-soluble salts of nitrate and sulfate, usually in aqueous droplets. During the "on"-half cycle of the IR heating beam, the particle rapidly heats until it reaches a new equilibrium temperature determined by I_0, C_{abs}, and thermal conduction. As the particle warms, water is evaporated from its surface until a new equilibrium droplet size is established that is inversely proportional to the droplet temperature. For small changes, a reduction in particle size can be induced that is proportional to its absorption cross-section. During, the "off" cycle of the IR source, the particle rapidly returns to its ambient temperature and recondenses to its original size. Thus C_{sca} oscillates between two values. The theory describing this case is quite involved (Arnold, Murphy, and Sageev, 1985). We outline here an approximation that appears to give good results. We assume that when exposed to a constant light field, the particle heats so quickly (within milliseconds) to an equilibrium temperature that little evaporation initially takes place. Thereafter, the particle modifies its size by evaporation but with negligible temperature change. This approximation was justified by Baker (1975), who investigated the effect of radiation on volatile particles by solving the complete hydrodynamic equations. An approximate expression for the equilibrium temperature change ΔT is given by Chan (1975), $\Delta T = IQ_{abs}a(4k_a)^{-1}$ (see Section II.B.). For a 0.3-μm diameter ammonium sulfate water droplet illuminated by 35 W/cm^2 of 9.2-μm radiation, ΔT is about 0.1°C. At the elevated temperature, water evaporates from the particle until an equilibrium water-to-salt ratio is established, and the water-vapor pressure above the particle surface just equals that of the surrounding air. Tang (1976), by using thermodynamic arguments, has derived an expression relating the equilibrium radius of a solution droplet containing nonvolatile multicomponent electrolytes to the temperature T, relative humidity, and various solution parameters. For particles of > 0.05 μm, the droplet's behavior can be described by Raoult's law. If the temperature change is small, the approximation

$$\frac{\Delta a}{a} = -\frac{1}{3}\left(\frac{\Delta T}{T}\right)(1 - X)^{-1} \tag{21}$$

appears justified. Here X is the mole fraction of water in the droplet, and T is the temperature in degrees Celsius. Thus, heating a submicrometer particle by 0.1°C would typically result in a radius change of 0.8% and a Mie scattering cross-section reduction in the visible of 2.4%. Interpreting the PMMS signal from a single particle is straightforward, whereas interpretation from an aerosol cloud requires more care. The absolute magnitude of the photothermal modulation amplitude will be a strong function of the exact aerosol size distribution encountered, particles in the 0.1- to 2-μm size range standing to contribute more to the scattering process in the visible. However, an important approximation is that for vanishingly small induced changes from clouds an expression for the modulation amplitude can be justifiably separated into multiplicative factors, one of which, the absorption, depends solely on the wavelength. For example, a good approximation for a monodisperse distribution of submicrometer sized particles is

$$\lim_{I \to 0} \frac{\Delta I_{sca}}{I_{sca}} = -2A_a(\nu)\left[\frac{I\rho_p a^2}{9(1 - X)k_a T}\right]. \qquad (22)$$

Here $A_a(\nu)$ is the specific absorption coefficient, I_{sca} is the intensity of the scattered light in the visible in the absence of the heating (IR) illumination, ΔI_{sca} is the photothermally induced change, and other parameters are as previously defined. This factorability results from the assumption that the IR absorption cross-section per unit mass is independent of the particle size (Rayleigh regime). In general, a plot of the modulation amplitude vs. laser wavelength for most monodisperse or polydisperse aerosols will yield a good approximation to the relative absorption spectrum. In the single-particle case, there is sufficient information present in the scattered profile to determine the particle's size, index of refraction, and absorption spectrum.

Historically, the cloud case was the first to be conducted experimentally (Campillo, Dodge, and Lin, 1981; Lin and Campillo, 1985). Campillo, Dodge, and Lin (1981) performed PMMS on a cloud of ammonium sulfate and ammonium bisulfate aqueous submicrometer droplets. Use of particles in the Rayleigh regime made interpretation of their spectra straightforward. The aerosol was generally a polydisperse log-normal size distribution with 0.1-μm geometric mean diameter and a geometric-mean standard deviation of 2 and was generated from aqueous solution using a TSI model 3076 constant-output atomizer. An impactor was employed to remove particles of > 1 μm and the resultant aerosol introduced at a constant rate (0.1-2 l/min) into a central inlet of an open-ended 20-cm-long, 4-mm-i.d. glass capillary tube, which served as sample cell. Linearly polarized 633-nm light emitted by a Tropel 100 He–Ne laser was directed through the sample and elastically scattered off the particles where the horizontally polarized component was detected at 20° with respect to the forward-transmitted beam by using a 1P28 photomultiplier. A grating tunable CO_2 laser beam, superimposed spatially on the He–Ne laser beam in the sample, was used to modify the visible Mie-scattered light. Depending on experimental conditions, modulation amplitudes as high as 40% were observed. Although not useful for diagnostic purposes at this level, the effect was easily visible to the eye at sub-Hertz chopping rates. While taking spectra, the intensity of the CO_2 laser was kept constant at a level where the modulation amplitude was well below 1% to ensure that the modulation amplitude is linearly proportional to C_{abs}. A lock-in amplifier was employed to synchronously detect the time-varying component of the scattered light. Figure 10 shows the 920–1080-cm^{-1} absorption of ammonium bisulfate aerosols obtained

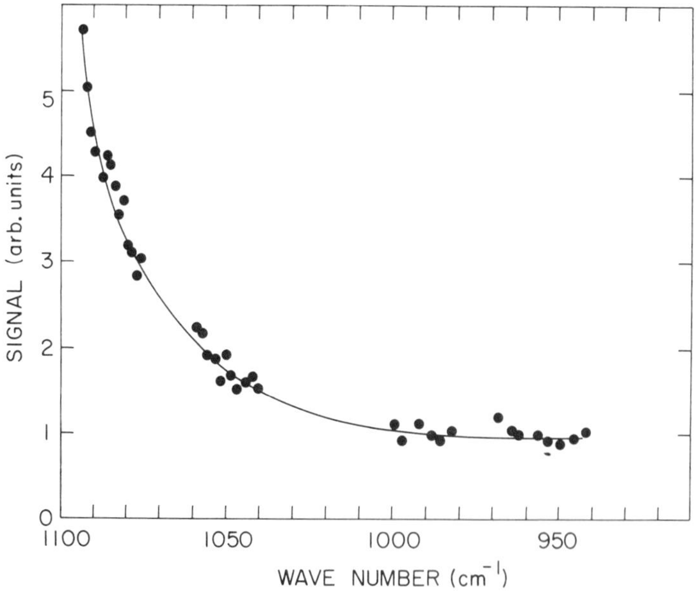

FIG. 10. Absorption spectrum of NH_4HSO_4 aerosol stream obtained by using photothermally modulated Mie scattering. (From Lin and Campillo, 1985).

using this method. These data agree well with additional data obtained using PFLOH spectroscopy and support the hypothesis that the PMMS technique yields a valid particle-relative absorption spectrum. Previous IR spectra of ammonium bisulfate crystals show a strong $SO_4^{=}$ ion band at 1100 cm^{-1}, as in Fig. 10. Assuming the absorption of 920 cm^{-1} to be caused totally by particulate water, we estimate that the mass fraction of water and sulfate in the droplets to be $1/3$ and $2/3$, respectively. These values are consistent with those expected from wet droplets partially dried in a 60% relative humidity environment. Had a He–Ne 3.39-μm laser source also been used for aerosol heating, the fraction of $(NH_4)^+$ ions could also have been determined. Precise chemical composition information, in this case the relative amounts of sulfate and attached cations as a function of relative humidity and the presence of other gases, would be useful in understanding the chemistry of acid rain.

Single-particle PMMS has recently been demonstrated (Arnold, Neuman, and Pluchino, 1984; Arnold, Murphy, and Sageev, 1985). In this case, somewhat larger particles (α = 2–10 μm) are utilized and suspended in a levitator (see Section VI), and use is made of surface wave resonances. This is accomplished by using a dye laser as the scattering probe and by tuning its wavelength so that the droplet is very near one of the resonances shown in Fig. 1. When operating on the steep slope of a scattering resonance, the effect of particle-size change is greatly magnified. In practice, monolayer-size changes may be detected (Ashkin and Dziedzic, 1981). Although the surface wave resonance also enhances the effect of various noise sources such as background thermal fluctuations and so does not improve the signal–noise ratio, it does have one very practical advantage in allowing the use of a low-power broad-band IR source such as that obtained from a blackbody source filtered by a monochromator. This allows spectra to be obtained over most of the molecular fingerprint region in the mid-IR (1–15 μm) and overcomes limitations inherent in discrete frequency sources such as CO_2 lasers. The use of a dye laser is not mandatory. Lin and Campillo (1985) demonstrated that a CO_2 laser may be used to shrink the size of the droplet to a point where it is in near resonance with a He–Ne laser.

B. Photothermal Modulation of Inelastic Scattering

Sharp spectral features resulting from particle surface wave resonances have also been observed in inelastic scattering processes such as fluorescence (Brenner *et al.*, 1980; Tzeng *et al.*, 1985), spontaneous Raman scattering (Turn and Kiefer, 1985), coherent Raman processes (Snow, Qian, and Chang, 1985; Qian and Chang, 1986; Qian, Snow, and Chang, 1985), and droplet lasing (Tzeng *et al.*, 1984; Lin *et al.*, 1986). Because such phenomena are sensitive to particle properties, they may be photothermally modulated as well. Unfortunately, to date, this possibility has not been widely exploited although such an approach has a number of attractive features. Figure 11 shows results of Tzeng *et al.* (1985) in which they used the wavelength of the surface fluorescence resonances to monitor photothermally induced shape and size changes in particles upon passing through a focused CW laser beam. Here the particles were generated using a Berglund–Liu vibrating orifice generator (Berglund and Liu, 1973), providing a linear stream of falling 42.2-μm diameter droplets of ethanol containing 10^{-4} M rhodamine 6G. The spectra from several falling particles are simultaneously displayed, with the distance downstream corresponding to the time following passage of a droplet through two counterpropagating Ar^+ laser beams (514.5 nm, 8 mW each) at $t = 0$. For times $t < 0$, the features at 596.2 and 597.6 nm are observed from the unperturbed spherical droplet. At $t = 0$, the counterpropagating beams distort the spherical droplets into prolate spheroids with corresponding shifted and degraded resonances. As the droplets pass out of the heating beam, they mechanically

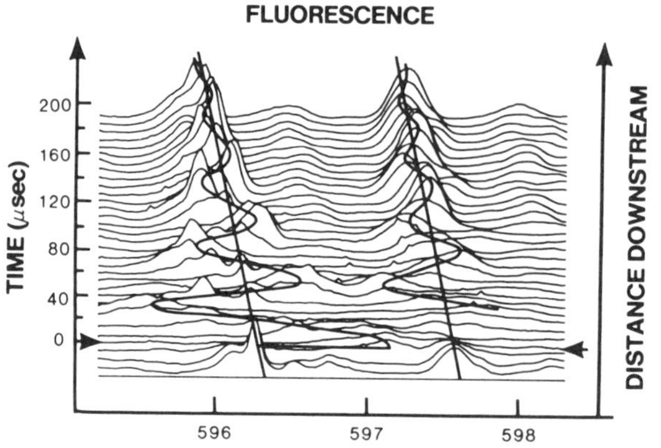

FIG. 11. Fluorescence spectra from laser-perturbed droplets. The distance downstream is plotted as time delay after the argon-ion laser perturbation occurring at $t = 0$ (see arrows). (From Tzeng *et al.*, 1985).

oscillate with small amplitudes (5 parts in 10^4) between prolate and oblate spheroids, with the observed damped oscillations in the spectra providing information on the surface tension and viscosity of the ethanol droplets.

Lin, Eversole, and Campillo (unpublished) have monitored the photothermally induced changes in fluorescence and lasing quantum efficiencies rather than making use of spectral shifts. In their work, they probed fluorescein water droplets with 533-nm, 20-n sec laser pulses. Fluorescein is a weak absorber at 533 nm under ambient conditions. As the particle is heated in a CO_2 laser beam, the red wing absorption edge broadens, thereby increasing the absorption and the resulting observed fluorescence. The enhancement in fluorescence and lasing efficiency increased exponentially with temperature, with a 40°F temperature rise above ambient resulting in a factor-of-five enhancement.

VI. Photophoresis

When spherical particles are subjected to directed light illumination from a laser or an arc lamp, some are observed to move away from the light source while others are observed to move towards it. Nonspherical particles may be observed to travel in closed orbits or along irregularly shaped paths.

This phenomenon, discovered by Ehrenhaft (1917), is called photophoresis and arises from a radiometric force (Kerker, 1969; Preining, 1966; Deguillon, 1950). The force may be towards (negative photophoresis) or away (positive photophoresis) from the light source and is quite distinct from radiation pressure. It results from momentum transfer between a fluid (gas or liquid) and an unevenly illuminated (heated) surface following light absorption. The key to this force is the asymmetry in the surface temperature resulting from the particle acting as an optical cavity. Thus, for example, if the particle is composed of a strong absorber and the light is largely absorbed at the front surface, the hottest side (front of particle) would be preferentially pushed as rebounding gas molecules collide, extract energy, increase their velocity, and impart momentum. In this case, a positive photophoretic force results. The effect is proportional to the incident light intensity and the particle absorption coefficient as well as size, shape, and complex index of refraction. Small changes in the size of an evaporating droplet may cause it to reverse its direction. Because photophoresis is intimately related to the absorption properties of a particle, it can be an important analytical tool as was first demonstrated by Pope, Arnold, and Rozenshtein (1979), and since by others (Arnold and Amani, 1980; Arnold, Amani, and Obrenstein, 1980; Lin, 1985). A practical photophoretic spectrometer (Lin, 1985) is shown in Fig. 12. The theory of the radiometric force is quite complex (Kerker and Cooke, 1982; Yalamov *et al.*, 1976; Lin, 1975; Akktaruzzaman and Lin, 1981) because it involves (1) calculation of the internal source function of the particle and (2) solution of the heat transfer equation. The latter is strongly dependent on the Knudson number. Single-particle experimental photophoretic spectroscopy, however, is fairly straightforward and extremely sensitive. Typically, the particle is given a charge and levitated in an electric quadrupole cell as shown to Fig. 12. Application of appropriate electrical fields to the quadrupole maintains the charged particle at the center of the cell or null point. Suspension of a charged particle is normally affected by parameters such as excess particle charge, density and mass, and electrical fields and frequencies. The cell of Fig. 12 is composed of two hyperboloidal end electrodes (endcaps) placed above and below a central toroidal electrode having a bihyperbolic cross-section. An ac voltage is applied to the toroidal electrode to form a hyperbolic trapping field, whereas a dc voltage is applied to the endcaps to compensate for the gravitational pull on the particle. With this design, particles in the 5- to 50-μm size range in diameter were stably suspended with 200-Hz ac operation at ambient pressure. Particles outside this range may be suspended by adjusting both the field strength and the frequency. In Fig. 12, a CW CO_2 laser, tunable between

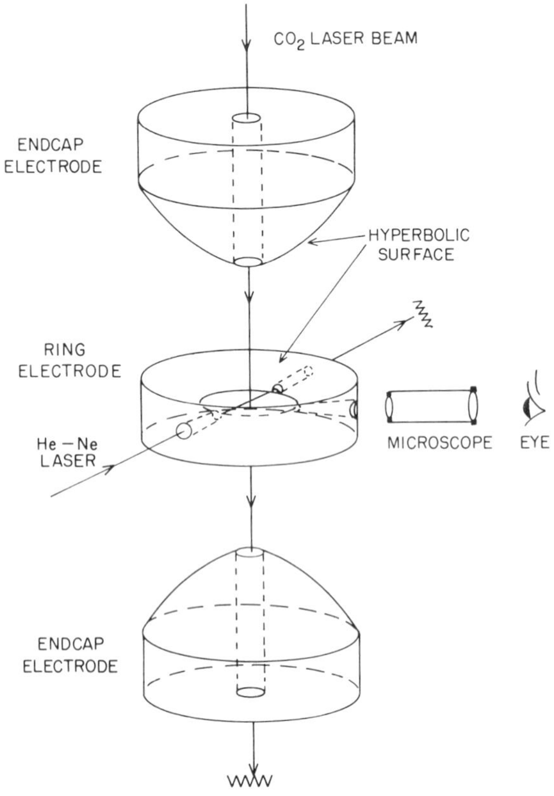

FIG. 12. Experimental schematic depicting geometry of electrodes and optics of the electrical quadrupole chamber used for single-aerosol-particle suspension and photothermal experiments.

9.2 and 10.8 μm is brought in through the endcaps to heat the particle. When exposed to CO_2 laser radiation, the particle tends to be displaced from its previous equilibrium position and requires an offset dc voltage to restore the particle to the previous null position. The balance between the gravitational and electrical field can be described by $(V/d)q = mg$. Here V is the dc voltage between endcaps, d is the separation of the endcap electrodes, q is the charge on the particle (typically ca. 10^5 e) of mass m, and g is the gravitational constant. The photophoretic force F_p is exerted on the particle along a gravitational field line:

$$mg + F_p = \left(\frac{V + \Delta V}{d}\right) q, \tag{23}$$

and from the previous expressions it follows that

$$\frac{F_p}{mg} = \frac{\Delta V}{V}. \tag{24}$$

In a typical experiment, ΔV and V are observables and may be related to F_p by Eq. (24). It should be cautioned that Eq. (24) breaks down whenever intense laser radiation causes surface ionization, photoinduced chemistry, or particle evaporation. However, this phenomenon is sufficiently sensitive so that low-power blackbody radiation or low-power IR laser irradiation appear safe for use under almost all conditions.

The characteristics of IR photophoresis for 8.4- and 11.7-μm ammonium sulfate particles was described in Lin (1985). A linear function of 9.2-μm irradiation power vs. photophoretic force was observed up to 500 W/cm^2. The pressure dependence of F_p is shown in Fig. 13. Initially, positive photophoresis is observed for sizes investigated at low pressures. As the pressure is gradually increased, the magnitude of the positive photophoresis decreases until a reversal in the direction of the force occurs. On sign reversal, the photophoretic force increases monotonically with pressure up to 1 atm. The negative photophoresis is consistent with theoretical predictions for this size particle (Kerker and Cooke, 1982). In general, large ($a \gg \lambda$) transparent particles act like microlenses and partially focus the input radiation near the rear of the particle, leading to a hotter back surface and a force directed back towards the laser. At low pressure, photophoretic effects should disappear due to the absence of a fluid. The small positive force observed at low pressure is probably due to direct radition pressure, a phenomenon which is independent of surrounding fluid.

Figure 14 shows typical photophoretic action spectra between 920 and 1080 cm^{-1} for 10- and 100-μm single suspended particles of ammonium sulfate (Lin and Campillo, 1985). Also shown for comparison are an absorption curve obtained by using photothermal modulation of Mie scattering and one by using Fourier-transform infrared (FTIR) absorption spectroscopy on ammonium sulfate powder pressed into KBr pellets. The photophoretic action spectrum for the 10-μm particle is in very good agreement with both the photothermally modulated Mie spectrum obtained on submicrometer polydispersed clouds and with that obtained using PFLOH spectroscopy. The photophoretic action spectra mimic the absorption quite well in this frequency range, and photophoresis appears to be an attractive spectroscopic tool for such measurements. Although both the particle size and the CO_2 laser wavelengths are ca. 10 μm, there do not appear to be any resonances in the absorption cross-section (i.e., similar to

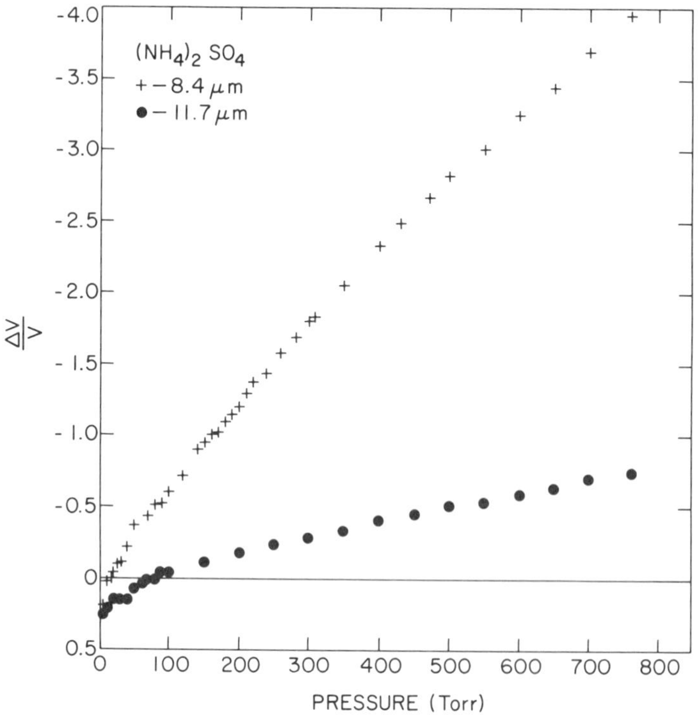

FIG. 13. Plot of normalized photophoretic signal ($\Delta V/V$) vs. chamber gas pressure for an 11.7-μm (●) and an 8.4-μm (+) ammonium sulfate crystalline particle. The normalized signal is a measure of the quantity F_p/mg and is inversely proportional to the particle radius. The photophoretic force for this size particle is negative (toward the laser). However, at low pressure, direct radiation pressure is stronger than the fluid-dependent photophoretic force, and the total force is positive (away from the laser). (From Lin and Campillo, 1985).

that shown in Fig. 1), because the imaginary component of the refractive index is quite large. The 100-μm particle data does not agree with data obtained by using the other photothermal schemes but agrees well with those using FTIR spectroscopy on coarse powder pressed into KBr pellets. This difference appears to be due to spectral broadening caused by saturation of the absorption when the characteristic absorption depth is much less than the particle diameter.

In principle, one does not have to measure the photophoretic force as above but can devise geometries to couple photophoretic motion information directly onto a visible probe beam. For example, in the geometry shown in Fig. 12, by employing a He–Ne visible beam having a Gaussian cross-section and a chopper to modulate the CO_2 laser, a periodic variation

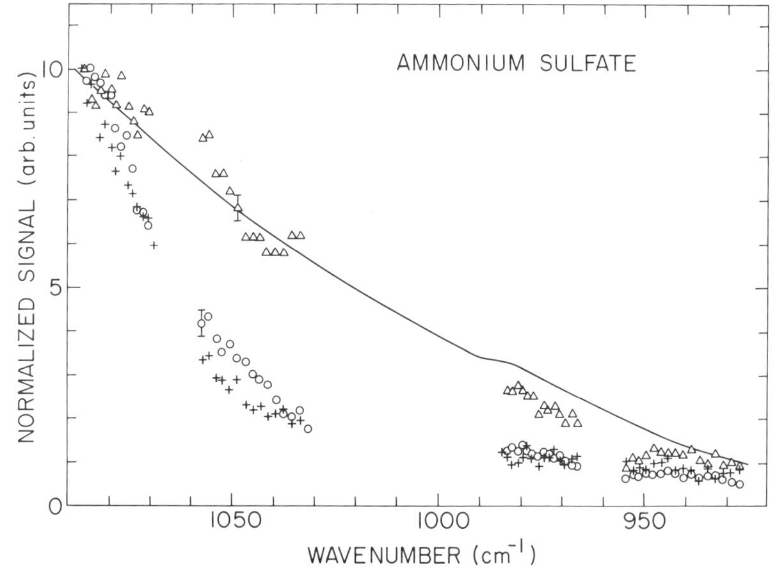

FIG. 14. Ammonium sulfate aerosol-absorption spectra: (○) and (△) correspond to photophoretic action spectra of single suspended crystalline particles of 10- and 100-μm diameter, respectively; (+) corresponds to data obtained using photothermal modulation of Mie scattering on submicrometer polydisperse aqueous aerosols. The solid curve corresponds to a FTIR absorption spectrum of coarse ammonium sulfate powder pressed into KBr pellets. The photophoretic, Mie, and PFLOH data (see Fig. 7) for small particles are in excellent agreement. The photophoretic 100-μm data and the FTIR data are also in agreement and indicate that the 1100-cm^{-1} absorption bandwidth is particle-size dependent. (From Lin and Campillo, 1985).

in the elastic scattered He–Ne light will be observed as the particle moves out of and back into the null position. Although this and other geometries that can be devised for freely floating or flowing particles appear to be very attractive, feasibility studies have not yet been reported in the literature. Recently, however, a somewhat different approach has been explored by Eversole and Lin (unpublished) that appears to be exceptionally promising. A quadrupole suspension apparatus similar to that of Fig. 12 is utilized. The magnitude of the quadrupole field was adjusted to balance the particle against gravitation, and the particle was stabilized at a null point through the use of appropriate field strengths and ac frequency. It is a characteristic of this type of levitator that even at the null point there is a small component of oscillatory motion at the ac frequency of the toroidal electrode. At the null point, the particle sinusoidal motion is too small to be detected by eye even through a microscope. Any deviation from the null point results in the establishment of a new equilibrium position and increases the amplitude of the oscillatory motion in proportion to the displacement. Eversole and Lin probe the peak velocity of this oscillatory motion by utilizing a He–Ne Doppler velocimeter. An IR heating source is brought in horizontally through a slot in the ring electrode, and the particle is displaced towards or away from the source, thereby increasing the amplitude of the oscillatory motion in proportion to the size of the photophoretic force. This approach appears to be very sensitive and compares favorably to other forms of photophoretic spectrometers.

VII. Conclusion

Although many of the photothermal spectroscopies discussed in other chapters of this volume may be used to measure the combined gas–particle absorption properties of aerosol clouds, we have chosen to limit discussion to photothermal schemes that are inherently particle specific (that is, yielding a signal that is to first order free of the much larger gaseous absorption contribution) and that display sufficient sensitivity to characterize single micrometer-sized particles. These schemes rely to some extent on the peculiar heat patterns developing near the illuminated particle surface or on the heat-induced changes in the particles or upon the unusual optical properties of such fine particles.

Several of the more promising spectroscopies were discussed. Many more can be visualized. Of those covered, three seem very promising: (1) photothermal interferometry, (2) photothermal modulation of Mie scattering and (3) photophoretic spectroscopy. Variations of photothermal interferometry appear to have the greatest potential. The sensitivity is sufficient for single-particle analysis, and the scheme may be applied to particles regardless of shape. Feasibility studies of each of the above three spectroscopies have demonstrated that broadband absorption data may be obtained from single particles. Various laboratories have obtained such spectra by using low-power blackbody sources as well as lasers. The use of blackbody sources is especially significant because spectroscopy over exceptionally broad wavelength ranges may now be performed. This will no doubt be a boon to aerosol chemists, who previously were unable to conduct *in situ* aerosol absorption studies for quantitative analysis.

Developments in the photothermal spectroscopy of aerosols have been both dramatic and sudden. As the field is still in its infancy, much remains to be done both experimentally and theoretically. The ultimate sensitivities of the above three schemes have yet to be achieved or elucidated. Currently, sensitivity is largely limited by convective currents set up by the heating laser. As designs are improved, sensitivity limits may approach photon noise and Brownian motion limits. The future of photothermal spectroscopy in aerosol science looks bright indeed.

ACKNOWLEDGEMENTS

The authors gratefully acknowledge useful discussions with Dr. Jay Eversole of NRL. The authors' work is funded through the Office of Naval Research.

REFERENCES

Akktaruzzaman, A. F. M., and Lin, S. P. (1981) *J. Colloid Interface Sci.* **61**, 170–182.

Armstrong, R. L. (1984). *Appl. Opt.* **23**, 148–155.

Arnold, S., and Amani, Y. (1980). *Opt. Lett.* **5**, 242–244.

Arnold, S., Amani, Y., and Orenstein, A. (1980). *Rev. Sci. Instrum.* **51**, 1202–1204.

Arnold, S., Neuman, M., and Pluchino, A. B. (1984). *Opt. Lett.* **9**, 4–6.

Arnold, S., Murphy, E. K., and Sageev, G. (1985). *Appl. Optics* **24**, 1048–1053.

Ashkin, A., and Dziedzic, J. M. (1981). *Appl. Opt.* **20**, 1803–1814.

Baker, M. E. (1975). *Atmos. Environ.* **10**, 241–248.

Berglund, R. N., and Liu, B. Y. H. (1973). *Environ. Sci. and Tech.* **7**, 147–153.

Bohren, C. F., and Huffman, D. R. (1983). *Absorption and Scattering of Light by Small Particles*. Wiley, New York.

Brenner, R. E., Barber, P. W., Owen, J. F., and Chang, R. K. (1980). *Phys. Rev. Lett.* **44**, 475–478.

Brock, J. R. (1982). In *Aerosol Microphysics I: Particle Interactions* (W. H. Marlow, ed.), pp. 15–88. Springer-Verlag, New York.

Bruce, C. W., and Pinnick, R. G. (1977). *Appl. Opt.* **16**, 1762–1765.

Campillo, A. J., and Lin, H.-B. (1983). "Method and Apparatus for Aerosol Particle Absorption Spectroscopy," U.S. Patent #**4,415,265**, (filed 25 June 1981).

Campillo, A. J., Dodge, C. J., and Lin, H.-B. (1981). *Appl. Opt.* **20**, 3100–3102.

Campillo, A. J., Lin, H.-B., Dodge, C. J., and Davis, C. C. (1980). *Opt. Lett.* **5**, 424–426.

Campillo, A. J., Petuchowski, S. J., Davis, C. C., and Lin, H.-B. (1982). *Appl. Phys. Lett.* **41**, 327–329.

Carslaw, H. S., and Jaeger, J. C. (1959). *Conduction of Heat in Solids*, 2nd ed. Clarendon, Oxford.

Chan, C. H. (1975). *Appl. Phys. Lett.* **26**, 628–630.

Chylek, P, Pendleton, J. D., and Pinnick, R. G. (1985). *Appl. Opt.* **24**, 3940–3942.

Davis, C. C. (1980). *Appl. Phys. Lett.* **36**, 515–517.

Davis, C. C., and Petuchowski, S. J. (1981). *Appl. Opt.* **20**, 2539–2554.

Davis, E. J. (1983). *Aerosol Sci. and Tech.* **2**, 121–144.
Deguillon, F. (1950). *Comp. Rend.* **231**, 274–275.
Ehrenhaft, F. (1917). *Phys. Z.* **18**, 352–368.
Fluckiger, D. U., Lin, H.-B., and Marlow, W. H. (1985). *Appl. Opt.* **24**, 1668–1681.
Friedlander, S. K. (1977). *Smoke, Dust, and Haze: Fundamentals of Aerosol Behavior*. Wiley, New York.
Fuchs, N. A. (1964). *The Mechanics of Aerosols*. Macmillan, New York.
Hidy, G. M., and Brock, J. R. (1970). *The Dynamics of Aerocolloidal Systems*. Pergamon, Oxford.
Kerker, M. (1969). *The Scattering of Light and Other Electromagnetic Radiation*. Academic, New York.
Kerker, M., and Cooke, D. D. (1982). *J. Opt. Soc. Am.* **72**, 1267–1272.
Lin, H.-B. (1985). *Opt. Lett.* **10**, 68–70.
Lin, H.-B., and Campillo, A. J. (1985). *Appl. Opt.* **24**, 422–433.
Lin, H.-B., Gaffney, J. S., and Campillo, A. J. (1981). *J. Chromatog.* **206**, 205–214.
Lin, H.-B., Huston, A. L., Justus, B. L., and Campillo, A. J. (1986). *Opt. Lett.* **11**, 614–616.
Lin, S. P. (1975). *J. Colloid Interface Sci.* **51**, 66–71.
Liu, B. Y., and Lee, K. W. (1975). *Am. Ind. Hyg. Assoc. J.* **36**, 861–865.
Mie, G. (1908). *Ann. Phys.* **25**, 377–455.
Pendorf, R. B. (1958). *J. Phys. Chem.* **62**, 1537–1542.
Pope, M., Arnold, S., and Rozenshtein, L. (1979). *Chem. Phys. Lett.* **62**, 589–591.
Preining, O. (1966). In *Aerosol Science* (C. N. Davies, ed.), pp. 111–135. Academic Press, London.
Qian, S.-X., and Chang, R. K. (1986). *Phys. Rev. Lett.* **56**, 926–929.
Qian, S.-X., Snow, J. B., and Chang, R. K. (1985). *Opt. Lett.* **10**, 499–501.
Rayleigh, Lord (1914). *Phil. Mag.* **27**, 100–109.
Sageev, G., and Seinfeld, J. H. (1984). *Appl. Opt.* **23**, 4368–4374.
Snow, J. B., Qian, S.-X., and Chang, R. K. (1985). *Opt. Lett.* **10**, 37–39.
Tang, I. N. (1976). *J. Aerosol Sci.* **7**, 361–371.
Terhune, R. W., and Anderson, J. E. (1977). *Opt. Lett.* **1**, 70–72.
Thurn, R., and Kiefer, W. (1985). *Appl. Opt.* **24**, 1515–1519.
Tzeng, H.-M., Wall, K. F., Long, M. B., and Chang, R. K., (1984). *Opt. Lett.* **9**, 499–501.
Tzeng, H.-M., Long, M. B., Chang, R. K., and Barber, P. W. (1985). *Opt. Lett.* **10**, 209–211.
Wagner, P. E. (1982). In *Aerosol Microphysics II: Chemical Physics of Microparticles* (W. H. Marlow, ed.), pp. 129–178, Springer-Verlag, New York.
Yalamov, Yu L., Kutukov, V. B., and Shchikin, E. R. (1976). *J. Colloid Interface Sci.*, **57**, 564–571.
Yasa, Z., Amer, N. M., Rosen, H., Hansen, D. A., and Novakov, T. (1979). *Appl. Opt.* **18**, 2528–2530.

Index

A

Acoustic
  absorption, 21
  transit time, 161, 180, 181, 184, 185
  waves, 155, 157, 159, 160, 166, 170, 175, 181, 182, 186
Aerosol(s), 309–342
  ammonium bisulfate, 332
  ammonium sulfate, 324–326, 329–332, 337–339
  carbon, 321, 327
  generators, 325, 333
  laser propagation through, 310, 320, 321
  mass concentration, 325
  monodisperse distribution, 330, 331
  photoacoustic spectroscopy of, 321
  polydisperse distribution, 320, 321, 331, 332, 339
    log normal, 325, 331
  water, 319–321, 329–333
Amino acid, detection of, 139
Analytical detection, 17
Azulene ($C_{10}H_8$), 175–179

B

Boundary conditions, 166, 209
Bragg scattering, 22

C

Calorimetry, 4, 5
Carrier diffusion, 63
Chromatography
  gas, 131
    thermal lens detector, 131–133
  liquid, 127
    thermal lens detector, 133–140
  thin layer, 127, 145–153
CO_2, 174, 175
Combustion, 249–267
Concentration
  gradients, 276, 277, 285, 290, 292
  measurements in flames, 260
  profile, 291
Contact thermal resistance, 29
Cooling, kinetic, 173
Copper
  electrodeposition, 289
  electrodes, 288
  film, 208
  layer, 275, 277
  plate, 275, 285
Corrosion, iron, 271, 285–290
Cracks, 193
  brittle fracture, 196
  fatigue, 194, 196
  inside bolt hole, 197
  planar, 193
CS_2, 18
Cyclic photoacoustic and voltammetric measurements, 275
Cyclic voltammetry, 293
Cyclopropane, 179, 180
Cyclobutanone, 179

D

Deflection angle, probe beam, 11, 13, 15, 290, 293
Delayed heat, 2, 14, 18
Detection of thermal image, 93
Detection of thermal lens, 113
Differential
  deflection, 292
  photoacoustic reflection absorption (PARAS), 271
  reflectance, 288

Diffraction
Fraunhofer, 165
Fresnel, 165
Diffusion, 45, 54, 65, 72, 159–161, 170, 175, 176, *see also* Thermal diffusion
Diffusion constant, 256, *see also* Thermal diffusivity
Dye, 302–306
rug, detection of, 139

E

Electrochemical
interface reactions, 277–285
interface sensitized, 302–306
reaction, 276, 285, 292
species diffusion, 290–293
Electrode(s)
back-surface illumination, 274
characterization, 271–277
deflection, parallel to plane of, 291
passive, 273–275
surface species, 290
Electrophoresis, 129, 141–145
Energy conversion efficiency, 294, 295, 300–302
Energy equation, 224, *see also* Heat conduction equation
Energy transfer, 155, 157, 160, 172, 180, 185
physics, 293–306
vibration-to-translation, 178
vibration-to-vibration, 178
Excited states
electronically, 157
vibrationally, 296

F

Flames, 249–267
Flow
in liquid chromatography, 134–135
injection analysis, 140–141
velocity, 11, 17, 213–248, 252–253, 264
Fluid flows, 84, 213–248
Fluid velocimetry, 213–248
Focal length of thermal lens, 116
Fractal, 69–76
Freon, detection of, 131–133

G

GaAs, 50
Gas dynamic effects, 161
Grating, 235–240
Greens function, 12, 15

H

Hadamard transform, 147–153
Heat conduction equation, 224, 316–320
Heterogeneous media, 75
Hydrodynamic equations, 166, 167

I

Imaging, 17, 232–235
Impulse approximation, 90, 107
Index of refraction, 162, 163, 179, 286
gradient, 7–22
Interface
electrochemical, 301
electrode-electrolyte, 279, 288, 297, 299
photoelectrochemical, 293–306
powder electrode, 278
Interferometer, Mach–Zehnder, 322–325
Interferometry, 26, 39, 94, 321

L

Laminar flow, 221, 228
Layered structures, thickness, 29, 204
Lensing, see thermal lens
Light polarization modulation, 277

M

Microphone, 167, 170, 181, 186
Mirage detection, 40–42, 62, 64, 191, 198, *see also* Photothermal deflection theory, 200, 209
Multiplexing, 76
spatial, 76–78, 147–153

N

NH_2, measurement of, 262
Nitroanaline, detection of, 135–138
NO_2, 18, 171, 172, 186, 187
Nonradiative
  deexcitation, 297–300
  quantum efficiency, 278
  recombination, 294

O

OH, 257–260, 266
Optical band gap, 298
Optical reflection coefficient, 287
Oxide
  film, 289
  growth, 276, 288
Oxygen atoms, 186, 187

P

Particles, 309–342
  complex index of refraction, 311–314
  cross section, 311
    absorption, extinction, and scattering, 311, 316, 329–331
    efficiency, 311
      absorption, 311–312, 318, 324, 327, 330
      extinction, 311
      scattered, 311–312, 316, 329
  electric dipole levitator, 327, 335–336, 339
    null point, 335–336, 339
  lasing, 313, 333–334
  resonances, 313, 332–334, 337
  size-parameter, 311–312, 314
  surface waves, 313, 332–334, *see also* Whispering gallery waves
  simulated effects, 313, 333
Particle photothermal spectroscopies, 309–342
  photophoresis, 316, 327–329, 334–340
  photothermal Fraunhofer diffraction, 315
  photothermal interferometry, 315, 320–328, 339–340
  photothermal modulation of inelastic scattering, 333–334
  photothermal modulation of Mie scattering (PMMS), 316, 329–334, 339–340
  survey, 315–316
Passivation, 281, 289
PDS, *see* Photothermal deflection spectroscopy
PDV, *see* Photothermal deflection velocimetry
P^2ES, *see* Photopyroelectric spectroscopy
Phase fluctuation optical heterodyne (PFLOH) spectroscopy, 315, 320–328, 339, 340
Photoacoustic
  calorimetry, 184
  deflection spectroscopy, 220, 229, 260
  generation, 5
  measurement, 6
Photophoresis, 316, 327–329, 334–340
Photopyroelectric spectroscopy (P^2ES), 270, 283, 285
Photothermal deflection spectroscopy (PDS), 16, 17, 61–63, 159, 213–248
  flowing medium, 101, 217
  pressure dependence, 245
  solids, 216
  static fluids, 216
  temperature dependence, 245, 253–257
  transverse, 61, 145–153
  velocimetry, 107, 213–248
    2-D, 242–243
    equipment, 246–247
    liquids, 243–244
    multipoint measurements, 232–235
    relaxation rate method, 221–228
    time-of-flight, 228–235
    transient thermal grating, 235–240
    turbulent flows, 240–242
Photothermal detection, 4, 37
Photothermal displacement spectroscopy, 23–25
Photothermal effects, 3, 4
  interferometric detection, 94
Photothermal imaging, 17
Photothermal interferometry, 26, 39, 94, 321
Photothermal lensing, flowing medium, 112, 120–125
Photothermal phase-shift spectroscopy, 96–101
Photothermal radiometry, 27–29, 38, 74

Photothermal spectroscopy (PTS), 273–277, 289, 294, 300, *see also* Photothermal deflection spectroscopy
Porphyrins, 302–306
Pressure change, 5–7
Probe beam refraction, 8–15
Propane, 266
Prompt heat, 2, 14
Proteins, electrophoresis of, 141–145
Pulsed excitation, 67, 104

Q

Quantum efficiency, 290, 299

R

Radiation pressure, 335–337
Radiometry, *see* Photothermal radiometry
Rate constant, 161
Ray equation, 101, 163
Reaction(s), 184, 277–285
  rate, 155
Refractive index, *see* Index of refraction
Refractive index gradient, 7–22, 276, 291–293
Rough surface, 69

S

Scattering, 310–314
  elastic, 310, 329
  inelastic, 333–334
  Mie, 310–314, 316, 329–331
  Rayleigh, 312
Self-defocusing, 8
Semiconductor, 281, 293–306
SF_6, 179, 180, 184, 186
Si, 68
Spectroscopy, 16–17
Spectroscopy, in flames, 265
Superlattice, 206
Surface and subsurface defects in solids, 191
Surface deformation, 22–25
Surface temperature
  absorbing layer, 50
  thick samples, 49

T

Temperature
  measurements in flames, 252, 260, 263
  profile, 12, 15
  rise, 4–5, 44–61, 215
Thermal blooming, 8, 320–321, *see also* Thermal lens
Thermal
  conductivity, 159–161, 170, 172–176
  diffusion, 66, 84, 290
  diffusivity, 198
    of compound semiconductors, 203
    of elements, 202
  interferometry, 287
  lens, 8, 42, 220, 329
    CW, intracavity, 132
    CW, modulated, 136–145
    CW, time-resolved, 136, 140
    focal length, 116
    pulsed, 131–132
  properties, 204
    of layered structures, 204
    of superlattice materials, 206
  relaxation time, 11, 14
  wave interference, 286
Thermistor, 274, 275, 289
TDTL, *see* Time dependent thermal lens
Time resolved optoacoustics (TROA), 156–157
Thin films, 285–290
Time dependent thermal lens, 155, 157
Transient lensing, 162
Transient thermal grating, 235–242
Transient thermal reflectance, 30
TROA, *see* Time resolved optoacoustics
Turbulence, 240–242

V

Velocity, measurements, 107, 213–248, 264

W

Whispering gallery waves, 313, 332–334